ИЗОТОПНЫЙ ОБМЕН
И ЗАМЕЩЕНИЕ ВОДОРОДА
В ОРГАНИЧЕСКИХ СОЕДИНЕНИЯХ

IZOTOPNYI OBMEN
I ZAMESHCHENIE VODORODA
V ORGANICHESKIKH SOEDINENIYAKH

ISOTOPIC EXCHANGE
AND THE REPLACEMENT OF HYDROGEN
IN ORGANIC COMPOUNDS

ISOTOPIC EXCHANGE
AND THE REPLACEMENT OF HYDROGEN
IN ORGANIC COMPOUNDS

by

A. I. Shatenshtein

Authorized translation from the Russian
by C. Nigel Turton, B.Sc., Ph.D. and Tatiana I. Turton, B.A.

CONSULTANTS BUREAU
NEW YORK
1962

The Russian text was published by the USSR
Academy of Sciences Press in Moscow in 1960.

Александр Исаевич Шатенштейн

Изотопный обмен и замещение водорода
в органических соединениях

Library of Congress Catalog Card Number 62-12859
©1962 Consultants Bureau Enterprises, Inc.
227 West 17th St., New York 11, N.Y.

Printed in the United States of America

I dedicate this book to
the memory of my brother
David Isaevich Shatenshtein

PREFACE

Hydrogen replacement reactions play an outstanding part in organic chemistry. They include such industrially important processes as halogenation, nitration, sulfonation, alkylation, and direct amination and metallation of organic substances. It is no accident that the data obtained in the investigation of these reactions have been used extensively in the solution of a series of fundamental problems in theoretical organic chemistry and this tendency still exists.

Isotopic exchanges of hydrogen also belong to this group of chemical conversions. They have attracted considerable attention as they are the simplest and, to some extent, models of hydrogen replacement reactions, which are convenient for elucidating the mechanism of hydrogen replacement and the effects of interaction between atoms in organic molecules.

The mechanisms of hydrogen replacement reactions are examined in this monograph and the most space is allocated to hydrogen isotope exchanges. The mechanisms of all the given reactions are interpreted from the point of view of the acid–base properties of the reagents. In this connection, a large section of the book is devoted to a thorough substantiation of the thesis that hydrocarbons participating in the given reactions act as acids or bases. Problems of acid–base catalysis and the mechanism of acid–base interaction are also discussed. Thus, a considerable part of the book is concerned with problems of the theory of acids and bases.

In general, the contents of this book fall within the field of physical organic chemistry, which is evolving very rapidly nowadays. Most of the phenomena described lie at the meeting point of three sections of chemistry: the study of the structure and reactivity of organic compounds, the chemistry of isotopes, and the theory of acids and bases. It is most tempting to bridge the gap between them and thus make it possible to use the conclusions reached in the study of some phenomena for interpreting the nature of others and also to promote a critical attitude toward some existing generalizations from a new point of view.

Definite opinions are put forward on the nature and mechanism of heterolytic hydrogen exchange in solutions and the mechanisms of acid–base reactions as well as hypotheses on the effect of the medium on the reactivity of organic substances and on the manifestation of the interaction between atoms in their molecules. These hypotheses required a multiplicity of facts for a thorough and objective substantiation. In the hope of stimulating the discussion of the problems considered in the book, the author has tried to group the material in such a way that concrete scientific conclusions evolve from it and these are formulated at the end of each chapter and section.

Soviet literature includes many books on problems in isotope chemistry, the mechanism of organic reactions, and the theory of acids and bases. It is sufficient to mention the valuable works of A. I. Brodskii, "The Chemistry of Isotopes" (1957), S. Z. Roginskii, "Theoretical Bases of Isotopic Methods of Studying Chemical Reactions" (1956), O. A. Reutov, "Theoretical Problems in Organic Chemistry" (1956), and N. A. Izmailov, "Electrochemistry of Solutions" (1959). I should also mention my previous monograph, "Theories of Acids and Bases." It seemed useless to duplicate the contents of these books. The reader can refer to them for further information. The present monograph summarizes and generalizes data obtained by Soviet and foreign authors and published in journals mainly in the decade up to and including 1958. Some work from 1959 is mentioned. *

The book contains numerous cross references, which help to associate the material in the different sections.

The monograph is the final outcome of investigations of the isotopic exchange of hydrogen in nonaqueous solutions under the general direction of the Scientific Council of the "Theory of Chemical Structure, Kinetics, and Reactivity" of the Division of Chemical Sciences of the Academy of Sciences USSR. The author is very grateful to the chairman of the Scientific Council Academician V. N. Kondrat'ev for his interest in these investigations and assistance in the publication of the book. The author considers it his pleasant duty to thank the Responsible Editor M. I. Kabachnik for much valuable advice and fruitful discussion and Academician Acad. Sci. Ukr.SSR A. I. Brodskii and Corresponding Member Acad. Sci. Ukr. SSR N. A. Izmailov for friendly help in the form of a critical review of the manuscript and valuable comments. All remarks as to the shortcomings of the book will be gratefully received.

Moscow,
May, 1959

---

* The translation includes supplementary material written by the author, which takes into consideration work published in 1959-1961.

CONTENTS

THE FOLLOWING SOVIET JOURNALS CITED IN THIS BOOK
ARE AVAILABLE IN COVER-TO-COVER TRANSLATION

Russian Title	English Title	Publisher
Doklady Akademii Nauk SSSR	Proceedings of the Academy of Sciences USSR	Consultants Bureau (chemistry sections)
Kinetika i kataliz	Kinetics and Catalysis	Consultants Bureau
Optika i spektroskopiya	Optics and Spectroscopy	American Institute of Physics
Izvestiya Akademii Nauk SSSR, Otdelenie khimicheskikh nauk	Bulletin of the Academy of Sciences USSR, Division of Chemical Sciences	Consultants Bureau
Zhurnal fizicheskoi khimii	Journal of Physical Chemistry USSR	The Chemical Society (London)
Zhurnal obshchei khimii	Journal of General Chemistry USSR	Consultants Bureau

INTRODUCTION

The history of any science is an absorbing account of the discovery of new phenomena and their generalization, the rise and development of concepts, hypotheses, and theories, and the interaction of ideas.

In chemistry there are many concepts that have occupied the thoughts of scientists for centuries and these include an element, valence, chemical affinity, etc. The concepts of "acid" and "base" have an even longer history as the peculiar properties of fermented grape juice and an infusion of wood ash have been known since remote antiquity. An attempt to follow the successive changes in the ideas on acids and bases was made in the monograph, Theories of Acids and Bases [1].

The hydrogen theory of acids originated at the beginning of the last century. Davy, Ampère, and Gay-Lussac discovered that hydrochloric, hydrofluoric, and prussic acids contain hydrogen. The idea that acids must contain hydrogen was expressed particularly clearly by Dulong. An important stage in the history of the hydrogen theory is associated with the name of Liebig. He defined acids as hydrogen compounds in which the hydrogen may be replaced by a metal. This definition is still of some value. The success of Arrhenius' theory of electrolytic dissociation stimulated a reexamination of the definition of acids and alkalis. The acidity and alkalinity of substances were related to their dissociation in aqueous solution to form hydrogen ions (H^+) and hydroxyl ions (OH^-), respectively.

To counterbalance Arrhenius' physical theory of solutions, D. I. Mendeleev put forward a chemical theory of solutions and the Russian physical chemists D. P. Konovalov, I. A. Kablukov, and others participated actively in substantiating it.

Hantzsch, Franklin, Kraus, P. I. Walden, V. A. Plotnikov, and many other scientists helped to develop the chemistry and physical chemistry of nonaqueous solutions. The study of nonaqueous solutions showed that the properties and reactions of aqueous solutions of acids and alkalis are similar to those of solutions which not only contain no H^+ and OH^- ions, but do not even conduct an electric current.

In 1923 Brønsted [2] and Lowry [3] simultaneously and independently defined acids and bases as donors and acceptors of protons. The new definition agreed well with all experimental material. Brønsted's orderly quantitative theory of acid–base equilibrium and catalysis has had a great influence on the investigations of chemists in different countries for several decades. Nonetheless, in the summary of the results achieved during the twenty-five years that Brønsted's theory has existed given in the monograph Theories of Acids and Bases [1] it was noted that there were already serious deviations from the theory and its limitations were beginning to be felt, while in the last decade so many facts have accumulated which do not agree with Brønsted's theory that a reexamination of it has become imperative. N. A. Izmailov [4] has established experimentally a theory of the dissociation of acids and bases, which allows for considerably greater complexity of the dissociation process than follows from Brønsted's theory. The equations of Brønsted's qualitative theory were found to be a particular case of N. A. Izmailov's equations.

The transfer of a proton, to which Brønsted reduced the reaction between an acid and a base, is actually only the final stage of an acid–base interaction. It occurs only under favorable conditions.

Brønsted greatly extended the scope of acids and bases, but still did not cover the whole range of acid–base processes. This became evident with the application of methods that were more sensitive than classical ones, namely, potentiometry, conductimetry, and electronic spectrophotometry. Measurements of infrared spectra, dipole moments, and proton magnetic resonance have made it possible to not only detect, but even characterize quantitatively much weaker interactions of the same type as normal acid–base reactions.

One extremely sensitive method utilizes the isotopic exchange of hydrogen in nonaqueous protophilic (basic) and protogenic (acidic) media. This was used in the Isotopic Reactions Laboratory of the L. Ya. Karpov Physicochemical Institute over the last ten years for a systematic study of the acidic and basic properties of various hydrocarbons.

Isotopes play a predominant part in contemporary science. They help investigators to elucidate the details of chemical reaction mechanisms, to obtain much important information on the reactivity and structure of substances, etc. Exchange reactions in which atoms of the same element participate can be followed readily by the use of isotopes. Such reactions were predicted by D. I. Mendeleev [5] in 1886: "In a mass of molecules there must or, at least, there may be exchange movement of the atoms. If we have the molecules AB and AB, then A from the former may transfer to the latter molecule and back."

Among the stable isotopes, the heavy isotope of hydrogen, deuterium, which was discovered by Urey, Brickwedde, and Murphy in 1932 [6], has been of the greatest value in chemical investigations. Gaseous deuterium was obtained by fractionation of liquid hydrogen. Washburn and Urey [7], and Lewis and MacDonald [8] soon proposed a simpler method of separating hydrogen isotopes by electrolysis of water; "heavy water" was obtained in this way. Even in the first work with it, Lewis [9], Hall [10], Bonhöffer [11], Klar [12], and other scientists discovered that it was then possible to make a detailed study of the isotopic exchange of hydrogen for deuterium or, more briefly, deuterium exchange. In the USSR, A. I. Brodskii [13] prepared water enriched in heavy isotopes of hydrogen and oxygen and used it successfully for many valuable studies in isotopic chemistry.

The hydrogen bound by O−H, N−H, S−H, and Hal−H bonds exchanges very rapidly with heavy water, but it is not usually possible to effect this exchange with C−H bonds, which is most promising for the solution of many problems in theoretical organic chemistry. It is readily observed that the exchange of hydrogen for the deuterium of heavy water occurs only with the C−H bonds of substances which are weak acids or weak bases in aqueous solutions and the exchange is catalyzed by strong acids and bases. Ingold [14] was able to replace hydrogen in some hydrocarbons by deuterium by the action of deuterosulfuric acid. These facts suggest the acid−base nature of hydrogen exchange in solutions and from this it follows that if the acidic or basic properties of hydrocarbons and their derivatives are increased by the use of appropriate solvents and catalysts, hydrogen exchange in C−H bonds becomes a normal phenomenon [15]. Knowledge of the rules of acid−base equilibrium and catalysis in nonaqueous solutions has helped in the search for solvents and catalysts which have made it possible to extend the range of hydrogen exchange reactions considerably. In addition, knowledge of the rules of acid−base interaction has made it possible to predict the factors which must affect hydrogen exchange.

Studies on hydrogen exchange in nonaqueous solutions were begun in the Isotopic Reactions Laboratory of the L. Ya. Karpov Physicochemical Institute on the basis of Brønsted's theory, which was found to be of inestimable help in their development [16]. At the same time, the results of these studies have made the limitations of the theory obvious as was also the case in the development of other aspects of acid−base interaction. The accumulation of knowledge of acids and bases has made it necessary to extend the concept of protolytic reactions and to determine their mechanism more accurately.

The isotopic exchange of hydrogen has much in common with chemical hydrogen replacement reactions.* Ingold was the first to note this when he compared deuterium exchange between aromatic hydrocarbons and sulfuric acid with reactions of the former involving strong acids (nitration and sulfonation). However, isotopic reactions are much simpler as the molecular structure of the substance remains almost unchanged during the replacement of hydrogen by its isotope. If we neglect the small change in free energy caused by the differences in the zero point energies of bonds with tritium, deuterium, and protium, then it may be assumed that there is no thermal effect during isotopic exchange and the reaction is determined solely by kinetic parameters. In most cases the kinetics of isotopic exchange can be described quite accurately by a simple first-order equation. By measuring the rate of the exchange of deuterium for ordinary hydrogen in various monodeutero derivatives of the same compound in different solvents, one can readily evaluate quantitatively the reactivities of nonequivalent atoms in the molecule, examine the fine effects of the interaction of atoms, and observe the active role of the medium.

The book consists of five sections. In the first section the classification of reagents and reactions is described and it is shown that only acids and bases may participate in heterolytic hydrogen exchange reactions. In the second section there is a comparison od data on the catalysis of hydrogen exchange by acids and bases, which demonstrates the nature of this catalysis and provides a key to the understanding

---

*In this book, hydrogen replacements which differ from isotopic reactions are arbitrarily termed chemical. Isotopic reactions essentially are also chemical.

of the mechanism of exchange reactions. The third section consists of a detailed review of the literature on the properties of hydrocarbons as acids and bases. This material demonstrates the need for extending the field of acids and bases beyond the limits set by Brønsted's theory and provides further proof for the thesis on the acid–base nature of hydrogen exchange in solutions. The fourth section contains a discussion of the mechanism of acid–base interaction and a new definition of acids and bases is formulated. Finally, in the fifth and last section of the book the mechanisms of hydrogen replacement in heterolytic chemical reactions are examined. A knowledge of them, together with an understanding of the mechanism of acid–base reactions, is essential for an accurate appraisal of the mechanism of hydrogen exchange in solutions.

The principles of methods of preparing deuterium compounds are examined in the Appendix.

LITERATURE CITED

1. A. I. Shatenshtein, Theories of Acids and Bases [in Russian] (Goskhimizdat, Moscow, 1949).
2. J. N. Brønsted, Rec. trav. chim Pays-Bas 42, 718 (1923).
3. T. M. Lowry, Chem. and Ind. 42, 43 (1923); J. Chem. Soc. 123, 822 (1923).
4. N. A. Izmailov, Zhur. Fiz. Khim. 24, 321 (1950); 30, 2164 (1956).
5. D. I. Mendeleev, Zhur. Russ. Fiz. Khim. Obshch. 18, 8 (1886).
6. H. C. Urey, F. G. Brickwedde, and G. M. Murphy, Phys. Rev. 39, 164 (1932).
7. E. W. Washburn and H. C. Urey, Proc. Nat. Acad. Sci. U.S. 18, 496 (1932).
8. G. N Lewis and R. MacDonald, J. Chem. Phys. 1, 341 (1933).
9. G. N. Lewis, J. Am. Chem. Soc. 55, 3502 (1933).
10. N. F. Hall, E. Bowden, and T. Jones, J. Am. Chem. Soc. 56, 750 (1934).
11. K. F. Bonhöffer and G. W. Brown, Z. phys. Chem. 23B, 171 (1933).
12. R. Klar, Z. phys. Chem. 24B, 335 (1934).
13. A. I. Brodskii, Chemistry of Isotopes [in Russian] (Izd. AN SSSR, 1957); Zhur. Fiz. Khim. 6, 1431, 1437 (1935); 9, 417, 755 (1937).
14. C. K. Ingold and C. L. Wilson, Z. Elektrochem. 44, 62 (1938).
15. A. I. Shatenshtein, Uspekhi Khim. 21, 914 (1952); 24, 377 (1955); 28, 3 (1959).
16. A. I. Shatenshtein and Yu. P. Vyrskii, Doklady Akad. Nauk SSSR 70, 1029 (1950); Zhur. Fiz. Khim. 25, 1206 (1951).

TYPES OF REAGENTS AND REACTIONS

In order to be clear on the place of hydrogen isotope exchange reactions among other reactions, it is essential to be familiar with the existing classification of reagents and reactions [1-7].

1. OXIDIZING AND REDUCING AGENTS

When an atom is converted into an ion, it donates or adds one or several electrons. A substance whose atoms act as electron donors is called a r e d u c i n g a g e n t and one whose atoms act as electron acceptors, an o x i d i z i n g a g e n t.

An oxidation–reduction process is represented by equations (1) and (2):

$$\text{reducing agent}_1 \rightleftharpoons \text{oxidizing agent}_1 + e^- \tag{1}$$

$$\text{oxidizing agent}_2 + e^- \rightleftharpoons \text{reducing agent}_2 \tag{2}$$

where e^- is an electron.

An oxidation--reduction reaction actually occurs only when an electron donor and an electron acceptor are present at the same time. Summing equations (1) and (2) gives the equation of an oxidation--reduction reaction:

$$\text{reducing agent}_1 + \text{oxidizing agent}_2 \rightleftharpoons \text{oxidizing agent}_1 + \text{reducing agent}_2 \tag{3}$$

Examples of the reactions are the mutual ionization of sodium and chlorine atoms (4) and the formation of a hydride from atomic sodium and hydrogen (5):

$$\text{red}_1 \rightleftharpoons \text{ox}_1 + e^-; \quad \text{ox}_2 + e^- \rightleftharpoons \text{red}_2; \quad \text{red}_1 + \text{ox}_2 \rightleftharpoons \text{ox}_1 + \text{red}_2$$

$$\text{Na} \rightleftharpoons \text{Na}^+ + e^-; \quad \text{Cl} + e^- \rightleftharpoons \text{Cl}^-; \quad \text{Na} + \text{Cl} \rightleftharpoons \text{Na}^+ + \text{Cl}^- \tag{4}$$

$$\text{Na} \rightleftharpoons \text{Na}^+ + e^-; \quad \text{H} + e^- \rightleftharpoons \text{H}^-; \quad \text{Na} + \text{H} \rightleftharpoons \text{Na}^+ + \text{H}^-. \tag{5}$$

Oxidation–reduction relationships extend to reactions between an atom and an ion (6), between ions (7) and (8), between an atom or an ion and a molecule (9) and (10), and between molecules (11).

1

Reactions involving atoms and ions:

$$red_1 + ox_2 \rightleftharpoons ox_1 + red_2$$

$$Fe + Cu^{++} \rightleftharpoons Fe^{++} + Cu \tag{6}$$

$$Cu^+ + Fe^{+++} \rightleftharpoons Cu^{++} + Fe^{++} \tag{7}$$

$$Sn^{++} + 2Fe^{+++} \rightleftharpoons Sn^{++++} + 2Fe^{++}. \tag{8}$$

Reactions involving molecules:

$$Na + H_2O \rightarrow Na^+ + OH^- + H \tag{9}$$

$$H^- + H_2O \rightarrow OH^- + H_2 . \tag{10}$$

Reactions between molecules:

$$H_2O + Cl_2 \rightleftharpoons 2H^+ + Cl^- + OCl^-$$

$$2H^+ + 2H_2O \rightarrow 2H_3O^+,$$

or summing:

$$Cl_2 + 3H_2O \rightarrow 2H_3O^+ + Cl^- + OCl^-. \tag{11}$$

In reactions (9) and (10) a water molecule acts as an oxidizing agent and in reaction (11) as a reducing agent. In oxidation–reduction reactions involving molecules, chemical bonds are broken and formed and such reactions are often complex, especially in organic chemistry.

2. ACIDS AND BASES

In 1923 Brønsted [8] and Lowry [9] defined a c i d s as substances which denote a proton and b a s e s as substances which add a proton, i.e.,

$$acid_1 \rightleftharpoons base_1 + H^+ \tag{12}$$

$$base_2 + H^+ \rightleftharpoons acid_2 . \tag{13}$$

Brønsted pointed out the analogy between acid–base and oxidation–reduction reactions: both reactions involve elementary particles, electrons in oxidation–reduction and protons in acid–base reactions. In accordance with the schemes presented above, in acid–base reactions (as in oxidation–reduction reactions) the substances formed have charges different from those of the starting materials as the reactions involve the transfer of a charged particle, namely, a proton (or an electron in the case of oxidation–reduction reactions). The simple acid–base equilibria (12) and (13) do not exist independently as is also the case with the simple oxidation–reduction equilibria(1) and (2). In actual fact, two pairs of oxidizing and reducing agents and two pairs of acids and bases, respectively, always participate in the reactions:

$$acid_1 + base_2 \rightleftharpoons base_1 + acid_2, \tag{14}$$

for example:

$$\text{acid}_1 \rightleftharpoons \text{base}_1 + \text{H}^+; \quad \text{base}_2 + \text{H}^+ \rightleftharpoons \text{acid}_2; \quad \text{acid}_1 + \text{base}_2 \rightleftharpoons \text{base}_1 + \text{acid}_2$$

$$\text{HF} \rightleftharpoons \text{F}^- + \text{H}^+; \quad \text{H}_2\text{O} + \text{H}^+ \rightleftharpoons \text{H}_3\text{O}^+; \quad \text{HF} + \text{H}_2\text{O} \rightleftharpoons \text{F}^- + \text{H}_3\text{O}^+ \tag{15}$$

$$\text{H}_2\text{O} \rightleftharpoons \text{OH}^- + \text{H}^+; \quad \text{NH}_3 + \text{H}^+ \rightleftharpoons \text{NH}_4^+;$$
$$\text{H}_2\text{O} + \text{NH}_3 \rightleftharpoons \text{OH}^- + \text{NH}_4^+ \tag{16}$$

$$\text{H}_2\text{O} \rightleftharpoons \text{OH}^- + \text{H}^+; \quad \text{H}_2\text{O} + \text{H}^+ \rightleftharpoons \text{H}_3\text{O}^+;$$
$$\text{H}_2\text{O} + \text{H}_2\text{O} \rightleftharpoons \text{OH}^- + \text{H}_3\text{O}^+. \tag{17}$$

A water molecule is a base toward HF (Equation 15), but water reacts as an acid in the presence of NH_3 (Equation 16). Finally, two water molecules may act as an acid and a base toward each other. Equation (17) represents the self-ionization of water.

The other acids and bases appearing in the equations given may also react with each other in pairs, for example, the acid from (15) and the base from (16):

$$\text{HF} + \text{NH}_3 \rightleftharpoons \text{F}^- + \text{NH}_4^+ \tag{18}$$
$$\text{H}_3\text{O}^+ + \text{OH}^- \rightleftharpoons \text{H}_2\text{O} + \text{H}_2\text{O}. \tag{17a}$$

Equation (17a) corresponds to the neutralization of an acid and a base ionized in aqueous solution and is the reverse of the self-ionization of water (17). In the latter there are formed a l y o n i u m   i o n (the cation obtained by the addition of a proton to a solvent molecule) and a l y a t e   i o n (the anion formed by the elimination of a proton from a solvent molecule*). In the solvents NH_3, N_2H_4, $\text{C}_2\text{H}_5\text{OH}$, $\text{CH}_3\text{CO}_2\text{H}$, H_2SO_4, and HF, the lyonium ions are NH_4^+, N_2H_5^+, $\text{C}_2\text{H}_5\text{OH}_2^+$, $\text{CH}_3\text{CO}_2\text{H}_2^+$, H_3SO_4^+, and H_2F^+ and the lyate ions are NH_2^-, N_2H_3^-, $\text{C}_2\text{H}_5\text{O}^-$, CH_3CO_2^-, HSO_4^-, and F^-.

Some oxidation — reduction reactions, for example, (9) and (10), may be represented as acid — base conversions if it is assumed that the water molecule acts as a proton donor in them. The lattice of metallic sodium (and other metals) contains free electrons together with metal ions ($\text{Na} \rightarrow \text{Na}^+ + e^-$) and consequently it is permissible to write:

$$\text{acid}_1 + \text{base}_2 \rightarrow \text{base}_1 + \text{acid}_2$$

$$\text{H}_2\text{O} + e^- \rightarrow \text{OH}^- + \text{H} \tag{9a}$$
$$\text{H}_2\text{O} + \text{H}^- \rightarrow \text{OH}^- + \text{H}_2. \tag{10a}$$

At first glance this way of writing the reactions appears to be formal: metallic sodium is not normally regarded as a Brønsted base nor molecular hydrogen as a Brønsted acid. However, there is not merely a formal correspondence to Brønsted's definitions. In his work it was emphasized [11] that the power to react with metals with the liberation of hydrogen has long been regarded as one of the most important and obvious criteria of acids. In actual fact, all the substances Brønsted designated as acids react with metals, for example, the molecules H_2SO_4, HCl, CH_3COOH, $\text{C}_6\text{H}_5\text{OH}$, H_2O, $\text{C}_6\text{H}_5\text{NH}_2$,

---

* Terms proposed by N. Bjerrum [10].

and NH_3, the cations H_3O^+, NH_4^+, $C_6H_5NH_3^+$, and $Co(NH_3)_5H_2O^{+++}$, and the anions $H_2PO_4^-$, HPO_4^{--}, etc. Even some hydrocarbons react with alkali metals with the liberation of hydrogen (see p. 83) and their properties and reactions as acids will be discussed in detail in Section III. The following are examples of such reactions:

$$\overset{acid_1}{} \qquad \overset{base_2}{} \qquad \overset{base_1}{} \qquad \overset{acid_2}{}$$

$$HC \equiv CH + e^- (+ Na^+) \rightleftarrows HC \equiv C^- + H(+ Na^+) \qquad (19)$$

$$(C_6H_5)_3CH + e^- (+ Na^+) \rightleftarrows (C_6H_5)_3C^- + H(+ Na^+). \qquad (20)$$

The elimination of a proton from a hydrocarbon molecule forms a c a r b a n i o n, i.e., an ion with trivalent, negatively charged carbon (see pp. 88 - 91 for more details).

Below we examine reactions confirming the acidity of molecular hydrogen, which is observed in its reactions with organoalkali metal compounds (p. 84) and in deuterium exchange with bases (p. 66). *

The list of acids shows that they may be molecules, cations, and anions. The same is true of bases. First of all, let us list the bases conjugate with the acids mentioned above: HSO_4^-, Cl^-, CH_3COO^-, $C_6H_5O^-$, OH^-, $C_6H_5NH^-$, NH_2^-, H_2O, NH_3, $C_6H_5NH_2$, $Co(NH_3)_5OH^{++}$, HPO_4^{--}, PO_4^{---}, etc.

The reactions in which hydrocarbons act as bases are examined in detail in Section III. By adding a proton they are converted into c a r b o n i u m i o n s with trivalent, positively charged carbon.

$$\overset{base_1}{} \qquad \overset{acid_2}{} \qquad \overset{acid_1}{} \qquad \overset{base_2}{}$$

$$(C_6H_5)_2C = CH_2 + H_2SO_4 \rightleftarrows (C_6H_5)_2CCH_3^+ + HSO_4^- \qquad (21)$$

$$C_{10}H_{14} + HF \rightleftarrows C_{10}H_{15}^+ + F^- \qquad (22)$$

(see pp. 130-136 for more details).

Carbanions obtained by acid ionization of hydrocarbons, for example, $HC \equiv C^-$, $(C_6H_5)_3C^-$, $C_6H_5^-$, CH_3^-, etc., are among the strongest bases.

3. ACIDLIKE SUBSTANCES

In addition to hydrogen acids, aprotic substances are capable of participating in equilibrium reactions with bases. They are called a c i d l i k e s u b s t a n c e s [13,14].†

---

* Correctly speaking, a hydrogen atom may be an acid (Equations 19 and 20). Hydrogen also has an affinity for a proton (see [12]), i.e., is also a base.

† Bjerrum [15-17] subsequently named them a n t i b a s e s, while Bell [18, 19] calls them a c c e p t o r s. Other workers [20-23] subsequently arrived at the same conclusion as the author of this book on the need to distinguish between acidlike substances and hydrogen acids and, in particular, Ingold [1] holds this view. In the literature, especially American literature, acidlike substances are designated as L - a c i d s, i.e., Lewis acids, which is not quite accurate as Lewis himself [24] combined aprotic and hydrogen-containing substances in the class of acids. Hydrogen acids are often called Brønsted acids and designated as B - a c i d s.

This name emphasizes their similarity to acids and simultaneously marks the difference from the latter.

Like acids, acidlike substances react with bases by a donor-acceptor mechanism, accepting the electron pair of the base. The structure of the complex compounds formed is therefore similar to that of the products from reactions between acids and bases. This is reflected, for example, in the similarity of their spectra [25-27]. Some acidlike substances catalyze the same reactions as acids [28] (see also [15]).

Despite the considerable similarity in the reactions of a series of aprotic acidlike substances and hydrogen acids, which led Lewis to combine them in the class of "acids," the separate classification of hydrogen acids nevertheless is fully substantiated by the fact that hydrogen occupies a special place in the Periodic System. The ionization of a hydrogen atom forms an elementary particle, namely, a proton. This ion, which has no electron shell, is extremely small and hence is distinguished by a high electrical potential and a high mobility in chemical reactions.

Acidlike substances very often react specifically with different bases. It is therefore very difficult to predict their relative strength in any actual reaction. This circumstance particularly interested Lewis [24].

Boron trifluoride, BF_3, is a typical representative of acidlike substances. It reacts with bases, for example, the bases appearing in Equations (15) and (16), i.e., with the F^- ion and water and ammonia molecules. Reactions (23)-(25) proceed as a result of the coordinational unsaturation of the trivalent boron atom in the trifluoride. The boron is thereupon converted into the tetravalent state as the bases reacting with it

$:\ddot{F}:^-$, $H:\ddot{O}:H$, and $:\overset{H}{\underset{H}{\overset{..}{N}}}:H$, which fills the free electron orbital of the boron atom:

$$F:^- + BF_3 \rightarrow F : BF_3^- \qquad (23)$$

$$H_2O: + BF_3 \rightarrow H_2O : BF_3 \qquad (24)$$

$$H_3N: + BF_3 \rightarrow H_3N : BF_3 . \qquad (25)$$

The bond between the fluorine ion and the boron atom is equivalent to the bonds of the latter with the other fluorine atoms and the complex ion has the composition BF_4^-.

In addition to electrically neutral acidlike substances (BF_3, $AlCl_3$, SO_3, Br_2, etc.), there are acidlike ions. These include the ions Ag^+, NO_2^+, SO_3H^+, Br^+, CH_3^+, etc. The silver ion, for example, combines with an ammonia molecule. It also forms a compound with a benzene molecule, which acts as an electron-donor in this reaction:

$$Ag^+ + NH_3 \rightarrow Ag (NH_3)^+ \qquad (26)$$

$$Ag^+ + C_6H_6 \rightarrow Ag (C_6H_6)^+ . \qquad (27)$$

Let us consider the reaction of a nitronium ion, NO_2^+, with a benzene molecule. If nitric acid is dissolved in concentrated sulfuric acid it adds a proton, i.e., reacts like a base:

$$HNO_3 + H_2SO_4 \rightleftarrows H_2NO_3^+ + HSO_4^- . \qquad (28)$$

$$\text{base}_1 \quad \text{acid}_2 \quad \text{acid}_1 \quad \text{base}_2$$

In excess sulfuric acid, the ion $H_2NO_3^+$ decomposes and the water liberated is converted into a hydroxonium ion with the simultaneous formation of a nitronium ion:

$$H_2NO_3^+ + H_2SO_4 \rightleftarrows NO_2^+ + H_3O^+ + HSO_4^- . \qquad (29)$$

The nitronium ion adds even to such a very weak base as benzene. The formation of a complex between them is the first stage in nitration and is frequently written as follows: *

$$(30)$$

The proton then is transferred to some base in the reaction mixture and nitrobenzene is obtained:

$$+ H^+ . \qquad (31)$$

For example:

$$C_6H_6NO_2^+ + HSO_4^- \rightarrow C_6H_5NO_2 + H_2SO_4 . \qquad (32)$$

$$\text{acid}_1 \quad \text{base}_2 \quad \text{base}_1 \quad \text{acid}_2$$

Consequently, in this stage of the reaction the complex is an acid.

The sulfonation (with the SO_3H^+ ion), bromination (by Br^+), and alkylation (by CH_3^+ and $C_2H_5^+$ ions) of benzene proceed analogously (see also Section V for the mechanisms of these reactions). Carbonium ions like CH_3^+ and $C_2H_5^+$ are formed, for example, in the reaction:

$$RHal + AlHal_3 \rightarrow R^+ + AlHal_4^- . \qquad (33)$$

Molecules of organic substances which react with bases without the transfer of a proton are also acidlike. The first stage of the amination of pyridine by sodamide (Chichibabin reaction) may be treated in this way:

---

* The formulas in brackets indicate that the electron density is low in the ortho- and para-positions.

$$\text{(structure)} + NH_2^- \rightleftharpoons \left[\text{(structure)} \leftrightarrow \text{(structure)} \right]. \qquad (34)$$

The carbon atom in the α-position (and in part in the γ-position) of the pyridine molecule acts as an acceptor of the electron pair of the amide ion. A hydride ion(H^-) is then eliminated and an aminopyridine is formed:

$$\left[\text{(structure)} \leftrightarrow \text{(structure)} \right] \rightarrow \text{(structure)} + H^-. \qquad (35)$$

Sodium hydride (Na^+H^-) reacts with the aminopyridine, which is an acid, the proton of whose amino group adds to the hydride ion and gaseous hydrogen is liberated:

$$\text{(structure)} NH_2 + H^- \rightarrow \text{(structure)} NH^- + H_2. \qquad (36)$$

acid₁   base₂   base₁   acid₂

We will return to the discussion of the mechanism of such reactions in Section V.

4. NUCLEOPHILIC AND ELECTROPHILIC REAGENTS

The formal analogy between oxidation—reduction and acid—base reactions was noted above (p. 2). Representatives of these two groups of reagents are similar in that reducing agents and bases are e l e c t r o n   d o n o r s, while oxidizing agents, acids, and acidlike substances are e l e c t r o n   a c c e p t o r s.

Ingold [1, 29] combined electron donors in the class of n u c l e o p h i l i c   r e a g e n t s or n u c l e o p h i l e s and electron acceptors were called e l e c t r o p h i l i c   r e a g e n t s or e l e c t r o p h i l e s. The latter term requires no special explanation. The word "phileo" means "I love" in Greek, i.e., an electrophilic substance readily adds electrons. There is a complete transfer of one or several electrons of a reducing agent to an oxidizing agent, while an atom of an acid or acidlike substance shares it free electron pair with an atom of a base. According to Brønsted's scheme, in the reaction of an acid there is at the same time rupture of the chemical bond between the hydrogen atom and the acid residue and the proton adds to the electron pair of the base. In a reaction of an acidlike substance, for example, boron trifluoride, the electrons of the base make up the electron deficit at the boron atom and a covalent bond is formed.

"Nucleo" is the Greek work for "nucleus," i.e., nucleophilic substances have an affinity for an atomic nucleus. The simplest atomic nucleus is a proton. Bases have an affinity for a proton, i.e., are p r o t o p h i l i c and, consequently, nucleophilic. There are good grounds for applying this term to reactions between bases and acids or acidlike substances, but, as was noted by Luder and Zuffanti [5], it is less suitable when there is a transfer of the electrons of a reducing agent to an oxidizing agent. These authors, therefore, combined reducing agents and bases in the class of electrodote reagents and avoided the term "nucleophilic." Nonetheless, we will use the latter term in order to retain the connection with the generally accepted division of reactions into electrophilic and nucleophilic substitutions.

Ingold's classification of reagents and reactions has much in common with the previous classifications of Lapworth [30] and Robinson [31]. They distinguished between cationoid and anionoid reagents. The former behave like cations and the latter react similarly to anions. These terms are less appropriate than those adopted by Ingold as they unnecessarily emphasize the role of the charge of the reagent, which might be a molecule or an atom and not only an ion.

Examples of nucleophilic and electrophilic reagents are given in Table 1.

TABLE 1. Classification of Reagents
_____

I. Electron Donors, Nucleophilic Reagents

1. REDUCING AGENTS

    a. Atoms: Na, Fe
    b. Molecules: H_2, H_2O
    c. Cations: Cu^+, Fe^{++}
    d. Anions: S^{--}, $Fe(CN)_6^{----}$

2. BASES

    a. Molecules: H_2, H_2O, NH_3, C_5H_5N, C_6H_6, C_2H_4, $C_6H_5NH_2$, HNO_3
    b. Cations: $Co(NH_3)_5OH^{++}$
    c. Anions: H^-, OH^-, NH_2^-, $C_6H_5NH^-$, $H_2PO_4^-$, HPO_4^{--}, PO_4^{---}, NO_3^-, $CH_3CO_2^-$,
       $(C_6H_5)_3C^-$, $C_6H_5^-$, CH_3^-

II. Electron Acceptors, Electrophilic Reagents

1. OXIDIZING AGENTS

    a. Atoms: Cl
    b. Molecules: Cl_2, O_3, H_2O
    c. Cations: Cu^{++}, Fe^{+++}
    d. Anions: MnO_4^-, $Fe(CN)_6^{---}$

2. ACIDS

    a. Atoms: H
    b. Molecules: H_2, H_2SO_4, HNO_3, HCl, H_3PO_4, CH_3COOH, H_2O, NH_3, $C_6H_5NH_2$,
       C_6H_6, $HC \equiv CH$
    c. Cations: H_3O^+, NH_4^+, $H_2NO_3^+$, $C_5H_5NH^+$, $C_6H_7^+$, CH_5^+, $Co(NH_3)_5OH_2^{+++}$
    d. Anions: $H_2PO_4^-$, HPO_4^{--}

3. ACIDLIKE SUBSTANCES

    a. Molecules: BF_3, $AlBr_3$, SO_3, Cl_2, C_5H_5N
    b. Cations: NO_2^+, Br^+, SO_3H^+, CH_3^+, Ag^+
_____

A classification which provides for the existence of electron donors and electron acceptors is applicable to substituents as well as reagents. *

The first group contains e l e c t r o p o s i t i v e   s u b s t i t u e n t s: $-NH_2, -N(CH_3)_2$, $- OH, - OCH_3$, etc., while the second group contains e l e c t r o n e g a t i v e substi- tuen ts: $- NO_2, - SO_3H, - CN, - COOC_2H_5, - COCl$, etc.

The in troduction of a substituent into a molecule strongly changes its properties and reactions. This is readily demonstrated by a comparison of the substances obtained by substituting NH_2 and NO_2 groups for a hydrogen atom in water, methane, and ben- zene molecules:

$$H_2O \; NH_2OH \; - \text{hydroxylamine}$$
$$NO_2OH \; - \text{nitric acid}$$
$$CH_4 \; NH_2CH_3 \; - \text{methylamine}$$
$$NO_2CH_3 \; - \text{nitromethane}$$
$$C_6H_6 \; NH_2C_6H_5 \; - \text{aniline}$$
$$NO_2C_6H_5 \; - \text{nitrobenzene}$$

With the replacement of hydrogen by an electron-donor group, the substance be- comes more strongly basic, while an electron-acceptor substituent increases the acid- ity of the substance.

The introduction of a substituent into an aromatic nucleus changes the distribution of electron density at the carbon atoms of the ring (primarily at the ortho- and para- atoms). The density is reduced in the presence of an electron-acceptor group, while an electron-donor group increases it. Between a substituent and an aromatic ring there is a peculiar donor-acceptor interaction (c o n j u g a t i o n   e f f e c t). The dissimilar polarizations of aromatic rings of substances containing substituents of opposite types are the reason for the interaction between the rings with the formation of a complex, which in the case of nitroaniline and nitrobenzene, for example, is shown by the ap- pearance of a color when they are mixed [32, 33].

Electron-donor and electron-acceptor atoms and groups replacing hydrogen in molecules of aliphatic hydrocarbons displace electrons in opposite directions. This i n d u c t i v e   d i s p l a c e m e n t falls off rapidly with distance from the substituent.†

The conjugation and inductive effects are of great importance in reactions with nucleophilic and electrophilic reagents, including hydrogen isotope exchange reactions.

5. WIDER DEFINITIONS OF ACIDS AND BASES

Together with the definitions of acids and bases proposed by Brønsted and Lowry (p. 2), there are other points of views on acids and bases, which are formulated in

---

* This classification of substituents is conditional in the sense that some substituents which are electron donors under some conditions are electron acceptors in other reac- tions. For example, substituents of the first group are normally electropositive, but sometimes show electronegativity as the oxygen or nitrogen atoms in them have an affinity for an electron (p. 245).

† See p. 277 for an examination of the inductive displacement of electrons in aromatic molecules and its role in hydrogen isotope exchange with bases.

wider theories of acids and bases. We will not dwell on the theories themselves here
as they have been presented in detail and analyzed critically [13, 14, 34]. We will
only discuss the relation between Ingold's classification of reagents and the wider definitions of acids and bases of Lewis [14, 24] and M. I. Usanovich [35] (see Table 2).

TABLE 2. Classification of Reagents

Electrophilic reagents (electron acceptors)			Nucleophilic reagents (electron donors)	
Oxidizing agents	Brønsted acids	Acidlike substances	Reducing agents	Brønsted bases
—	Lewis acids		—	Lewis bases
Usanovich acids			Usanovich bases	

According to Lewis, bases are donors and acids acceptors of an electron pair.
Consequently, Lewis bases coincide with Brønsted bases, while Lewis acids include
aprotic acidlike substances in addition to Brønsted hydrogen acids. Nowadays only
aprotic substances are usually called Lewis acids, though this does not correspond to
the views put forward in his original work [24]. Lewis's theory contains serious defects
and contradictions. For example, in applying it to hydrogen acids it would be necessary to assume the divalence of hydrogen (the product from the reaction between $H:Cl$
and NH_3 would have to be represented by the formula $Cl:H:NH_3$).

M. I. Usanovich [35] defines b a s e s as anion donors and cation acceptors and
a c i d s as cation donors and anion acceptors. In the particular case when the cation
is a proton (H^+), we are dealing with the normal acids and bases of Brønsted. If an
electron is regarded as an "anion," as is done by M. I. Usanovich, the definition covers
oxidation–reduction reactions, which, in his opinion, also constitute an inseparable
section of acid–base processes. M. I. Usanovich also designates as acid–base reactions
reactions between bases and acidlike substances which do not involve the transfer of
electrically charged particles and consequently are not covered by the definitions he
proposes. This has naturally provoked serious critical comments [13, 14, 34].

Thus, "acids" and "bases" by M. I. Usanovich's definition correspond to electrophiles and nucleophiles of Ingold's classification. The latter is preferable to M. I.
Usanovich's classification as it lacks the contradictions of the former and quite clearly
defines the similarities and differences between oxidants, acids, and acidlike substances
on the one hand and reducing agents and bases on the other.

M. I. Usanovich subsequently [36] rejected his definitions of acids and bases,
stating that the mechanisms of acid–base interaction are so diverse that acidity and
basicity are properties inherent to all substances and therefore he considers that there
is no point in debating the definitions of acids and bases (see [34] for a criticism of
this work).

6. RELATIVE AND APPROXIMATE NATURE OF THE CLASSIFICATION OF REAGENTS

In using a classification of reagents one must always bear in mind the fact that,
depending on the partner in the reaction, the same substance may be either a nucleophile or an electrophile and may belong to different groups within the limits of these

classes of reagents. Returning to the above equations we see that a water molecule is able to react as an oxidizing agent [Equations (9) and (10)], a reducing agent (Equation 11), a base (Equation 15), or an acid (Equation 16). Moreover, in self-ionization, one water molecule behaves as a base toward another water molecule, which behaves as an acid (Equation 18).

Amphoteric character in the wide sense of this term, i.e., with reference to oxidation—reduction as well as acid—base processes, is inherent in comparatively few substances. Others are predominantly oxidizing agents (e.g., O_3), reducing agents (Sn^{++}), bases (CH_3^-), or acids (H_3O^+).

A wider knowledge of the reactions of a given substance sometimes reveals its functions as a reagent that are unexpected at first glance. For example, it is found that sodium hydroxide may react not only as a base, but also as a Brønsted acid. This is the case in its reaction with metallic sodium [37]:

$$NaOH + Na \rightarrow Na_2O + H \tag{37}$$

or

$$OH^- + e^- \rightarrow O^{--} + H . \tag{37a}$$

$$acid_1 \quad base_2 \quad base_1 \quad acid_2$$

The acidity of the hydrogen molecule was mentioned above. However, as it has an affinity for a proton (70 kcal) and is converted into the ion H_3^+, the hydrogen molecule shows the properties of a base:

$$H_2 + H^+ \rightarrow H_3^+ . \tag{38}*$$

Any classification reflects natural phenomena only approximately. The classification of reagents described above is no exception. It covers either the complete transfer of electrons from one atom to another or the formation of a new covalent bond between two atoms by coordination of the electron pair of one of them, though in addition to heteropolar (ionic, electrovalent) and homopolar (molecular, covalent) bonds there are bonds of a transitional type. Exchange forces cannot be neglected completely even in ionic crystals, while valence bonds in organic molecules are more or less polar.

The whole variety of actual forms of donor-acceptor interaction cannot be reduced merely to the formation of a covalent coordination bond through an unshared pair of electrons. This is readily demonstrated by the numerous reactions listed in Mulliken's article [38], for example. This author listed about forty varieties of donor-acceptor interaction. However, it is sufficient to recall the reactions mentioned above. Thus, although there is undoubted similarity between reactions (26) and (27), the former involves the free pair of p-electrons of the nitrogen atom, while the latter involves the whole π-electron system of the benzene ring. (For more details see the section on the donor-acceptor interaction on p. 12).

---

* This reaction occurs in the ionization chamber of a mass spectrometer [12].

Brønsted's scheme of acid–base interaction is accurate only in the first approximation as the reaction between an acid and a base is not always effected by the complete transfer of a proton. It is necessary therefore to change the definitions of an acid and a base proposed by Brønsted (p. 236).

There are completely different views on acid–base reactions. For example, according to Ya. K. Syrkin [39] neutralization in an aqueous medium is not effected by the transfer of the proton of the acid to the base, but by an electron transfer with a molecule of the solvent, namely, water, acting as the hydrogen source. The rupture of bonds between the atoms and the formation of new bonds by electron transfers are accomplished in one act in a cyclic complex, whose formation is energetically favored. For example, the neutralization of pyridine by acetic acid is written in the following way by Ya. K. Syrkin:

$$\text{(structure)} \rightleftharpoons \text{(structure)} \overset{+}{N}H + CH_3COO^- + H_2O. \qquad (39)$$

This direction of the reaction is due to the considerable affinity of the electronegative oxygen atom for an electron. One of the electrons of the free electron pair of the nitrogen atom is transferred to it, while the second electron participates in a bond with the hydrogen atom. The nitrogen is converted into a tetravalent, positively charged ion. Ya. K. Syrkin [40] also noted the role of cyclic complexes in hydrogen exchange.

In recent years the idea of the participation of cyclic complexes in the activated state has acquired increasing popularity. Several hundred examples of organic reactions were analyzed in a recent review [41]. The authors of the review noted that too strict adherence to nucleophilic or electrophilic classification of the electronic mechanism of chemical reactions often hampers their interpretation.

Swain and E. A. Shilov [42] came to the conclusion that together with reactions proceeding by a donor-acceptor mechanism between two molecules, there are many organic reactions in which a molecule is attacked simultaneously by an electrophilic and a nucleophilic reagent and the transition complex is termolecular.

We should note yet another important circumstance. The classification of reagents is concenred with their electronic structure in the initial state and not in the activated complex. All arguments based on data on the electron density distribution in a nonreacting molecule, which may be deduced, for example, from the sign and magnitude of the dipole moment, etc., are therefore conditional.

7. DONOR–ACCEPTOR INTERACTION

Acids and acidlike substances react with bases by means of a donor-acceptor interaction. It has already been mentioned above that the latter is not simply the formation of a new covalent bond with the use of the electron pair of the donor. More than ten years ago in the monograph Theories of Acids and Bases [14, p. 259] it was stated that a thorough study of the properties and characteristics of various acidlike compounds in their reactions with bases and an extension of knowledge of the structure of molecular compounds will make it possible in the future to replace the empirical definition of an acidlike substance by a new classification with a theoretical and experimental foundation.

One of the first attempts at classifying reactions of the donor-acceptor type was made by Mulliken [38]. Good donors are molecules with a low ionization potential, while typical acceptor molecules have a high affinity for an electron. The former category also includes negative ions, while the latter includes ions with a positive charge.

Mulliken distinguishes between the following classes of donors: *

I. Donors with a readily "ionized" free electron pair:
    A. Molecules (amines, alcohols, ethers, nitriles, etc.).
    B. Anions ($CH_3CO_2^-$, OH^-, NH_2^-, etc.).
II. Molecules containing π-electrons (aromatic and unsaturated hydrocarbons and their derivatives with electron-donor substituents).
III. Molecules whose participation in a donor-acceptor interaction is accompanied by the rupture of a σ-bond (alkyl halides, ethers, etc.; example of the reactions: $RF + BF_3 \rightarrow R^+ + BF_4^-$). These reactions normally occur in a polar medium and the rupture of the bond is energetically favored by solvation of the ion .

In their turn, acceptors are also divided into several classes.

I. Acceptors with a vacant electron orbital:
    A. Molecules ($AlCl_3$, BF_3, $SnBr_4$).
    B. Cations (Ag^+, NO_2^+, R^+, etc.).
II. Aromatic and unsaturated hydrocarbons with electrophilic substituents.
III. Molecules whose participation in a donor-acceptor interaction is accompanied by rupture of a σ-bond (hydrogen acids).

Illustrative examples of twenty-five types of donor-acceptor interaction and series of subtypes are given by Mulliken's article [38]. The actual reactions mentioned by the author, which by no means exhaust all the forms of donor-acceptor interaction, give an idea of their great variety.

The problem of reactions of electron-donor and electron-acceptors was discussed by Mulliken [43] in a report to the Symposium on Molecular Physics in Tokyo in 1954.

The author noted that the interaction between substances proceeds with different degrees of intensity. It is weakest with like molecules and normally may be represented as a physical interaction (electrostatic forces and quantum mechanical dispersion forces). Chemical interaction proceeds in definite stoichiometric ratios and is most often described in the form of a donor-acceptor interaction.

There are intermediate and mixed types of interaction. The formation of neutral molecular complexes involves both "chemical" and "physical" forces. The wave function (ψ) of a complex consisting of a donor and acceptor is written as a linear combination of two functions. The first expresses the physical and the second the chemical forces. In the latter case the electrons of the donor are transferred to the acceptor and

---

* The classification described below is considerably simplified in comparison with that given in the original, which had a complex symbolism.

covalent bond is formed:

$$\psi(D \ldots A) = a\psi(D \ldots A) + b\psi(D^+ \cdot A^-). \tag{40}$$

The coefficients a and b characterize the fraction of physical and chemical forces, respectively. For example, in the reaction between ammonia and boron trifluoride (Equation 25) $b^2 \gg a^2$ *, but $b^2 \approx a^2$ in the reaction between Ag^+ and C_6H_6.†

Mulliken examined the role of the relative orientation of the molecules in a reaction.

A weak molecular compound, which Mulliken called an "outer" or "loose" complex is often formed first. Under energetically favorable conditions it changes in a strong "inner" complex. This stage of the reaction, which is accompanied by much more complete transfer of charge, may be completed by an extensive chemical conversion.

In the absence of a polar medium a high potential barrier must be overcome for the change to an inner complex. This is shown by the form of the potential curve a in Fig. 1. The potential energy is plotted along the ordinate axis and the reaction coordinate along the abscissa. An "outer" complex therefore has a greater or lesser stability in the gas phase and in nonpolar solvents. It is readily formed in a polar medium. The gain in energy as a result of solvation of the complex by polar molecules facilitates the formation of an ion pair and its separation into free ions (curve b in Fig. 1).

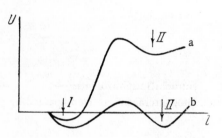

Fig. 1. Potential curve of a donor-acceptor interaction. U) Energy; l) reaction coordinate; a) reaction in a nonpolar medium; b) reaction in a polar medium. I) "outer" complex; II) "inner" complex.

Similar ideas have been put forward by other authors, for example, Brown [44, 45]. He stated that from the chemical point of view, reactions between acids and bases may have much in common with reactions between bases and other acceptors and the strength of the complex (reaction product) depends on the strength of the base. Thus, molecular iodine and hydrogen chloride react with an aromatic hydrocarbon to form an unstable compound with a 1:1 composition:

$$ArH + I_2 \rightleftarrows ArH \ldots I - I \tag{41}$$

$$ArH + HCl \rightleftarrows ArH \ldots H-Cl. \tag{42}$$

With a somewhat stronger base such as ethyl ether, a more stable molecular compound is obtained.

---

* If the whole wave function is equated to unity, $a^2 + b^2 = 1$. According to Mulliken, b^2 conveys the fraction of "dative" character in the total wave function better than the coefficient b.

† White crystals with the composition $C_6H_6 \cdot AgClO_4$ may be isolated from a solution of silver perchlorate in benzene.

$$(C_2H_5)_2O + I_2 \rightleftarrows (C_2H_5)_2O \cdot I - I \qquad (43)$$

$$(C_2H_5)_2O + HCl \rightleftarrows (C_2H_5)_2 O \cdot H - Cl. \qquad (44)$$

However, even here the reaction does not proceed to ionization.

Finally, if a still stronger base such as pyridine is used, ions are formed:

$$C_5H_5N + I_2 \rightleftarrows C_5H_5NI^+ \cdot I^- \qquad (45)$$

$$C_5H_5N + HCl \rightleftarrows C_5H_5NH^+ \cdot Cl^-. \qquad (46)$$

Brown made a detailed examination of the case of donor-acceptor interaction when the electron donor is an aromatic ring. In his opinion the whole system of π-electrons participates in a weak interaction (reactions 41 and 42) and the electron cloud is changed comparatively little. With an increase in the electrophilicity of the reagent, the cloud of π-electrons is deformed and a new σ-bond is formed between one of the carbon atoms of the aromatic ring and the reagent atom so that the π-complex is converted into a σ-complex. For example, this conversion occurs on addition of aluminum chloride, which strongly polarizes the H—Cl bond in the hydrogen chloride molecule:

$$ArH + HCl + AlCl_3 \rightleftarrows ArH_2^+ \cdot AlCl_4^-. \qquad (47)$$

The π-complex corresponds to the "outer" and the σ-complex to the "inner" complex of Mulliken. They may be represented by the following formulas:*

π-complex σ-complex

The term π-complex was introduced into science by Dewar [46-48], but he applied it to the product of any interaction between an aromatic nucleus and an electrophilic reagent.

The electrophilic replacement of hydrogen in an aromatic ring (Equations 29-32) has already been discussed above (p. 6). Let us reexamine it in the light of Mulliken's interpretation [43].

When a nitronium ion NO_2^+ adds to a benzene ring, an "outer" complex is formed first. As in any other π-complex, the NO_2^+ ion is not localized at any particular carbon atom of the benzene molecule. The change to an "inner" complex involves overcoming the activation barrier. This is promoted by the polarity of the medium, which contains different ions (NO_3^-, H_3O^+, HSO_4^-, and $H_3SO_4^+$). The charge of the nitronium ion is transferred to a considerable extent to the benzene and the NO_2 group is firmly bound to the latter. The NO_2 group is thereupon deformed and a bond is formed be-

---

* The formulas in brackets indicate that the electron density is low in the ortho- and para-positions.

tween the nitrogen atom and the carbon atom in the carbonium ion (σ-complex). In the wave function of this complex, $b^2 \gg a^2$.

The next step in the reaction is the attack of a donor molecule to which the proton is transferred (Equation 32). The intermediate complex is converted into the reaction product, i.e., nitrobenzene. This also requires the expenditure of energy and the potential curve as a whole has the form shown in Fig. 2.

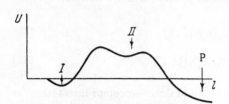

Fig. 2. Potential curve of an irreversible reaction proceeding through a donor-acceptor interaction stage. The symbols are the same as in Fig. 1. P denotes the reaction product.

Thus, a chemical reaction proceeds through several stages. The determining stages can often be established. We will return to this question again in Section V.

8. HETEROLYTIC AND HOMOLYTIC REACTIONS

A covalent bond between atoms, which is normally represented by a line or a pair of electrons, may be ruptured in different ways in a reaction.

In heterolytic reactions, the pair of electrons is transferred to one of the two atoms:

$$A \vdots B \rightarrow A + :B \qquad (48)$$

or

$$A : \vdots B \rightarrow A: + B . \qquad (49)$$

Such reactions are also called polar [2] or ionic, though, as we shall see later, they are not always completed by the formation of ions.

If a chemical bond is ruptured in such a way that one electron remains with each of the atoms:

$$A : B \rightarrow A \cdot + B \cdot , \qquad (50)$$

then radicals are formed. Each of them has an atom with an unpaired electron. Such reactions are called homolytic or radical reactions.

Heterolysis of a bond by scheme (48) or (49) is the opposite of complex formation with the production of a covalent coordination bond. For example, when heated, the molecular compound of boron trifluoride and water dissociates into its components:

$$H_2O : BF_3 \rightarrow H_2O : + BF_3. \qquad (51)$$

Two molecules were obtained from the molecular compound and a heterolytic reaction occurred, though ions were absent.

Heterolytic reactions in which the bond is broken as a result of the displacement of one of the components of the complex by a reagent of a similar type are somewhat more complicated:

$$C: + A \,\vert\, : B \rightarrow A : C + : \overset{..}{B} \qquad (52)$$

or

$$C + A : \,\vert\, B \rightarrow A : C + B \qquad (53)$$

for example,

$$H_3N: + H_2O : BF_3 \rightarrow H_3N : BF_3 + H_2O \qquad (54)$$

$$BF_3 + H_3N : B\,(CH_3)_3 \rightarrow H_3N : BF_3 + B\,(CH_3)_3 \;. \qquad (55)$$

These schemes also include reactions in which the bond is broken heterolytically in a simple molecule (or ion) rather than a molecular compound. For example, the following reactions proceed according to scheme (52):

$$H_3N : + H : OH \rightarrow H_3\overset{+}{N} : H + : OH^{-} \qquad (56)$$

$$H_3N : + R : I \rightarrow H_3\overset{+}{N} : R + : I^{-} \qquad (57)$$

$$H_3N : + H_3\overset{+}{O} \rightarrow H_3\overset{+}{N} : H + : OH_2 \qquad (58)$$

$$R_3N : + R\overset{+}{S}R_2 \rightarrow R_3\overset{+}{N} : R + : SR_2 \qquad (59)$$

$$H\overset{-}{O} : + H : I \rightarrow H : OH + : I^{-} \qquad (60)$$

$$H\overset{-}{O}{}^{\cdot} + R : I \rightarrow R : OH + : I^{-} \qquad (61)$$

$$H\overset{-}{O} : + H : C\,(C_6H_5)_3 \rightarrow H : OH + : \overset{-}{C}\,(C_6H_5)_3 \;. \qquad (62)$$

The following reaction proceeds according to scheme (53):

$$BF_3 + Cl : C\,(C_6H_5)_3 \rightarrow F_3\,B: Cl + \overset{+}{C}\,(C_6H_5)_3. \qquad (63)$$

Reactions (56), (58), (60), and (62) belong to the acid−base class. Acid−base processes are heterolytic reactions involving a proton.

Reactions (57), (59), and (61) are similar to those just mentioned and this induced M. I. Usanovich [35] to consider them also as acid−base reactions, though this classification is inexpedient for the reasons given above (p. 10).

A further complication of the scheme of a heterolytic reaction occurs in the case of reactions in which bonds between the atoms of both reagents are broken and there is a double replacement, for example:

$$C : D + A : B \rightarrow A : D + C : B \qquad (64)$$

$$H : OH + R : I \rightarrow R : OH + H : I \qquad (65)$$

or ionization

$$C : D + A : B \rightarrow A + C : B + D: \tag{66}$$

$$HOSO_2OH + (C_6H_5)_3COH \rightarrow (C_6H_5)_3C^+ + HOH + OSO_2OH^-. \tag{67}*$$

These are examples of reactions in which a chemical bond is broken heterolytically.

9. PLACE OF HYDROGEN EXCHANGE AMONG OTHER REACTIONS

We are now quite ready to answer the question posed at the beginning of this section, namely, that of the place of hydrogen isotope exchange among other reactions.

First of all, it is clear that the general classification of reagents and reactions is also applicable to hydrogen exchange reactions. Consequently, it is necessary to distinguish between homolytic and heterolytic hydrogen exchange. In this book, we will not consider at all the vast region of homolytic hydrogen exchange, whose development is associated with the names of Farkas, Taylor, Polanyi, Steacie, V. V. Voevodskii, and many other authors. Radical hydrogen exchange occurs predominantly between substances in the gas or vapor state in heterogeneous catalysis on metal catalysts. Radical exchange reactions in solution have been studied little as yet (G. A. Razuvaev).

Heterolytic hydrogen exchange reactions occur largely in solution and in order to establish their place among other reactions, it is necessary to turn again to Ingold's classification, which we reproduce in the following form (Table 3).†

TABLE 3. Ingold's Classification of Reagents

Electrophilic reagents			Nucleophilic reagents	
Oxidizing agents	Acidlike substances (Lewis acids)	Acids	Bases	Reducing agents
		Region of hydrogen exchange reactions		

It is evident that oxidation–reduction processes, which consist of electron transfers are not related directly to hydrogen exchange. Reducing agents are therefore eliminated from the class of nucleophiles and bases remain. Among electrophiles, in addition to oxidizing agents, acidlike substances (for example, NO_2^+ and Cl_2) do not participitate in the exchange as they do not contain hydrogen, which is essential for hydrogen exchange, whereas chemical electrophilic replacements (nitration and chlorination) proceed with the active participation of precisely acidlike substances, i.e., aprotonic electron acceptors. Consequently, only acids and bases may participitate in heterolytic hydrogen exchange reaction.

---

*The following reactions then occur: $HI + H_2O \rightarrow H_3O^+ + I^-$ (65a) and $H_2O + H_2SO_4 \rightarrow$ $\rightarrow H_3O^+ + HSO_4^-$ (67a).

†Brønsted acids and bases appear in Ingold's classification. Later, in Section IV in particular, it will be shown that Brønsted's scheme does not completely reflect the mechanism of the reaction between nucleophile-bases and hydrogen-containing electrophilic reagents.

10. SUMMARY

1. Homolytic and heterolytic reactions differ in the manner in which the covalent bond is broken. In reactions of the first type, to each of the atoms is transferred one electron from the electron pair forming the bond. In reactions of the second type, the electron pair is transferred to one of the atoms.

2. Heterolytic reactions involve electron-donor (nucleophilic) and electron-acceptor (electrophilic) reagents. The former group includes reducing agents and bases, while the latter includes oxidizing agents, acids, and acidlike substances.

3. It is necessary to consider the relative and approximate nature of the classification of reagents.

4. A donor-acceptor interaction may be effected in several stages and it is sometimes possible to establish the separate stages. A donor-acceptor interaction leads to ionization only under conditions that are energetically favorable.

5. Only acids and bases may participate in heterolytic hydrogen exchange reactions.

LITERATURE CITED

1. C. K. Ingold, Structure and Mechanism in Organic Chemistry (New York, 1953).
2. J. Hine, Physical Organic Chemistry (New York, Toronto, London, 1956).
3. A. Remick, Electronic Interpretations of Organic Chemistry [Russian translation] (IL, Moscow, 1950).
4. O. A. Reutov, Theoretical Problems in Organic Chemistry [in Russian] (Izd. MGU, Moscow, 1956).
5. W. Luder and S. Zuffanti, Electronic Theory of Acids and Bases [Russian translation] (IL, Moscow, 1950).
6. W. Luder, Chem. Rev. 27, 547 (1940).
7. E. D. Hughes, Ionization in Organic Chemistry, The Royal Institute of Chemistry, Lectures, monographs and reports, No. 1 (1955).
8. J. N. Brønsted, Rec. trav. chim. Pays-Bas. 42, 718 (1923).
9. T. M. Lowry, Chem. and Ind. 42, 43 (1923); J. Chem. Soc. 123, 822 (1923).
10. N. Bjerrum, Chem. Rev. 16, 287 (1935).
11. J. N. Brønsted, J. Am. Chem. Soc. 53, 3624 (1931).
12. V. L. Tal'roze and E. L. Frankevich, Doklady Akad. Nauk SSSR 119, 1174 (1958).
13. A. I. Shatenshtein, Zhur. Obshchei Khim. 9, 1603 (1939).
14. A. I. Shatenshtein, Theories of Acids and Bases [in Russian] (Goskhimizdat, Moscow, 1949), ch. 17.
15. J. Bjerrum, Acta chem. scand, 1, 526 (1947).
16. J. Bjerrum, Naturwiss. 38, 461 (1951).
17. J. Bjerrum, Angew. Chem. 63, 527 (1951).
18. R. P. Bell, Quart. Rev. 1, 113 (1947).
19. R. P. Bell, Acids and Bases (London, New York, 1952).
20. J. Kolthoff, J. Phys. Chem. 48, 51 (1944).
21. J. Kolthoff, Ind. Eng. Chem., News Ed. 27, 835 (1949).
22. R. Ferriera, J. Chem. Phys. 19, 794 (1951).
23. M. M. Davis and H. B. Hetzer, J. Res. Nat. Bur. Stand. 46, 496 (1951).
24. G. N. Lewis, J. Franklin Inst. 226, 292 (1938).
25. A. I. Shatenshtein, Zhur. Fiz. Khim. 10, 766 (1937).

26. A. I. Shatenshtein and E. A. Izrailevich, Zhur. Fiz. Khim. 13, 1791 (1939).

27. V. N. Filimonov, D. S. Bystrov, and A. N. Terenin, Zhur. Optika i Spektroskopiya 3, 480 (1957).

28. R. P. Bell and B. G. Skinner, J. Chem. Soc. (1952) p. 2955.

29. C. K. Ingold, Chem. Rev. 15, 265 (1934).

30. A. Lapworth, Nature 115, 625 (1925).

31. R. Robinson, Solvay Reports (1931) p. 434.

32. R. E. Gibson and O. H. Loeffler, J. Am. Chem. Soc. 62, 1324 (1940).

33. J. Landauer and H. McConnell, J. Am. Chem. Soc. 74, 1221 (1952).

34. A. I. Shatenshtein, Vestn, Akad. Nauk Kaz.SSR, No. 12 (105), pp. 32, 43 (1953).

35. M. I. Usanovich, Zhur. Obshchei Khim. 9, 182 (1939).

36. M. I. Usanovich, What Are Acids and Bases? [in Russian] (Izd. An Kaz.SSR, Alma-Ata, 1953).

37. M. LeBlanc and L. Bergmann, Ber. 42, 4744 (1909).

38. R. S. Mulliken, J. Phys. Chem. 56, 801 (1952).

39. Ya. K. Syrkin, Ukr. Khim. Zhur. 22, 52 (1956); Izvest. Akad. Nauk SSSR, Otdel. Khim. Nauk 238, 600 (1959).

40. Ya. K. Syrkin, Doklady Akad. Nauk SSSR 105, 1018 (1955).

41. J. Mathieu and J. Valls, Bull. Soc. Chim. France, p. 1509 (1957); Uspekhi Khimii 28, 1216 (1959).

42. E. A. Shilov, Coll: "Problems in Chemical Kinetics, Catalysis, and Reactivity" [in Russian] (Izd. AN SSSR, Moscow, 1955), p. 749.

43. R. S. Mulliken, Symposium on Molecular Physics (Tokoyo, 1954) p. 45.

44. H. W. Brown and J. D. Brady, J. Am. Chem. Soc. 74, 3570 (1952).

45. H. W. Brown, J. Phys. Chem. 56, 821 (1952).

46. M. J. S. Dewar, J. Chem. Soc. 406, 777 (1946).

47. M. J. S. Dewar, Bull. Soc. Chim. France 18, C71 (1951).

48. M. J. S. Dewar, Electronic Theory of Organic Chemistry (London, 1949).

ACID–BASE CATALYSIS OF HYDROGEN EXCHANGE

1. INTRODUCTION

In the discussion of the classification of reagents and reactions it was shown that only acids and bases may participate in heterolytic hydrogen-exchange reactions. Sensitivity to acids or bases is an important criterion of them and simultaneously serves as a convincing demonstration of the acid–base nature of these reactions. It is therefore convenient to use this criterion for classifying the reactions interesting us [1].

It is the same in principle if exchange reactions are classified in accordance with the protolytic function of the substrate or reagent (solvent or catalyst). We will call reactions induced by acids a c i d h y d r o g e n e x c h a n g e and reactions induced by basic reagents, b a s i c h y d r o g e n e x c h a n g e. If a reaction proceeds in an amphoteric or aprotic solvent it is necessary to know how its rate and other kinetic parameters change when an acid or a base is added. Hydrogen exchange can also occur between particles of identical or very similar structure, of which one acts as an acid and the second as a base. Such reactions, which are acid–base reactions, cannot be assigned to either of the classes mentioned and must be incorporated into the group of a m - p h o t e r i c h y d r o g e n e x c h a n g e.

Until recently, acid–base catalysis of hydrogen exchange was well known largely for CH bonds of organic compounds. Hydrogen exchange in CH bonds in amphoteric solvents (water and alcohol) in the absence of a catalyst was observed only in exceptional cases [1-5]. New methods of measuring the kinetics of very fast reactions made it possible to demonstrate experimentally the existence of acid–base catalysis of isotopic exchange of hydrogen bound to atoms of other elements. These facts are summarized in this section.

It is very significant that very powerful acid–base catalysts in nonaqueous solvents which themselves are very strong acids or bases (liquid hydrogen halides, anhydrous sulfuric acid, liquid ammonia, hydrazine, ethylene diamine, etc.) promote hydrogen exchange even with such inert substances as saturated hydrocarbons. It is shown below that the exchange rate is a function of the protolytic properties of the substrate, solvent, and catalyst. Deviations from a simple relation between the strength of the acid or base and the rate of hydrogen exchange can often be explained by the dual reactivity of the substance and also by its binding of the catalyst. Deviations may be produced also by steric factors. A consideration of concrete examples leads to the conclusion that these apparent deviations from the rule only provide additional proof of the conception of the acid–base nature of heterolytic hydrogen exchange. From it there also follow the experimentally confirmed relation of the hydrogen isotope exchange rate to the charge of the substrate and catalyst and to the dielectric constant of the medium [6] and the applicability of the normal Brønsted and Hammett relations for acid–base catalysis. Much attention has been paid to the latter in publications on acid hydrogen exchange and therefore we devote considerable space to it, the more so as the applicability of this relation is often regarded as a criterion of the probability of a reaction mechanism. It is also necessary to consider possible reasons for deviations from Hammett's relation.

The acid–base nature of hydrogen exchange is also confirmed by the occurrence of heterogeneous hydrogen isotope exchange on solid catalysts with acid or basic properties. These heterogeneous reactions have much in common with the corresponding reactions in solution.

In addition, in this section of the book we consider the problems of the catalytic activity of complexes of acidlike substances and the salt effect in hydrogen exchange.

Thus, on the whole, in this section we examine the most important aspects of phenomena which form the essence of hydrogen isotope exchange reactions in solution.

2. COMPARISON OF THE RATES OF HYDROGEN EXCHANGE WITH AMPHOTERIC AND PROTOPHILIC SOLVENTS

Only a few hydrocarbons (and a limited number of their derivatives) are able to react as acids and bases when dissolved in an amphoteric solvent (water or alcohol) and the exchange of hydrogen in CH bonds, which is most promising for determining the reactivity and structural characteristics of organic compounds, occurs comparatively rarely. The acid properties of substances are very much increased by solution in a protophilic solvent such as liquid ammonia. This was shown previously in work on acid catalysis in liquid ammonia and on the electrical conductivity of solutions in it and by other physicochemical measurements (see the review [7] of acids and bases in liquid ammonia). Acetic acid, hydrogen sulfide, and even p-nitrophenol become equal in strength to hydrochloric, nitric, and perchloric acids. This is understandable: all the acids listed are converted to their ammonium salts in liquid ammonia and ammonolysis is actually catalyzed by the same acid, namely, the ammonium ion. Such substances as urea and acetamide, which are practically neutral in water, are partly ionized in liquid ammonia and converted into the ions $CO(NH_2)NH^-$ and CH_3CONH^-. These substances catalyze ammonolysis and react with alkali metals with the liberation of hydrogen. In ammonia solution, potassamide (a strong base) neutralizes weak acids such as indene, fluorene, triphenylmethane, diphenylmethane, etc., to form colored anions of hydrocarbons:

$$(C_6H_5)_3CH + K^+NH_2^- \rightarrow (C_6H_5)_3C^-K^+ + NH_3$$
$$(C_6H_5)_2CH_2 + K^+NH_2^- \rightarrow (C_6H_5)_2CH^-K^+ + NH_3 .$$

Toluene is not ionized appreciably under these conditions. The degree of ionization of the hydrocarbons listed is also very low in the absence of potassamide. However, the increase in their acidity in ammonia solution is indicated by the very considerable acceleration of hydrogen exchange in comparison with water or alcohol solutions.

The rates of hydrogen isotope exchange of CH bonds in several organic substances with ethanol and with liquid ammonia were compared directly in work of the Isotopic Reactions Laboratory [8]. Two of these substances were hydrocarbons (indene and fluorene) and two were ketones (acetophenone and β-naphthyl methyl ketone). The same hydrogen atoms, namely, those in the methylene groups of the hydrocarbons and the methylene groups of the ketones, exchanged with both solvents in the substances named. This was demonstrated in the following way: the deuterium introduced into the substance by isotopic exchange with heavy alcohol was again replaced by protium when the substance was dissolved in liquid ammonia of normal isotopic composition.

Table 4 gives the rate constants of the exchange reactions. As everywhere, where it is not specifically stated, the rate constants were calculated from a first-order equation and expressed in sec^{-1}. The exchange reactions were carried out in ethanol, liquid ammonia, and a solution of potassium alcoholate ($C_2H_5O^-K^+$) in ethanol.

TABLE 4. Comparison of Rate Constants of Deuterium Exchange with Alcohol and with Liquid Ammonia (Temperature is given in brackets)

Substance	C_2H_5OD	ND_3	$C_2H_5O^-K^+ +$ $+C_2H_5OD$
Indene	10^{-6}—10^{-7} (150°)	$1 \cdot 10^{-4}$ (10°)	$0.05N$ $3 \cdot 10^{-5}$ (0°)
"	—	$4 \cdot 10^{-4}$ (0°)	$1N$ $7 \cdot 10^{-4}$ (0°)
"	—	$9 \cdot 10^{-4}$ (10°)	—
Fluorene	$1 \cdot 10^{-7}$ (180°)	$2 \cdot 10^{-4}$ (25°)	$1N$ $6 \cdot 10^{-4}$ (25°)
Acetophenone	$2 \cdot 10^{-7}$ (120°)	$1.1 \cdot 10^{-5}$ (0°)	$1N$ $4 \cdot 10^{-5}$ (0°)
"	—	$6.5 \cdot 10^{-5}$ (25°)	—
"	—	$1.3 \cdot 10^{-4}$ (40°)	—
β-Naphthyl methyl ketone	10^{-6}—10^{-7} (120°)	$1.5 \cdot 10^{-5}$ (0°)	—

As Table 4 shows, the exchange of hydrogen in hydrocarbons and ketones with liquid ammonia proceeds at a much lower temperature and much more rapidly than with ethanol. For example, one hydrogen atom in fluorene exchanges with liquid ammonia after 1 hr at 25° and with ethanol after 2000 hr at 180°. By using approximate values of the activation energy for hydrogen exchange in acetophenone and fluorene (12 and 11 kcal), we established that the rate of exchange of the most labile hydrogen atoms with liquid ammonia is 4-6 orders greater than with ethanol.

As regards the solution of potassium ethoxide in alcohol, at a concentration of 1 N the rate of the exchange reaction was higher than in liquid ammonia as a result of the high affinity of the $C_2H_5O^-$ ion for a proton.

A proton of the substrate is added to the ethoxide ion. Deuterium is absent from this ion. Consequently, at least three particles are involved in the catalytic reaction, namely, those of substrate, catalyst, and solvent.

A solution of KND_2 in ND_3 catalyzes hydrogen exchange much more strongly than a solution of the alcoholate in alcohol in accordance with the higher affinity for a proton of ND_2^- and ND_3 in comparison with $C_2H_5O^-$ and C_2H_5OD. This is illustrated by experiments with quinaldine [8]. The rates of exchange of hydrogen in the methyl group of quinaldine with C_2H_5OD and ND_3 are comparatively close ($k = 2 \cdot 10^{-6}$ and $8 \cdot 10^{-8}$ at 120°*). With the addition of base to the two solvents, the same rates of exchange ($k = 10^{-4}$ sec^{-1}) could be attained, but under quite different conditions, namely, at 120° for 0.1 N solution of alcoholate in alcohol and at −30° for a 0.02 N solution of the amide in ammonia. The actual concentration of amide ions was even much less as the amide was neutralized by quinaldine according to the equation:

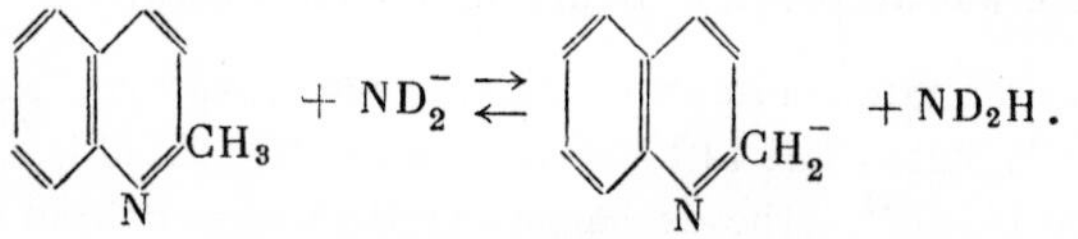

We can also give an example of other work in the Isotopic Reactions Laboratory [9] where the substance was such a weak acid that neutralization of the amide could be neglected (naphthalene). If naphthelane was dissolved in ammonia containing about 100% of deuterium and the solution heated at 120° for 100 hr, the deuterium concentration in the water from combustion of the naphthalene after the experiment did not reach 0.05 at. %. Consequently, $k_{120°} < 10^{-9}$ sec^{-1}. The rate constant of exchange of the hydrogen in naphthalene catalyzed by a 0.01 N solution of KND_2 is given by the equation:

$$k = 10^{7} e^{-14200/1.98\, T}.$$

Assuming that the rate constants is proportional to the catalyst concentration, which is not quite accurate, we find $k = 10^1$ sec^{-1} for 1 N solution of KND_2. Consequently, KND_2 accelerates the reaction by a factor of not less than 10 orders. These approximate estimates are in agreement with conclusions drawn by Wilmarth [10] by comparing the rates of the exchange reactions between molecular hydrogen and a solution of potassium hydroxide in water and a solution of potassmide in liquid ammonia (of equal concentration). The reaction in the latter proceeds at −53° with a rate constant that is 10^4 times greater than that found for the exchange with an aqueous solution of potassium hydroxide at 100°. With an activation energy for the exchange of molecular hydrogen with aqueous alkali of E = 24 kcal/mole, the approximate exchange rate constant is a factor of 10^{14} less than the rate constant of the analogous process with amide ions in liquid ammonia as the catalyst.

---

*See p. 35 for the reasons why the rate of exchange of hydrogen in the methyl group of quinaldine with alcohol is higher than with ammonia.

A solution of potassamide catalyzes hydrogen exchange so strongly that with long experiments (200-1500 hr) and a considerable amide concentration (1-8 N) it was possible to effect the partial exchange of hydrogen in saturated hydrocarbons [11,12] (p. 113).

3. COMPARISON OF THE RATES OF HYDROGEN EXCHANGE WITH AMPHOTERIC AND PROTOGENIC SOLVENTS

While protophilic solvents reduce the differences in the strengths of acids and, at the same time, many neutral substances become acids on solution in them, in acid, protogenic solvents, the differences in the strengths of bases are reduced and many substances show basic properties [13-16]. Even if the solvent is such a comparatively weak acid as acetic acid, almost identical titration curves are obtained for H_2SO_4 with all bases that are stronger than aniline in water. There is still greater increase in the basic properties of substances dissolved in formic and trifluoroacetic acids and even more so in anhydrous sulfuric acid and liquid hydrogen halides. Only a very few substances, for example, $HClO_4$, remain acids when dissolved in liquid hydrogen fluoride. Such compounds as CH_3COOH and HNO_3, which are typical acids, are capable of reacting as bases in this extremely acid solvent and ionize to form the cations $CH_3CO_2H_2^+$, and $H_2NO_3^+$:

$$CH_3CO_2H + HF \rightleftarrows CH_3CO_2H_2^+ + F^-$$
$$HNO_3 + HF \rightleftarrows H_2NO_3^+ + F^-$$

Urea and acetamide, which are acids in liquid ammonia and ionize to form the anions $CO(NH_2)NH^-$ and CH_3CONH^-, on solution in it, are converted into the cations $CO(NH_2)NH_3^+$ and $CH_3CONH_3^+$, by glacial acetic acid. Many aromatic hydrocarbons such as toluene, mesitylene, anthracene, pyrene, etc., are capable of adding a proton and ionizing like bases in liquid hydrogen fluoride. When boron trifluoride is added to a solution of benzene in this solvent, the equilibrium of benzene ionization (C_6H_6 + + HF + $BF_3 \rightarrow C_6H_7^+ + BF_4^-$) is practically completely displaced to the right.

Let us attempt to estimate the acceleration of an exchange reaction when water is replaced by various acid solvents. According to Small and Wolfden [17], heating a solution of phenol in heavy water at 100° results in the exchange of two hydrogen atoms of the aromatic ring after 400 hr, i.e., calculated on three hydrogen atoms, $k = 10^{-7}$ sec^{-1}. According to the measurements of Koizumi [18], two hydrogen atoms are exchanged after two hours in 0.5 N hydrochloric acid solution at 100° ($k = 10^{-5}$ sec^{-1}). Work of the Isotopic Reactions Laboratory [19] has shown that with liquid deuterium bromide, four hydrogen atoms in phenol are exchanged immeasurably rapidly even at room temperature. The reaction between DBr and anisole is equally rapid [19], while, according to the measurements of Brown [20], anisole does not exchange hydrogen with glacial acetic acid even on heating (90°) for two days. The addition of sulfuric acid (0.1 N solution in glacial acetic acid) results in the exchange of more than two hydrogen atoms under the same conditions. Hydrogen exchange with liquid DF proceeds even more energetically than with DBr [21]. For example, not more than six hydrogen atoms in the diphenyl ether molecule are exchanged with liquid DBr even after 150 hr, while all ten hydrogen atoms are replaced by deuterium with liquid DF even after 3 hr [22]. Table 5 gives the number of hydrogen atoms (n) exchanging in a molecule of the substance after the time τ. We assume that the activation energy of hydrogen exchange between C_6H_5OH and D_2O equals 30 kcal/mole [18] and that the rate constants of exchange between C_6H_5OH and DBr $k < 10^{-2}$ sec^{-1}. In this case, the rates of the reactions in the two solvents named differ by at least 6-8 orders.

Hydrogen exchange with liquid deuterium bromide is very strongly catalyzed by aluminum bromide.* At a concentration corresponding to $\sim 10^{-3}$ mole of AlBr$_3$ per 1000 g of DBr, the hydrogen of benzene is exchanged immeasurably rapidly at room temperature and very rapidly at low temperatures, as is indicated by the fact that the time to cool the solution and evaporate the solvent (5-7 min) was sufficient for the establishment of the equilibrium distribution of deuterium, corresponding to a temperature from −50 to −70° [23, 24]. With an aluminum bromide concentration equal

TABLE 5. Comparison of Rates of Deuterium Exchange with Water and with Acid Solvents

Substance	Solvent	Catalyst	Temp., °C	τ, hr	n
Phenol	D_2O	—	100	400	2
"	D_2O	HCl(0.5 N)	100	2	2
"	DBr	—	20	0.2	3
"	DBr	—	20	72	3
"	DF	—	20	0.1	3.3
"	DF	—	25	1	4.5
Anisole	CH_3CO_2D	—	90	48	0
"	CH_3CO_2D	D_2SO_4(0.1N)	90	47	2.4
"	DBr	—	20	0.2	3
"	DF	—	20	0.05	3.5
"	DF	—	25	1	4.8
Diphenyl ether	DBr	—	20	0.2	6
" "	DBr	—	20	150	6
" "	DF	—	20	0.1	7
" "	DF	—	25	3	10

to 10^{-5} mole per 1000 g of DBr, it was possible to pass into the kinetic region (k = $= 10^{-5}$ sec^{-1} at 25°) [25] (see p. 42 for more details). The hydrogen in benzene exchanges extremely slowly in the absence of a catalyst: $k = 5 \cdot 10^{-8}$ sec^{-1} at 20° [26]. According to a very rough estimate, a normal solution of AlBr$_3$ in liquid DBr increases the rate constant by not less than ten orders in comparison with the rate constant of exchange with the same solvent without catalyst, while the rate of the latter reaction is at least 6-8 orders higher than that of the reaction of the same substance in aqueous solution. Thus, the use of solutions of AlBr$_3$ in liquid DBr makes it possible to increase the rate of an exchange reaction by almost twenty orders in comparison with its rate in an amphoteric solvent, water. The exceptionally high catalytic activity of aluminum bromide in liquid deuterium bromide makes it possible to effect hydrogen exchange

---

* See p. 42 for the reasons for this.

in alicyclic hydrocarbons [23, 27-29]. Its rate was measured in work of the Isotopic Reactions Laboratory [29]. At an AlBr$_3$ concentration of 1 mole per 100 moles of HBr, the rate constants of hydrogen exchange in cyclohexane and methylcyclopentane equals $1 \cdot 10^{-5}$, for methylcyclohexane k = $1 \cdot 10^{-5}$, and for cyclopentane, k = $6 \cdot 10^{-5}$ sec^{-1}. The rate of deuterium exchange in a saturated hydrocarbon is not less than four orders below the rate of hydrogen exchange in benzene at an equal concentration of aluminum bromide.

As already mentioned above, liquid hydrogen fluoride accelerates exchange reactions to an even greater extent than liquid hydrogen bromide [21] (see p. 169 for more details). Hydrocarbons with a tertiary carbon atom exchange hydrogen with liquid DF without a catalyst. The reaction rate is increased by the addition of BF$_3$, which is an acid catalyst in liquid deuterium fluoride (see p. 183). With catalysis by BF$_3$ (0.4 mole/1000 g of DF), the hydrogen is exchanged even in a saturated hydrocarbon, cyclopentane, which does not have a tertiary carbon atom and, as far as is known, is not isomerizable.

To sum up, it can be stated that the use of very strong basic and acidic catalysts in protophilic and protogenic solvents, respectively, makes it possible to accelerate hydrogen exchange by up to about twenty orders in comparison with the rate in amphoteric solvents without a catalyst. This has made it possible to make a thorough study of isotopic exchange of hydrogen in CH bonds of hydrocarbons and their derivatives [30].

4. EFFECT OF THE DIELECTRIC CONSTANT OF THE SOLVENT AND THE POLARITY OF ITS MOLECULES ON THE HYDROGEN EXCHANGE RATE

Above we compared the rates of exchange in an amphoteric solvent, on the one hand, and in a protophilic or protogenic solvent, on the other. The results of work in the Isotopic Reactions Laboratory have made it possible to compare the kinetics of exchange reactions in solvents that are similar chemically, but differ in physical properties [6, 12]. The dielectric constant of the solvent (DC) and dipole moment of its molecules (μ) were selected from among these properties. The former is of great importance in proton transfer processes [15]. As regards the dipole moment, the energy of an intermolecular interaction depends on its value. $K_i^{H_2O}$ is the ionization constant of the acid or base in water and K_S is the ionic product of the solvent (Table 6).

In Table 7 we compare the rate constants of the exchange of hydrogen k (sec^{-1}) at 120° in the methylene group of fluorine and diphenylmethane and in the methyne group of triphenylmethane with several protophilic solvents, namely, liquid deuteroammonia, anhydrous, deuterohydrazine, ethylenediamine, and ethanolamine [31].

Hydrogen exchange with deuterohydrazine and deuteroethylenediamine proceeds more rapidly than with deuteroammonia (Table 7). The results of experiments with hydrazine are particularly conclusive because triphenyl- and diphenylmethanes are sparingly soluble in hydrazine, in contrast to liquid ammonia. The higher rate of deuterium exchange with hydrazine, which is evidently a weaker base than ammonia, is explained by the fact that the DC of hydrazine is three times that of ammonia and its molecules are more polar.

In addition to the substances mentioned, the methyl group of the acetate ion exchanges its hydrogen more rapidly with anhydrous hydrazine than with liquid ammonia

($k_{120°} = 3 \cdot 10^{-7}$ and $2 \cdot 10^{-8}$ sec^{-1}) [32]. Sodium and potassium acetates are soluble in both solvents and do not react with them chemically.

The dielectric constants of liquid ammonia and anhydrous ethylenediamine are very similar. We are therefore inclined to ascribe the considerable acceleration of

TABLE 6. Constants of Solvents

Solvent	M.p., °C	B.p., °C	DC$_{25°}$	$\mu \cdot 10^{18}$	K_s	$K_i^{H_2O}$
NH$_3$	−77.7	−33.35	16	1.46	10^{-33}	$2 \cdot 10^{-5}$
NH$_2$NH$_2$	2	113.5	52	1.84	10^{-25}	$3 \cdot 10^{-6}$
NH$_2$CH$_2$CH$_2$NH$_2$	11	116.2	14	1.94	—	$1 \cdot 10^{-1}$
NH$_2$CH$_2$CH$_2$OH	10.5	172.2	38	2.27	10^{-6}	$3 \cdot 10^{-5}$
H$_2$O	0.0	100.0	80	1.84	10^{-14}	—
CH$_3$OH	−97.5	64.5	32.6	1.66	$10^{-16.8}$	—
C$_2$H$_5$OH	−114.5	78.3	24.3	1.68	$10^{-19.1}$	—
iso-C$_3$H$_7$OH	−89.5	82.4	18.3	1.68	—	—
tert- C$_4$H$_9$OH	25.7	82.4	10.9 (30°)	1.66	—	—
HCOOH	8.25	100.7	58.5 (16°)	1.19	10^{-6}	$2 \cdot 10^{-4}$
CH$_3$COOH	16.63	117.72	6	—	10^{-12}	$2 \cdot 10^{-5}$
CH$_2$ClCOOH	62.5	189.3	—	—	—	$2 \cdot 10^{-3}$
CHCl$_2$COOH	9.7	194.4	8.3 (20°)	—	—	$5 \cdot 10^{-2}$
CCl$_3$COOH	58	196.5	4.5 (61°)	—	—	$2 \cdot 10^{-1}$
CF$_3$COOH	−15.25	72.4	8.4 (20°) *	—	—	1.8
HF	−83.0	19.5	175 (−73°)	1.91	10^{-10}	$6 \cdot 10^{-4}$
HCl	−111	−85.0	9 (−90°)	1.04	—	$10^6 - 10^7$
HBr	−86	−67.1	6 (−80°)	0.79	—	$10^8 - 10^9$
HI	−50.8	−35.5	3 (−50°)	0.38	—	10^{10}
H$_2$SO$_4$	10.5	332	101	—	10^{-4}	—

* According to other data, DC = 39.5°.

TABLE 7. Rate Constants of Hydrogen Exchange with Protophilic Solvents

Substance	ND$_3$	N$_2$D$_4$	ND$_2$C$_2$H$_4$ND$_2$	ND$_2$C$_2$H$_4$OD
Fluorene*	$2 \cdot 10^{-4}$	—	$3 \cdot 10^{-3}$	$2 \cdot 10^{-4}$
Triphenylmethane	$2 \cdot 10^{-7}$	($2 \cdot 10^{-6}$)	$2 \cdot 10^{-6}$	$4 \cdot 10^{-7}$
Diphenylmethane	$7 \cdot 10^{-9}$	($2 \cdot 10^{-7}$)	$1 \cdot 10^{-7}$	—
Quinaldine**	$7 \cdot 10^{-8}$	—	$1 \cdot 10^{-6}$	$8 \cdot 10^{-6}$

* The rate constants were measured at 25° in experiments with fluorene.
** The experiments with quinaldine are discussed on p. 34.

exchange (by a factor of 10-15) with the latter solvent to its higher protophilicity and the higher polarity of its molecules. The "statistical factor," i.e., the presence of four deuterium atoms and two nitrogen atoms in hydrazine and ethylenediamine molecule as compared with three deuterium atoms and one nitrogen atom in ammonia, may also be of some importance.

The effect of the dielectric constant of the solvent and the polarity of its molecules it shown even more clearly by a comparison of the kinetics of hydrogen exchange in two acid solvents, namely, liquid hydrogen bromide and fluoride, whose constants are given in Table 6. Even a purely qualitative comparison of hydrogen exchange in phenol and its ethers (Table 5) indicates the much higher rate of exchange reactions with liquid hydrogen fluoride. There are also measurements [21, 26] confirming this conclusion. For example, the rate constants of deuterium exchange between benzene and HF are 4.5 orders higher than the constants obtained for the exchange with HBr ($k_{20°} = 1 \cdot 10^{-3}$ and $5 \cdot 10^{-8}$ sec^{-1}). In experiments with toluene, the same degree of exchange (n = 4.8) was reached after 6 and $9 \cdot 10^4$ min, respectively.

The effect of the dielectric constant of the solvent on deuterium exchange was also observed in experiments with solutions of dimethylaniline in anhydrous carboxylic acids [33]. The rates of the exchange reactions with formic and trifluoroacetic acids differ little though the latter is much more acid (the acidity functions H_0 equal -2 and -4, respectively; see p. 53), but on the other hand, the dielectric constant of formic acid is much higher (58.5 and 8.1, respectively, at 25°).

5. EFFECT OF THE SUBSTRATE AND CATALYST CHARGE ON THE HYDROGEN EXCHANGE RATE

The effect of the substrate charge on the rate of a catalytic reaction was demonstrated by Brønsted [34, 35] on the base-catalyzed decomposition of nitramide in water (DC = 80), m-cresol (DC = 13), and isoamyl alcohol (DC = 5.7) (Table 8, see also [15]).

The rate constant of the reaction in aqueous solution was arbitrarily taken as unity. With catalysis by the benzoate ion, the rate constants increased from water to m-cresol

TABLE 8. Comparison of Rates of Catalytic Decomposition of Nitramide in Various Solvents

Catalyst	H_2O	$m-C_6H_4(CH_3)OH$	$iso-C_5H_9OH$
	DC = 80	DC = 13	DC = 5.7
$C_6H_5COO^-$	1	28	90
$C_6H_5NH_2$	1	0.3	0.06

and then to isoamyl alcohol. According to Brønsted, this is explained by the fact that the polarization of the substrate molecule in the field of the anion is increased with a reduction in the dielectric constant of the solvent. On the other hand, with catalysis by the electrically neutral molecule aniline, the rate constants decrease in the same sequence; a high dielectric constant promotes the transfer of a proton from one electrically neutral molecule to another.

An analogous effect must be observed in hydrogen exchange [6]. It may be predicted that an exchange reaction catalyzed by a base will be retarded by a negative charge on the substrate, while in catalysis by an acid, a positive charge on the substance will interfere as it hampers the attraction and abstraction of a proton (deuteron) from the substrate in the first case and its addition to the substrate in the second case. It is also evident that like charges on the substrate and catalyst will obstruct the reaction.

In actual fact, from work in the Isotopic Reactions Laboratory [32] it follows that hydrogen in the acetate ion exchanges with heavy water (DC = 80) and with deutero-ammonia (DC = 16) at approximately the same rate, which is somewhat lower than the rate of exchange with anhydrous deuterohydrazine (DC = 52). The following values were obtained for $k_{120°}$: $6 \cdot 10^{-8}$, $2 \cdot 10^{-7}$, and $3 \cdot 10^{-7}$, respectively. There was hardly any difference in the exchange rate constants ($k_{120°} = 4 \cdot 10^{-5}$ and $3 \cdot 10^{-5}$) with the same concentration (1 N) of OD^- ions in D_2O and ND_2^- ions in ND_3, which catalyze the reactions. The strength of the effect of the substrate charge on the rate of a reaction catalyzed by ND_2^- ions is shown by the fact that in the absence of the catalyst, hydrogen exchange with liquid deuteroammonia in the acetate ion and the quinaldine molecule occurs at the same temperature (120°) and there is little difference in the rates ($k = 2 \cdot 10^{-8}$ and $8 \cdot 10^{-8}$). For the exchange rates to be similar ($k = 4 \cdot 10^{-5}$ and 10^{-4}) with catalysis by potassamide, the reaction with the acetate ion must be carried out at 120° with a potassamide concentration $C_{KND_2} = 1$ N and the reaction with the quinaldine molecule at −30° and $C_{ND_2} = 0.02$ N [8].

The decrease in the rate of an exchange reaction catalyzed by anions with a negative charge on the substance is also indicated by experiments on the exchange of hydrogen in the phenolate ion, which is catalyzed by a solution of KND_2 in liquid ND_3. At a potassamide concentration of 0.4 N and 50°, the exchange of hydrogen in the phenolate ion proceeds at a rate which is 3 orders less than the rate in benzene [36].

The retardation of an exchange reaction catalyzed by an acid with a positive charge on the substrate was confirmed experimentally by comparing the rates of the exchange of hydrogen in aromatic hydrocarbons.

The hydrogen in anthracene exchanges with liquid DBr 5-6 orders faster than in benzene [26]. As the rate of hydrogen exchange between benzene and DF is 4.3 orders greater than with DBr, it would seem that the exchange between anthracene and deuterium fluoride would be immeasurably fast. In actual fact, the hydrogen in anthracene exchanges with DF at least an order slower than the exchange of hydrogen between benzene and DF [21].

The facts presented can be explained if we consider the strong ionization of anthracene in liquid HF, which was first revealed by Klatt's measurements of the electrical conductivity and elevation of the boiling point of these solutions [37]. A proton (or deuteron) adds to an anthracene molecule almost exclusively in the meso-position [38]. The meso-atoms therefore exchange rapidly, while the exchange of the other hydrogen atoms in the ion ($C_{14}H_{10}D^+$) is hindered. (In addition to the electrostatic action of the charge, the change in the electron density distribution in the molecule of the substance produced by the ionization is also of great importance in the exchange reaction.)

According to Reid [39], the addition of BF_3 completely ionizes anthracene dissolved in liquid HF. Therefore, in accordance with the interpretation given, the presence of BF_3 not only does not promote, but even hinders the exchange of hydrogen in anthracene, limiting it to only the two meso-atoms. For example, in a solution containing BF_3 (0.4 mole/1000 g of DF), two hydrogen atoms (the meso-atoms) are exchanged after 2 hr and in the absence of BF_3, nine atoms are exchanged in the same time, while in benzene five atoms are exchanged after half an hour without catalyst and there is complete exchange of hydrogen in less than 2 min in the presence of BF_3 (0.1 mole/1000 g of DF).

An increase in the number of condensed rings leads to strengthening of the basic properties of aromatic hydrocarbons and an increase in the degree of their ionization in DF. Therefore, the rate of hydrogen exchange catalyzed by BF_3 falls in the series: benzene > naphthalene > anthracene (Table 9).

TABLE 9. Rate of Hydrogen Exchange in Benzene, Naphthalene, and Anthracene with Liquid DF Catalyzed by Boron Trifluoride

Substance	BF_3 conc., mole per 1000 g of DF	Time, min	% exchange
Benzene	0.1	2	100
Naphthalene	0.2	6	80
Anthracene	0.4	120	20

When a substance is dissolved in liquid DBr, where there is a very low concentration of free ions as a result of the low dielectric constant of DBr and electrically neutral hydrocarbon molecules actually participate in the exchange, the exchange of hydrogen accelerates in a sequence corresponding to the increase in the strengths of the hydrocarbons as bases, namely: benzene < naphthalene < anthracene [26].

Thus, the same factor, namely, the acid—base interaction of the solvent with aromatic hydrocarbons containing nonequivalent hydrogen atoms, may lead to opposite sequences of relative rates of hydrogen exchange.

It is shown below (p. 33) that the same result may be caused by the dual reactivity of a substrate whose molecule contains a heteroatom. The addition of a proton (deutron) to it converts the molecule into a positively charged ion and the exchange is inhibited. A clear example is provided by the behavior of dimethylaniline and triphenylamine in glacial acid. The hydrogen of the former exchanges three orders more rapidly than that of the latter. In liquid hydrogen bromide, the isotopic exchange of hydrogen in triphenylamine proceeds at a rate that is six orders higher as dimethylaniline is converted into a salt and the exchange with the dimethylanilinium ion is very slow [33].

Finally, the absence of deuterium exchange between the positively charged tropylium ion and strong acids, which was described by M. E. Vol'pin, D. N. Kursanov, et al. [40], is a phenomenon of the same type.

The rate of hydrogen exchange is changed in the same direction as by the total charge of an ion by a partial charge, i.e., a reduction or an increase in the electron density at the carbon atom to which the hydrogen atom being exchanged for deuterium is attached. It occurs when an electron-acceptor or electron-donor substituent (p. 9) is introduced into the hydrocarbon molecule. An increase in the electron density at the carbon atom ($\delta-$) must promote isotopic exchange with an acid (arbitrarily denoted by D^+) as a result of the electrostatic attraction of the unlike charges, while a reduction in the electron density ($\delta+$) must hinder exchange. The opposite result is obtained in exchange with a base (arbitrarily denoted by B^-) as in this case the reaction occurs as a result of the abstraction of the proton of the CH bond.

Substrate − acid (proton donor*) Reagent − base (proton acceptor*)	Substituent	Substrate − base (deuteron acceptor*) Reagent − acid (deuteron donor*)
$\overset{\delta+}{C}$ − H + B⁻ → acceleration	Electron-acceptor	$\overset{\delta+}{C}$ − H + D⁺ → retardation
$\overset{\delta-}{C}$ − H + B⁻ → retardation	Electron-donor	$\overset{\delta-}{C}$ − H + D⁺ → acceleration
$\overset{-}{C}$ − H + B⁻ → inhibition		$\overset{+}{C}$ − H + D⁺ → inhibition

*In talking of a proton, we also imply a deuteron and vice versa.

The effect of the charge of the substrate on hydrogen exchange is given schematically below. The negatively and positively charged ions are arbitrarily denoted by $\overset{-}{C}$ − H and $\overset{+}{C}$ − H (the charge may be on another atom in the molecule apart from carbon).

The effect of a substituent on deuterium exchange is illustrated by experiments with aromatic compounds containing an electron-donor methyl group or an electron-acceptor nitro group. Thus, the α-hydrogen atoms in the naphthalene molecule exchange almost completely after half an hour in liquid deuterium bromide and the exchange of the β-atoms only just begins; while all the hydrogen atoms in the aromatic nucleus of methylnaphthalene are replaced by deuterium in the same time and only two hydrogen atoms in nitronaphthalene are replaced even after two thousand hours [26]. On the other hand, when the solvent is liquid deuteroammonia and potassamide is used as the catalyst, the introduction of a methyl group into benzene reduces the exchange rate constant by a factor of four, while exchange with m-dinitrobenzene proceeds without the addition of a catalyst [41].*

Let us summarize what has been stated above. As was to be expected, in view of the acid−base nature of hydrogen exchange in solutions, in addition to depending on the protophilicity or protogenicity of the substrate, solvent, and catalyst, the rate of the exchange reactions depends on parameters such as the dielectric constant of the solvent, the degree of polarity of molecules of the latter, and the charge on the sub-

*See also the collection: Problems in Chemical Kinetics, Catalysis, and Reactivity [in Russian] (Izd. AN SSSR, Moscow, 1955), p. 699.

strate and the catalyst. These parameters have such a substantial effect on the rate of hydrogen exchange that more rapid exchange may occur in a solvent which is a weaker acid (or base), so that ignoring all these factors may lead to incorrect conclusions on the relative strength of acids and bases from a comparison of the rates of hydrogen exchange.

6. EFFECT OF DUAL REACTIVITY OF A SUBSTANCE ON HYDROGEN EXCHANGE

The dual reactivity of a substance is a very important factor in hydrogen exchange [42, 43]. In the same reaction system, such a substance may fulfill two protolytic functions (acid and base), which are interconnected. Thus, for example, the change in the valence state of a heteroatom in a molecule on elimination of a proton or, on the other hand, on its addition affects the proton-acceptor or proton-donor capacity of another group or atom in the same molecule. This explains the apparently paradoxical cases of catalysis of basic hydrogen exchange by acids and the acceleration of acidic hydrogen exchange by bases, examples of which are given below.* All this means that the same molecule may be a donor and an acceptor of a proton (deuteron). As a result there may be an unusual ratio between the rate of hydrogen exchange in amphoteric and protophilic (or protogenic) solvents, i.e., more rapid exchange in an amphoteric solvent. As is shown below (p. 35), this occurs, in particular, when the valence state of an atom in a molecule changes as a result of intermolecular reactions with only one of the solvents.

A. Acceleration of Acid Exchange by a Base

According to Ingold [44], the addition of alkali to a solution of phenol in heavy water produces an increase in the rate of deuterium exchange in the aromatic nucleus until half of the phenol is neutralized. A further increase in the amount of alkali decreases the exchange rate. This reaction was studied in more detail later by Japanese authors [45]. Alkali converts phenol into the phenolate ion, at whose ortho- and para-carbon atoms the electron density is considerably higher than in the un-ionized molecule. Therefore the deuteron of the hydroxyl group of even such a comparatively weak acid as phenol, C_6H_5OD, adds to the aromatic ring of the ion. The addition of more than the stoichiometric amount of alkali to the phenol solution reduces the number of phenol molecules relative to the number of phenolate ions and the rate of the exchange reaction falls.

B. Acceleration of Basic Exchange by an Acid

An example is provided by work in the Isotopic Reactions Laboratory [8] on hydrogen exchange in the methyl groups of quinaldine and α-picoline. If an alcohol solution of quinaldine is acidified, at a ratio of 0.2 mole of HCl to a mole of quinaldine the exchange rate constant is increased by more than an order. If an equimolecular amount of acid is added to the quinaldine, i.e., the base is converted completely into the hydrochloride, the rate constant is decreased by an order and then hardly changes with an increase in the amount of acid by a factor of five (Table 10).

---

*With a heteroatom present, a reaction with an acid may proceed as far as ionization with the formation of charged particles. Therefore, as was pointed out above (p. 29), it is necessary to allow for the effect of the charge on the substrate on the deuterium exchange rate.

When quinaldine is neutralized by an acid, the nitrogen of the heterocycle becomes tetravalent and positively charged. The electrons of the CH bond of the methyl

TABLE 10. Hydrogen Exchange in Quinaldine

Solvent	DCl added, mole per mole of substance	$k_{120°}$, sec^{-1}
C_2H_5OD	None	$2 \cdot 10^{-6}$
C_2H_5OD	DCl 0.2	$5 \cdot 10^{-5}$
C_2H_5OD	DCl 1.0	$4 \cdot 10^{-6}$
C_2H_5OD	DCl 5.0	$5 \cdot 10^{-6}$
C_2H_5OD	None*	$\sim 3 \cdot 10^{-3}$
ND_3	None	$7 \cdot 10^{-8}$

* C_2H_5OK (1 N) was added to the solution instead of acid.

group are displaced toward the nitrogen and as a result, the hydrogen atoms in it become much more labile than in the original molecule. This is well known from the
chemical reactions of methylated heterocycles [46]. For example, acids catalyze condensations. The latter also proceed much more readily when the quinaldine and picoline are converted into halcalkylates, in which the nitrogen is a tetravalent, positively
charged ion. As long as excess quinaldine molecules are present in solution, they act
as acceptors of the protons of methyl groups of other molecules (I):

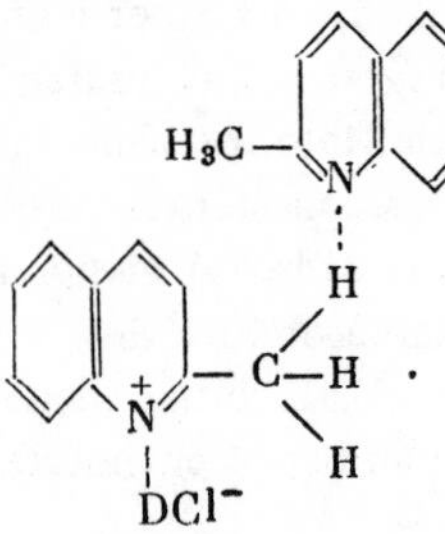

(I)

With complete neutralization of the quinaldine by acid, the molecules of the solvent, alcohol, i.e., a much weaker base than quinaldine, become the proton acceptors
(II):

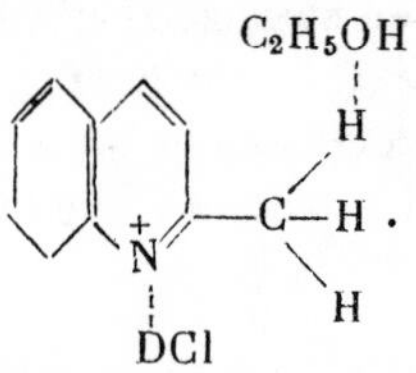

(II)

The exchange rate is therefore reduced.

The addition of a strong base, namely, an alcoholate (1 N) to the alcohol solution of quinaldine increases the reaction rate by more than three orders (Table 10). It is evident that a complex of type (III) is formed during the reaction:

$$\tag{III}$$

The accuracy of this explanation was confirmed later by the work of I. P. Gragerov et al. [47]. The authors found that the rate constant of deuterium exchange between ethanol and quinaldine is comparatively small ($k_{130°} = 3.6 \cdot 10^{-5}$ sec^{-1}), but that in a quaternary ammonium salt of quinaldine, 80% of the hydrogen of the CH_3 group is exchanged for deuterium in 8 min at 18°. The catalyst in these experiments was triethylamine.

C. Inversion of Relative Rates of Hydrogen Exchange with Amphoteric and Protophilic Solvents

The change in the valence state of the nitrogen in the quinaldine molecule and the corresponding increase in the lability of the methyl hydrogen are the reasons for the unusual ratio of the rates of deuterium exchange with amphoteric and protophilic solvents, namely, alcohol and ammonia [8]. Normally, the rate constant of an exchange with liquid ammonia is 4-6 orders higher than that of the reaction in alcohol solution (p. 22). However, in experiments with quinaldine (and also α-picoline) the rate of hydrogen exchange with alcohol was an order higher than with liquid ammonia. This may be explained by the formation of a hydrogen bond between the hydrogen atom of the alcohol and the nitrogen atom of the heterocycle. Its existence is reflected in the displacement of the frequency of the valence vibration of the alcohol hydroxyl group (p. 229). The formation of the hydrogen bridge is accompanied by partial protonization of the alcohol hydrogen with the result that the nitrogen atoms of the heterocycle tends toward a tetravalent positive ion with a consequential displacement of the electrons of the methyl group CH bond. This facilitates exchange of the methyl hydrogens under the action of a base, namely, another molecule of the heterocycle (IV):

$$\tag{IV}$$

Thus, in alcohol solution, in contrast to ammonia, the reaction involves not a quinaldine molecule, but a polarized complex, in which the hydrogen atoms of the methyl group are protonized and replaced by deuterium more readily than in the starting quinaldine molecule. The effect of the high protophilicity of ammonia is suppressed and the reaction with alcohol proceeds more rapidly than with ammonia, contrary to the normal order of rates of deuterium exchange with protophilic and amphoteric solvents.

This explanation of the different deuterium exchange rates in alcohol and ammonia solutions of hydrocarbons and methylated heterocycles is confirmed by a comparison of isotopic exchange of hydrogen in these substances with ethanolamine and liquid ammonia [31]. It was found that the hydrogen of the methylene group of fluorine and the methyne group of triphenylmethane exchanges with the two solvents at the same rate, while the rate of exchange of quinaldine with ethanolamine is two orders higher than with ammonia (Table 11).

Ethanolamine, $H_2NCH_2CH_2OH$, is an amino alcohol. The presence of amino and hydroxyl groups in its molecule allows it to react in two ways: in the reaction with fluorene and triphenylmethane, the amino group acts as a proton acceptor, while a hydrogen bond may be formed between the nitrogen atom of the quinaldine molecule and the hydroxyl group of the ethanolamine with the consequences discussed in detail above. Another ethanolamine molecule acts as an acceptor of a proton of the quinaldine methyl group.

There is a further demonstration of the accuracy of the above explanation of the higher rate of hydrogen exchange in methylated nitrogen heterocycles with alcohol

TABLE 11. Comparison of Rate Constants of Deuterium Exchange with Ethanolamine and Ammonia

Substance	Ammonia k, sec^{-1}	Ethanolamine
Fluorene (25°)	$2 \cdot 10^{-4}$	$2 \cdot 10^{-4}$
Triphenylmethane (120°)	$2 \cdot 10^{-7}$	$4 \cdot 10^{-7}$
Quinaldine (120°)	$7 \cdot 10^{-3}$	$8 \cdot 10^{-6}$

than with ammonia. The exchange rate in alcohol solutions should be symbatic with the degree of acidity of the alcohol as the polarity of the hydrogen bond and the valence state of the nitrogen atom depend on the latter. There is actually a parallel relation between these values and the exchange rate: the sequence of rate constants of deuterium exchange with various alcohols is as follows: ethylene glycol > methanol > ethanol > isopropanol (see p. 287).

D. Inhibition of Acid Hydrogen Exchange by an Acid

The same phenomenon, namely, the conversion of a nitrogen atom into a tetravalent positive ion, is the reason for retardation of acid hydrogen exchange as well as acceleration of basic hydrogen exchange.

Aromatic amines show dual reactivity. The proton of an acid normally adds to the nitrogen atom, which has an unshared pair of electrons. However, the position of the attack of a reagent may also be the ortho- and para-atoms of the aromatic ring (or rings) as a result of their high electron density, which is caused by conjugation of the p-electrons of the nitrogen atom with the π-electrons of the aromatic system. In the terminology of A. N. Nesmeyanov [43], the reaction proceeds with transfer of the reaction center. Therefore, in an acidified alcohol solution, the corresponding hydrogen atoms in aromatic amines are replaced by deuterium [5, 48].

In the series dimethylaniline, methyldiphenylamine, and triphenylamine, there is an increasing number of phenyl groups to each pair of p-electrons with the result that triphenylamine is the weakest base. It may be assumed that the electronegativity of the aromatic rings of these amines will decrease in the same sequence. In actual fact, Brown [48] established that the hydrogen of triphenylamine exchanges more slowly than that of methyldiphenylamine in an alcohol solution of an acid. A detailed study was made in the Isotopic Reactions Laboratory [33] of deuterium exchange in aromatic amines dissolved in carboxylic acids and liquid hydrogen bromide. The results are summarized in Table 12. The rate constants (sec^{-1}) for exchange of the para-atom are given for dimethylaniline and methyldiphenylamine, while the rate constants for triphenylamine are mean values, which may partially reflect the participation of ortho-atoms.

TABLE 12. Rate Constants of Hydrogen Exchange of Aromatic Amines with Carboxylic Acids

Solvent	$C_6H_5N(CH_3)_2$	$(C_6H_5)_2NCH_3$	$(C_6H_5)_3N$
CH_3COOH	$2 \cdot 10^{-4}$ (90°)	$2 \cdot 10^{-5}$ (90°)	$1 \cdot 10^{-7}$ (90°)
$HCOOH$	$3 \cdot 10^{-5}$ (90°)	$1 \cdot 10^{-3}$ (90°)	$6 \cdot 10^{-4}$ (90°)
$CH_2ClCOOH$	$7 \cdot 10^{-5}$ (90°)	$4 \cdot 10^{-3}$ (65°)	$1 \cdot 10^{-3}$ (90°)
CCl_3COOH	$3 \cdot 10^{-5}$ (90°)	$2 \cdot 10^{-3}$ (65°)	—
CF_3COOH	$6 \cdot 10^{-6}$ (90°)	$2 \cdot 10^{-4}$ (65°)	$2 \cdot 10^{-3}$ (30°)
HBr	$< 10^{-8}$ (20°)		$> 10^{-2}$ (20°)

The sequence of the rate constants for deuterium exchange with glacial acetic acid are as follows:

$$C_6H_5N(CH_3)_2 > (C_6H_5)_2NCH_3 > (C_6H_5)_3N,$$

i.e., the rate of the exchange reaction falls in parallel with the decrease in the electronegativity of the aromatic rings. The absorption spectrum of a solution of dimethyl-aniline in glacial acetic acid [48a] indicates that even this amine, the strongest of the bases examined, forms a molecular complex with the acid and not a true salt. In a salt, the proton of the acid adds to the nitrogen atom of the amine and converts the base into an ion with a tetravalent, positively charged nitrogen atom. As a result of saturation of the free electron pair of the nitrogen atom by the proton, there is disruption of the p, π-conjugation, on which depends the possibility of the addition of a proton (or deuteron) to the aromatic ring that results in the exchange reaction. The amine is ionized more readily, the stronger it is as a base and the stronger the carboxylic acid is. As a result, there may be a change in not only the magnitude, but even the sequence of the rate constants of the exchange reactions. Thus, if trifluoroacetic acid is used as the solvent, the sequence of the amines is the opposite to that in experiments with acetic acid:

$$(C_6H_5)_3N > (C_6H_5)_2NCH_3 > C_6H_5N(CH_3)_2.$$

Particularly clear results were obtained for hydrogen exchange in liquid deuterium bromide [19]. After less than five minutes at room temperature, nine hydrogen
atoms, i.e., the ortho- and para-atoms, in the triphenylamine molecule exchanged.
About 1000 hr was required for the exchange of six hydrogen atoms of the aromatic
rings of diphenylamine, but even after such a long time there was practically no exchange of hydrogen in the aromatic ring of dimethylaniline. Increasing the acidity of
the triphenylamine solution by adding an equimolecular amount of aluminum bromide
retarded the exchange and only one hydrogen atom instead of nine was exchanged in
triphenylamine in five minutes. It is evident that even the weakest of the three bases
is converted into a saltlike substance in this solution.

The inhibition by an acid of hydrogen exchange which is normally catalyzed by
an acid, which was unexpected at first, is explained simply in view of the dual reactivity of aromatic amines.

7. EFFECT OF STERIC HINDRANCE ON HYDROGEN EXCHANGE

In the example just examined, the rate of hydrogen isotope exchange decreased
on disruption of the conjugation between the π-electrons of the aromatic ring and the
free electron pair of the nitrogen atom of the amine when the proton of an acid was
added to the latter. The same result could be produced by prevention of conjugation
in another way. For conjugation to occur, the axes of the electron clouds must be
parallel. For this it is essential that all the atoms in the molecule lie in one plane,
i.e., the molecule must be coplanar. The introduction of a large substituent into a
molecule may force out of the plane the atom whose electrons participate in the conjugation. The reactivity of the substance is then changed. This occurs, for example,
if an ortho-hydrogen atom in the N, N-dimethylaniline molecule is replaced by a
large atom or group of atoms. Such substances become inert with regard to electrophilic substitution of hydrogen in the aromatic ring.

The effect of such steric hindrance on acid-catalyzed hydrogen exchange was
discussed in the articles of Kharasch and Brown [5, 20, 20a, 48]. In the first work [5,
48], the way in which the rate of hydrogen exchange varied when one of the hydrogen
atoms in N,N-dimethylaniline was replaced by a fluorine, chlorine, or bromine atom
was determined. Deuteroalcohol was used as the solvent in these experiments and the
exchange was catalyzed by sulfuric acid. Of the ortho-isomers, only o-fluoro-N,N-di-
methylaniline underwent exchange and the rate was much lower than that with unsubstituted N,N-dimethylaniline.* There was practically no isotopic exchange of the hydrogen in o-chloro- and o-bromo-N,N-dimethylanilines even after 140 hr at 115°,
while exchange occurred if the halogen atom (even bromine) was in a meta- or para-
position relative to the dimethylamino group and therefore did not hinder conjugation.
The numbers of hydrogen atoms exchanging in halogen derivatives of N,N-dimethyl-
aniline are compared below:

	ortho	meta	para
F	0.96	—	—
Cl	0.05	2.43	1.73
Br	0.02	2.53	1.70

*For this substance, the number of hydrogen atoms exchanging after 24 hr at 60° is
1.73.

An analogous "ortho effect" was also observed in the naphthalene series with large substituents (a nitro group or a bromine atom):

$$n$$

Dimethylnaphthylamine-1 . . . . 1.73
8-Nitrodimethylnaphthylamine-1 0
8-Bromodimethylnaphthylamine-1 0

Retardation of hydrogen exchange as a result of disruption of the coplanarity of the molecule was also found [48] on comparing the acid hydrogen exchange in N-methylindoline (I), N-methyl-1,2,3,4-tetrahydroquinoline (II), and N-methylhomo-tetrahydroquinoline (III).

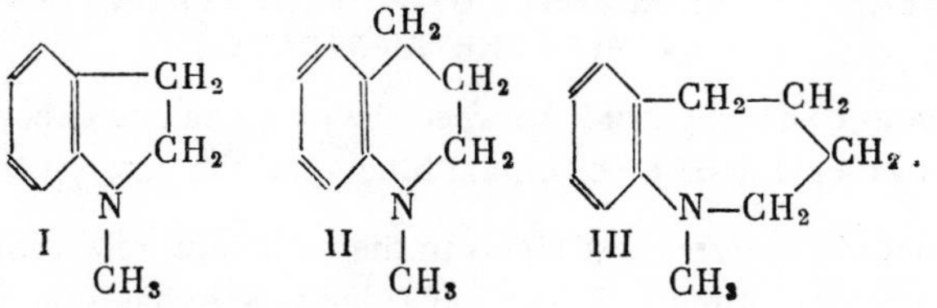

$$n = 1.58 \qquad n = 0.87 \qquad n = 0*$$

In the molecule of the first of these substances, all the atoms lie in one plane, while the second molecule is not quite planar and the third is not planar at all. Under each of the formulas is given the value of n from experiments with alcohol solutions of sulfuric acid at 60° (the first two values were obtained after 24 hr and the latter after 44 hr).

An analogous investigation was made with phenols and phenol ethers [20, 20a]. In this case, the free electron pair of the oxygen atom is conjugated with the π-electrons of the ring. The most interesting results were obtained with the following cyclic ethers of catechol:

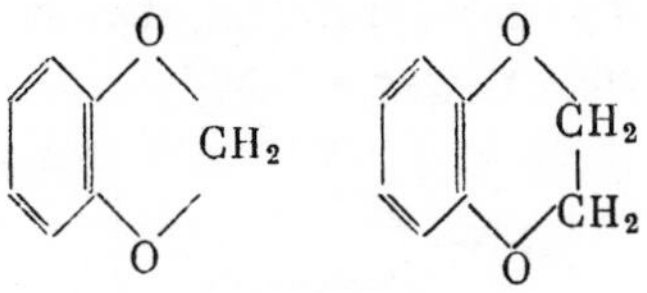

IV   V

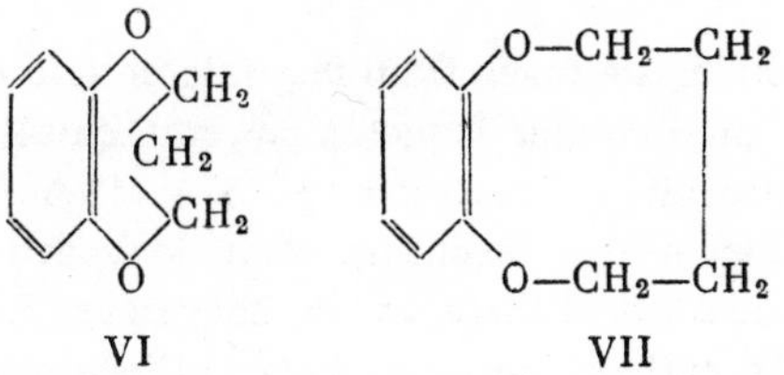

VI   VII

In view of certain side chemical processes which occur in alcohol solutions, glacial acetic acid was used as the solvent and sulfuric acid as the catalyst. The following values were obtained under comparable conditions (90°, 12 hr):

*After 70 hr at 120°, n = 0.85.

Substance	IV	V	VI	VII
n	1.30	0.79	0.23	0.93

The groups containing the oxygen atoms lie in the plane of the benzene ring in the cyclic ethers containing one and four methylene groups. The steric hindrance to conjugation is greatest in the trimethylene ether and therefore the exchange proceeds more slowly than in the other substances. Data on steric hindrance to hydrogen exchange were published recently by Satchell [J. Chem. Soc., 463 (1959)]. They concern the exchange of tritium for normal hydrogen in 1,2,3-trihydroxy-4,6-ditritiobenzene in aqueous sulfuric acid. As a result of steric hindrance, the rate of exchange is a factor of 60 lower than the rate calculated from partial rate factors (p. 258) for the same acid concentration.

8. DECREASE IN THE ACTIVITY OF A CATALYST ON ITS REACTION WITH THE SUBSTRATE

There are cases of discrepancies between the observed and expected exchange rates, which may be explained by chemical binding of the catalyst by the substrate.

As a rule, electronegative substituents in the molecules of organic substances activate isotopic exchange of the hydrogen in CH bonds with bases (p. 96). Therefore, alkali-catalyzed exchange proceeds at a considerable rate in nitromethane, acetone, and acetonitrile even in aqueous solution. The rate constants (sec^{-1}) of exchange at 25° in 1 N alkali solution are given in brackets [49]:

$$CH_3NO_2 \ (3 \cdot 10^1); \quad CH_3COCH_3 \ (2.5 \cdot 10^{-1}); \quad CH_3CN \ (3 \cdot 10^{-5}).$$

The hydrogen in m-dinitrobenzene exchanges with liquid ammonia without a catalyst [41] ($k_{50°} = 1 \cdot 10^{-6} \ sec^{-1}$). It might have been expected, therefore, that the reaction would proceed quite rapidly with benzonitrile and even more so, naphthoninitrile in liquid ammonia in the presence of potassamide. However, experiments did not confirm this surmise. With approximately 0.1 mole of catalyst per mole of nitrile, the fraction of exchange did not exceed 0.03-0.04 after an hour or even after a thousand hours at room temperature. Substances with the original constants were isolated from the reaction system. It is probable therefore, that the catalyst (NH_2^- ion) is bound by the nitrile in ammonia solution to give a complex

$$\left[H_2N-\left\langle \underline{} \right\rangle =CN \right]^-,$$

which is similar to those known for aromatic nitro compounds (p. 251).

The following examples are taken from the field of acid exchange. Aluminum bromide in liquid deuterium bromide is such a powerful catalyst that it even induces hydrogen exchange in alicyclic hydrocarbons (p. 182). Therefore, it was incomprehensible at first that there should be no exchange of the hydrogen in benzoic acid and nitrobenzene under conditions where there was exchange even with saturated hydrocarbons [25]. It was found that there was no exchange of hydrogen as long as the amount of aluminum bromide was less than equimolecular to the amount of the aromatic compound.

Consequently, there was no exchange as long as the substance bound the catalyst in a complex. It is well known that nitrobenzene forms complexes with both aluminum halides and strong protonic acids [50, 51]. Such a protonic acid could be the compound of aluminum bromide with hydrogen bromide, $HAlBr_4$, which does not exist in the free form (p. 159). Nitrobenzene reacts as a base either with an acidlike substance ($AlBr_3$) or with an acid ($HAlBr_4$), though the result is the same in this case. It is also known that benzoic acid can act as a proton acceptor [52] through the carbonyl group.

9. ACTIVATION OF A CATALYST BY A CHANGE OF SOLVENT

There has been little study of the change in the activity of a catalyst with a change in the solvent. It is understood that there is no reaction between them. For this reason, it is impossible to compare not only the behavior of an alcoholate in alcohol and in water as it reacts vigorously with the latter, but also the properties of a caustic alkali in aqueous and alcohol solutions as it is partly converted into an alcoholate in alcohol [53].

In Table 13 we compare the rate constants for the exchange of hydrogen in the methyne group of triphenylmethane with several protophilic solvents and basic catalysts [31].

It is assumed that the base conjugate with the ethylenediamine molecule is not formed when potassium ethylate or aminoethylate is dissolved in ethylenediamine. In other words, the following equilibria are not displaced appreciably to the right:

$$C_2H_5O^- + H_2NCH_2CH_2NH_2 \rightleftharpoons C_2H_5OH + H_2NCH_2CH_2NH^-$$

$$NH_2CH_2CH_2O^- + H_2NCH_2CH_2NH_2 \rightleftharpoons NH_2CH_2CH_2OH + H_2NCH_2CH_2NH^-$$

The data show that the rate constant (in sec^{-1}) of the exchange reaction with ethylenediamine is a factor of approximately five greater than that with ethanolamine, but when

TABLE 13. Rate Constants of Hydrogen Exchange in Triphenylmethane

Solvent	Catalyst	Temp., °C	k, sec^{-1}
Ethanolamine	–	120	$4 \cdot 10^{-7}$
Ethylenediamine	–	120	$2 \cdot 10^{-6}$
Ethanolamine	$NH_2C_2H_4OK \sim 0.01$ N	120	$3 \cdot 10^{-6}$
Ethanolamine	$NH_2C_2H_4OK \sim 0.1$ N	120	$2 \cdot 10^{-5}$
Ethylenediamine	$NH_2C_2H_4OK < 0.1$ N	25	$> 10^{-3}$
Ethanol	C_2H_5OK 1 N	120	$3 \cdot 10^{-5}$
Ethylenediamine	$C_2H_5OK < 0.1$ N	25	$> 10^{-3}$

the same catalyst ($NH_2C_2H_4OK$) is dissolved in them, the constants differ by not less than 4-5 orders with allowance for the difference in the temperature at which the experiments were carried out. A similar picture is observed on comparing the rates of the exchange reactions in ethanol and ethylenediamine catalyzed by alcoholate.

Potassium hydroxide in liquid ammonia is able to induce rapid isotopic exchange of deuterium in the methyl group of quinaldine [8].

It was also established later [53a] that the exchange of hydrogen for tritium in the CH_2 groups of indene and fluorene proceeds much more rapidly when catalyzed by OH^- ions in pyridine, containing tritiated water, than in tritiated ethanol in the presence of an alcoholate. Thus, for example, while the exchange reaction is 80% complete after 15 min in a pyridine solution of fluorene, there is only 7% exchange in an alcohol solution even after a day. This sharp acceleration of exchange reactions catalyzed by alcoholates and caustic alkalis in ethylenediamine, liquid ammonia, and pyridine is caused by vigorous solvation of the base cation by solvent molecules, as a result of which the bond between the cation and the anion becomes weaker than in alcohol solution and the electron-donor activity of the anion is increased [53b, c]. The proton-acceptor capacity of the base is also reduced in alcohol as a result of participation in the formation of a hydrogen bond.

10. CATALYTIC ACTIVITY OF COMPLEXES OF HYDROGEN ACIDS WITH ACIDLIKE SUBSTANCES IN HYDROGEN EXCHANGE

At the end of Section I it was concluded that only hydrogen acids and bases participate in heterolytic hydrogen exchange. At the same time, the facts presented on p. 26 indicate the exceptionally high catalytic activity in deuterium exchange of solutions of aluminum bromide in liquid deuterium bromide and solutions of boron trifluoride in liquid deuterium fluoride, i.e., aprotonic acidlike substances. However, there is no contradiction here. In ternary systems consisting of a hydrocarbon, a halide, and a hydrogen halide, complexes are formed because a coordinationally unsaturated halide has an affinity for a halogen ion. Thereupon, the bond between the halogen and deuterium (or hydrogen) atoms in the hydrogen halide molecule is polarized and the deuterium (hydrogen) atom is ionized more readily than in the original hydrogen halide molecule. A lower energy is required for rupture of the H—Hal bond and the exchange reaction is accelerated. Consequently, the aprotonic acidlike substance participates in the exchange not as itself, but in a complex with a hydrogen acid, increasing the strength of the latter. It was mentioned above (p. 25) that a hydrocarbon which is a sufficiently strong base is ionized by a similar reaction:

$$C_6H_6 + DF + BF_3 \rightleftarrows C_6H_6D^+ + BF_4^- .$$

However, polarized complexes are obtained most frequently, especially in a solvent with a low dielectric constant (liquid DBr).

$$C_6H_6 + DBr + AlBr_3 \rightleftarrows C_6H_6 . \overset{\delta+}{D}\text{---}\overset{\delta-}{Br}.AlBr_3.$$

In Section III (p. 157) there is a detailed examination of the properties of the above ternary systems, which are in agreement with the explanation given.

The high catalytic activity of aluminum chloride in deuterium exchange between DCl (gaseous) and benzene was reported first by Polanyi [54] in England and by Klit and Langseth [55] in Denmark. The first attempt at a quantitative comparison of the catalytic activities of halides in deuterium exchange between benzene and liquid deuterium bromide was made in the Isotopic Reactions Laboratory [25]. A special procedure for manipulating small samples of extremely hygroscopic substances without contact with atmospheric moisture [56] was used. It was found that aluminum bromide is the most active of the bromides (the activity of $TiBr_4$ was taken as unity).

$AlBr_3$	$GaBr_3$	$FeBr_3$	BBr_3	$SbBr_3$	$TiBr_4$
$5 \cdot 10^5$	$4 \cdot 10^5$	$1 \cdot 10^4$	30	6	1

In other work in the Isotopic Reactions Laboratory [57], catalysis by iodine of the exchange between isomers of monodueterotoluene and monodeuterobiphenyl and liquid hydrogen iodide was studied. In this case also, the catalysis is explained by the fact that in the ternary system aromatic hydrocarbon − hydrogen iodide − iodine, the H−I bond is polarized because the iodine ion tends to form the complex anion I_3^- with molecular iodine:

$$ArD + HI + I_2 \rightleftarrows ArD \cdot \overset{\delta+}{H} - \overset{\delta-}{I} \cdot I_2.$$

The function of molecular iodine is the same in the catalysis of deuterium exchange and bromination of an aromatic hydrocarbon. By polarizing the hydrogen halide molecule in the complex, iodine facilitates the transfer of a proton to the hydrocarbon and in bromination it promotes the conversion of Br_2 into Br^+.* In the same way, as a proton, the positive bromine ion attacks the electron-donor molecule of the aromatic compound (p. 267).

A similar explanation is given by Gold and Satchell [58, 59] for the catalysis of hydrogen exchange between p-deuteroanisole and a solution of stannic chloride or zinc chloride in glacial acetic acid. Complex acids are formed:

$$SnCl_4 + 2CH_3CO_2H \rightleftarrows H_2SnCl_4 (O_2CCH_3)_2$$
$$ZnCl_2 + 2CH_3CO_2H \rightleftarrows H_2ZnCl_2 (O_2CCH_3)_2.$$

By measurements with indicators it was shown that $H_2SnCl_4(O_2CCH_3)_2$ is similar in strength to sulfuric acid.

Thus, acidlike aprotonic substances increase the rate of hydrogen exchange by reacting with hydrogen acids to form complexes that are stronger acids than the original hydrogen acids.

11. BRØNSTED'S RELATION IN HYDROGEN EXCHANGE

A. Rate and Equilibrium of Protolyte Ionization

Let us first consider whether there is a connection between the rate of a reaction catalyzed by an acid or a base and the strength of the protolyte, expressed, for example, as the ionization constant, and whether there is a relation between the equilibrium constant of a protolytic reaction and the rate at which this equilibrium is established.

A relationship between the catalytic activity of a protolyte and its strength was established by Brønsted in 1924 in the form of the well-known equations $k_A = G_A K_A^\alpha$ for acid catalysis and $k_B = G_B K_B^\beta$ for basic catalysis, where k_A and k_B are the rate constants, K_A and K_B are the ionization constants of the acid and the base, and G_A, G_B, α, and β are constants characterizing the medium and temperature as well as the reaction. They also depend on the sign and magnitude of the charge on the catalyst. Brønsted's relation has been confirmed by extensive verification of many catalytic reactions in aqueous and nonaqueous solutions (see [60, p. 82] and [15, p. 181].

---

* Ionization does not actually occur, but the bromine molecule is polarized in the complex: $ArH + Br_2 + I_2 \rightleftharpoons ArH \cdot \overset{\delta+}{Br} - \overset{\delta-}{Br} \cdot I_2.$

As the substrate is also a protolyte, a similar relation describes the effect of a change in the strength of the substrate as a base or an acid with a given catalyst[61, 62].

The reasons for deviations from Brønsted's relation were discussed in Bell's article [62]. Strong changes in the molecular structure of the catalyst are usually accompanied by deviations. For example, in the acid catalysis of the dehydration of acetaldehyde hydrate by 45 different carboxylic acids and phenols, the dissociation constants of

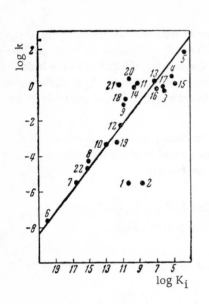

Fig. 3. Relation between the rate and equilibrium of acid ionization. k) ionization rate constant; K) ionization constant 1) CH_3NO_2; 2) $C_2H_5NO_2$; 3) $C_2H_5O_2CCH_2NO_2$; 4) $CH_3COCH_2NO_2$; 5) $CH_2(NO_2)_2$; 6) CH_2COCH_3; 7) CH_3COCH_2Cl; 8) $CH_3COOHCl_2$; 9) $CH_3COCH_2COC_2H_5$; 10) $CH_3COCH(C_2H_5)COOC_2H_5$; 11) $CH_3COCH_2COCH_3$; 12) $CH_3COCH(OH_3)COCH_3$; 13) $CH_3COCHBrCOCH_3$; 14) $CH_3COCH_2COC_6H_5$; 15) $CH_3COCH_2COCF_3$; 16) $C_6H_5COCH_2COCF_3$; 17) $COCH_2COCF_3$; 18) $COOC_2H_5$; 19) $COOC_2H_5$; 20) $CH_3COCH_2COOCH_3$; 21) $CH_2(CN)_2$; 22) $CH(C_2H_5)_5(COOC_2H_5)_2$.

which varied over ten orders, the deviations did not exceed 0.1 of a logarithmic unit on an average, but reached one or two logarithmic units when nitroparaffins or oximes were used as the catalyst. There is nothing remarkable in this in view of the fact that the ionization of protolytes depends on a large extent on specific characteristics of their structure, a phenomenon which will be discussed in detail in Section IV (p. 203). It is certain that deviations from Brønsted's relation may also be produced by steric hindrance which is observed in a reaction with some catalyst.

In his book "The Proton in Chemistry" [62a], Bell presents clear new examples of deviations from Brønsted's relation. Thus, although the acids examined (nitromethane, benzoylacetone, the trimethylammonium ion, and boric acid) are approximately the same in strength (pK_i from 10.2 to 9.1), the rates of proton transfer involving them differ by up to eight orders.

In 1935 Hammett [63] showed that Brønsted's equation is only a particular case of a more general relation between rate constants and equilibrium constants. This problem was considered by Leffler [64, p. 187]. He pointed out that a linear relation between the logarithms of the rate and equilibrium constants is equivalent to a linear relation between the free energy of a reaction and the free energy of its activation.

Many measurements have been made which make it possible to follow the relation between the ionization constants of acids and the ionization rate constants of acids.

Figure 3 shows that most points lie well on a straight line, i.e., there is a linear relation between the logarithms of the ionization rate constants k and the logarithms of the ionization constants K_i (the reasons for some deviations are discussed in Leffler's book [64, p. 188]).

TABLE 14. Ionization Constants $K_i^{H_2O}$ and Ionization Rate Constants k_i of Acids in Water at 25°

Acid	$K_i^{H_2O}$	k_i, min^{-1}
$CH_2(NO_2)_2$	$2.7 \cdot 10^{-4}$	50
$CH_3COCHBrCOCH_3$	$1 \cdot 10^{-7}$	1.4
$CH_3COCH(CH_3)COCH_3$	$1 \cdot 10^{-11}$	$5 \cdot 10^{-3}$
$CH_3COCH(C_2H_5)CO_2C_2H_5$	$2 \cdot 10^{-13}$	$4.5 \cdot 10^{-4}$
$CH_3COCHCl_2$	10^{-15}	$4.4 \cdot 10^{-5}$
$C_2H_5CH(CO_2C_2H_5)_2$	10^{-15}	$2 \cdot 10^{-5}$
CH_3COCH_2Cl	$3 \cdot 10^{-17}$	$2.3 \cdot 10^{-6}$
CH_3COCH_3	10^{-20}	$2.8 \cdot 10^{-8}$

Table 14 gives the rate and equilibrium constants of ionization of a series of weak acids, taken from Pearson's work [65, 66].

Table 15 gives the activation energies (E_a) and entropies ($\Delta S^{\neq}$) of ionization in aqueous solution of CH bonds in some compounds according to Pearson and Dillon's estimates [65].

TABLE 15. Activation Energy and Entropy of Ionization

Substance	E_a, kcal/mole	$\Delta S^{\neq}$, entropy units
Acetoacetic ester	14.2	− 26.2
α-Methylacetylacetone	18.0	− 16.0
Nitromethane	22.6	− 20.2
Nitroethane	22.9	− 18.4
Ethyl nitroacetate	16.0	− 15.3
Malononitrile	18.1	− 7.1

The rate of ionization of O−H and N−H bonds also depends, regularly on the strength of the protolyte, a measure of which may be, for example, the affinity for a proton (kcal/mole):

H_2O	NH_3	OH^-	NH_2^-
169	214	383	419

With an increase in the affinity of the reagent for a proton, there is an increase in the rate of the reactions listed in Table 16. On the other hand, the ammonium ion is a weaker acid than the hydroxonium ion and the rate of reactions No. 6 and 7 is less than that of reactions No. 8 and 9. Naturally this comparison is too simplified if only for the reason that it makes no allowance for possible differences in the mechanism of proton transfer from the molecules and ions mentioned.

B. Brønsted's Relation in Hydrogen Exchange

It appears that the rate of hydrogen exchange generally characterizes the strength of the protolytes involved. The rate of hydrogen exchange is a function of the rate of

TABLE 16. Ionization Rates

No.	Reaction	k, sec^{-1}	Literature reference[*]
1	$H_2O + H_2O \rightarrow H_3O^+ + OH^-$	$2 \cdot 10^{-5}$	1
2	$H_2O + NH_3 \rightarrow NH_4^+ + OH^-$	$5 \cdot 10^5$	1
3	$H_2O + OH^- \rightarrow OH^- + H_2O$	$1.3 \cdot 10^{10}$	1
4	$NH_3 + NH_3 \rightarrow NH_4^+ + NH_2^-$	Slowly	2
5	$NH_3 + NH_2^- \rightarrow NH_2^- + NH_3$	$4.8 \cdot 10^8$	2
6	$NH_4^+ + H_2O \rightarrow NH_3 + H_3O^+$	$2 \cdot 10^{-2}$	3
7	$NH_4^+ + OH^- \rightarrow NH_3 + H_2O$	$3 \cdot 10^{10}$	4
8	$H_3O^+ + H_2O \rightarrow H_2O + H_3O^+$	$5.7 \cdot 10^{10}$	5
9	$H_3O^+ + OH^- \rightarrow H_2O + H_2O$	$1.3 \cdot 10^{11}$	6

[*] 1. M. Eigen, Discuss. Farad. Soc. 17, 194 (1954).
2. R. A. Ogg, Discuss. Farad. Soc. 17, 215 (1954).
3. E. Grunwald, A. Löwenstein, and S. Meiboom, J. Chem. Phys. 27, 630, 1007 (1957).
4. M. Eigen and J. Schoen, Z. phys. Chem. (Neue Folge) 3, 126 (1955); Z. Elektrochem. 59, 483 (1955).
5. S. Meiboom, Z. Luz, and D. Gill, J. Chem. Phys. 27, 411 (1957).
6. M. Eigen and L. De Mayer, Z. Elektrochem. 59, 986 (1955).

rupture of the bond with a hydrogen atom, while the latter is symbatic with the ionization constant, which is normally a measure of acid strength. The ionization constants of CH bonds in weak carbo acids[*] in aqueous solutions given in Pearson's article [65] were calculated from the rate constants of isotopic exchange of the hydrogen in them with a normal alcohol solution in heavy water, taken from the work of Bonhöffer et al. [68, 69]. In Table 17, K_i is the ionization constant of the acid in water, k_i the rate constant of ionization (min^{-1}) in aqueous solution, and k_{H-D} the rate constant of hydrogen exchange.

Brønsted's relation holds in the exchange of tritium in 1,3,5-trimethoxybenzene with acids [69a] and in the acid catalysis of hydrogen exchange in azulene [69b].

[*] The term "carbo acids" is applied to substances whose acidic properties depend on the protonization of the hydrogen of a CH bond.

There is also the possibility of some more indirect comparisons. For example, there is a parallelism between the rate constants, k_{H-D} (in sec^{-1}), of hydrogen isotope exchange in aliphatic CH bonds of hydrocarbons with liquid deuteroammonia [8, 70] (recalculated from new data for a temperature of 120°) and the dissociation constants conditionally related to water from the determinations of Conant and Wheland [71] (K_1) and McEwen [72] (K_2) (p. 93) and this is shown in Table 18.

TABLE 17. Comparison of the Rate Constants of Ionization and Deuterium Exchange with the Ionization Constants of Carbo Acids

Substance	K_i	k_i	k_{H-D}
CH_3NO_2	$6 \cdot 10^{-11}$	$2.6 \cdot 10^{-6}$	$2 \cdot 10^3$
CH_3COCH_3	10^{-20}	$2.8 \cdot 10^{-8}$	$1.5 \cdot 10^1$
$(CH_3)_2SO_2$	10^{-22}	$2 \cdot 10^{-10}$	$6 \cdot 10^{-1}$
CH_3CN	10^{-25}	$4 \cdot 10^{-12}$	$2 - 3 \cdot 10^{-3}$

The graph taken from the work of Mackor [73] (Fig. 4) is very convincing. The logarithms of the ionization constants of polycyclic hydrocarbons in liquid hydrogen fluoride are plotted along the ordinate axis and the logarithms of the rate constants of deuterium exchange in CH bonds of deuterated hydrocarbons with a solution of H_2SO_4 in CF_3CO_2H are plotted along the abscissa axis [74] (see also p. 147).

TABLE 18. Comparison of Dissociation Constants of Carbo Acids and Deuterium Exchange Rate Constants

Hydrocarbon	k_{H-D}	K_1	K_2
Indene	$4 \cdot 10^1$	10^{-22}	10^{-21}
Fluorene	$2 \cdot 10^{-2}$	10^{-24}	10^{-25}
Triphenylmethane	$2 \cdot 10^{-7}$	10^{-29}	10^{-33}
Diphenylmethane	$7 \cdot 10^{-9}$	10^{-30}	10^{-35}

The relation between the rate of deuterium exchange and the strength of acids and bases participating in it may be estimated also from the kinetics of deuterium exchange between trifluoroacetic acid and alkylbenzenes.

Table 19 gives the logarithms of the basicity constants ($\log K_B$) of alkylbenzenes in liquid HF:

$$K_B = \frac{m_{ArH_2^+} \, m_{F^-}}{m_{ArH}} \frac{f_\pm^2}{f_{ArH}},$$

where m represents the molarities of the carbonium ion ArH_2^+, the F^- ion, and the hydrocarbon molecule; $f_\pm$ is the means activity coefficient of these ions for an infinitely dilute solution in HF [75]; K_i^{HF} is the ionization constant of the hydrocarbon in HF [76];

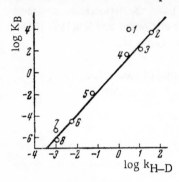

Fig. 4. Relation between "reduced" basicity constants of hydrocarbons and rate constants of deuterium exchange. K_B is the "reduced" basicity constant and k_{H-D} is the rate constant of deuterium exchange. 1) Perylene (3); 2) anthracene (9); 3) 1,2-benzanthracene (10); 4) pyrene (3); 5) chrysene (2); 6) naphthalene (1); 7) triphenylene (1); 8) biphenyl (2). The position at which deuterium exchange occurs is given in brackets.

k_{rel} is the relative rate constant of exchange of the p-deuterium atom in the hydrocarbon molecule in CF_3CO_2H with sulfuric acid added ($k_{C_6H_6} = 1$) [77].

Hine et al. [78] used the Brønsted relation and the rate constant of deuterium exchange between chloroform and water, catalyzed by methylamine, to calculate the

TABLE 19. Comparison of Strength of Alkylbenzenes as Bases and Deuterium Exchange Rate

Hydrocarbon	$\log K_B$	K_i^{HF}	k_{rel}
C_6H_5D	—9.4	$1 \cdot 10^{-8}$	1
$C_6H_4DCH_3$	—6.3	$2 \cdot 10^{-7}$	$3 \cdot 10^2$
1,3- $C_6H_3D\,(CH_3)_2$	—3.2	$8 \cdot 10^{-6}$	$2.5 \cdot 10^5$
1,3,5 - $C_6H_2D(CH_3)_3$	—0.4	$4 \cdot 10^{-3}$	$3 \cdot 10^7$

ionization constant of the latter ($K_i^{H_2O} = 7.27 \cdot 10^{-4}$), which was in excellent agreement with the literature data ($K_i^{H_2O} = 7.24 \cdot 10^{-4}$). The results obtained by Bigeleisen [79] in a study of isotopic exchange of hydrogen in phosphine with heavy water catalyzed by molecular acids are in agreement with Brønsted's equation (see also [117a]).

Accurate conclusions on the strength of acids and bases may be drawn from the rate constants of hydrogen exchange only if allowance is made for the other factors (apart from the protolytic properties of the reagents) on which the hydrogen exchange rate depends (sometimes very strongly) [6, 80]. Such factors include, for example, the effect of the charge of the substrate, the dual reactivity of the latter caused by the presence of a heteroatom, steric hindrance, etc., which were discussed above.

12. HAMMETT'S RELATION IN HYDROGEN EXCHANGE

A. Hammett's Acidity Function

a. ACIDITY FUNCTION H_0

Among the methods of comparing the acidity of solution proposed by Hantzsch (see [15]), the indicator method has become commonest and it was on the basis of this that Hammett [81, 82] formulated the concept of the acidity function, which may be used to express the acidity of concentrated solutions of strong acids quantitatively (see [15, ch. 12] and [83]).

When an electrically neutral base B is dissolved in an acid HA, the following equilibrium is established:

$$B + HA \rightleftarrows BH^+ + A^-. \tag{1}$$

The equilibrium state depends on the strength of the base and the acidity of the solution. The degree of ionization of the base is determined readily if ionization is accompanied by a change in color, i.e., if the base is an indicator. As a measure of the strength of the base it is possible to use the thermodynamic constant of the equilibrium:

$$BH^+ \rightleftarrows B + H^+ \tag{2}$$

$$K_{BH^+} = \frac{a_{H^+} a_B}{a_{BH^+}} = \frac{a_{H^+} f_B C_B}{f_{BH^+} C_{BH^+}}, \tag{3}$$

where a_{H^+} is the activity of the proton; f_B and C_B are the activity coefficients and concentration of the base B; f_{BH^+} and C_{BH^+} are the corresponding values for the conjugate acid BH^+.

The standard state is an infinitely dilute solution in water.

The range of solution acidities over which the color of any indicator changes is comparatively small and therefore it is normal to use a series of indicators, for which the values of K_{BH^+} are determined successively, beginning with the indicator that is a sufficiently strong base and ionizes in a dilute aqueous solution of the acid. It is postulated that the ratio of the activity coefficients of the ionized and un-ionized forms is constant for any electrically neutral basic indicators:

$$\frac{f_B}{f_{BH^+}} = \frac{f_{B_1}}{f_{B_1 H^+}} = \frac{f_{B_2}}{f_{B_2 H^+}}. \tag{4}$$

This condition is fulfilled quite well in solvents with a high dielectric constant (see [15, Fig. 8] and [83]).

The values of Hammett's acidity function H_0 are determined by the strength of a base capable of ionizing in the solution of the acid examined and the degree of its ionization. It equals the ratio of the concentration of the ionized and unionized forms of the indicator C_{BH^+}/C_B.

Let us adopt the form pK = $-\log$ K. From Equation (3)

$$pK_{BH^+} = -\log a_{H^+} + \frac{f_B}{f_{BH^+}} - \log \frac{C_B}{C_{BH^+}} \tag{3a}$$

or

$$-\log a_{H^+} \frac{f_B}{f_{BH^+}} = pK_{BH^+} - \log \frac{C_{BH^+}}{C_B} = H_0. \tag{5}$$

Hammett called the value H_0 the acidity function. In dilute aqueous solution, $f_B/f_{BH^+} = 1$ and the H_0 scale coincides with the pH scale:

$$H_0 = -\log a_{H^+} = pH. \tag{6}$$

The H_0 scale is a continuation of the pH scale into the region of concentrated solutions of acids. The value h_0 is often used and this is related to H_0 by the equation:

$$H_0 = -\log h_0; \quad h_0 = K_{HB^+} \frac{C_{HB^+}}{C_B} = a_{H^+} \frac{f_B}{f_{BH^+}}. \tag{7}$$

The acidities of aqueous solutions of mineral acids [83] at concentrations, C, from 0.1 to 10 mole/liter are compared in Table 20 (Fig. 5).

TABLE 20. Comparison of Acidities (h_0) of Aqueous Solutions of Acids

C	$HClO_4$	HCl	HNO_3	H_3PO_4
0.1	0.11	0.11	0.11	0.045
1.0	1.6	1.6	1.6	0.23
3.0	17.0	11.2	10.5	1.2
6.0	690	132	62	11
10.0	62000	4800	—	390

b. OTHER FORMS OF ACIDITY FUNCTION

The subscript "zero" denotes that the change of the base equals zero (electrically neutral molecule).

If the base is an anion, i.e.,

$$BH \rightleftarrows B^- + H^+ \tag{8}$$

or a cation

$$BH^{++} \rightleftarrows B^+ + H^+, \tag{9}$$

then it is possible to represent other acidity functions analogously to the acidity function H_0:

$$H_- = pK_{BH} - \log \frac{C_{BH}}{C_{B^-}} = -\log \frac{a_{H^+} f_{B^-}}{f_{BH}} \tag{10}$$

and

$$H_+ = pK_{BH^{++}} - \log \frac{C_{BH^{++}}}{C_{B^+}} = -\log \frac{a_{H^+} f_{BH^+}}{f_{BH^{++}}}. \tag{11}$$

Examples of indicators with different types of charge (B, B⁻, and B⁺) used for determining the functions H_0, H_-, and H_+ are given below. **B**: p-aminoazobenzene (2.77),* p-nitroaniline (0.99), 2,4-dinitroaniline (−4.55), 2,4,6-trinitroaniline (−9.41), nitrobenzene (−11.38), and 2,4-dinitrotoluene (2.78). **B⁻**: p-nitrophenolate, 2,4-dinitrophenolate, and picrate. **B⁺**: m-nitroanilinium ion.

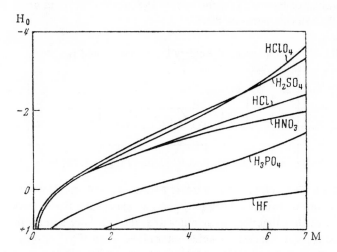

Fig. 5. Acidity functions H_0 of aqueous solutions of acids. M is the molarity of the solution.

In contrast to Hammett's acidity function H_0, a new acidity function was introduced by Westheimer and Kharasch [84] and this was determined by the concentration of the carbonium ion formed in the reaction between an arylcarbinol and the acid, for example:

$$(C_6H_5)_3\,COH + H_2SO_4 \rightleftarrows (C_6H_5)_3C^+ + H_2O + HSO_4^- \cdot$$

This reaction is represented by the scheme:

$$ROH + H^+ \rightleftarrows R^+ + H_2O, \tag{12}$$

while the following scheme is adopted for a normal acid−base equilibrium:

$$ROH + H^+ \rightleftarrows ROH_2^+ \cdot \tag{13}$$

Different authors (Gold, Brand, and Deno) have used different symbols for the new acidity function, namely, H_R, C_0, and J_0. Paul and Long [83] have recommended the use of the symbol J_0. The acidity function J_0 is determined by the equation:

$$J_0 = - pK_{ROH} - \log \frac{C_{R^+}}{C_{ROH}} = - \log \frac{a_{H^+} f_{ROH}}{a_{H_2O} f_{R^+}} \cdot \tag{14}$$

---

* The values of pK_{BH^+} are given in brackets.

where K_{ROH} is the equilibrium constant of reaction (12).

The relation between J_0 and H_0 is given by the equation:

$$J_0 = H_0 + \log a_{H_2O} - \log \frac{f_{BH^+} f_{ROH}}{f_B f_{R^+}}. \tag{15}$$

The values of H_0 and J_0 for the system $H_2SO_4 - H_2O$ [85] are given in Table 21 to illustrate the relation between them.

TABLE 21. Comparison of Acidity Functions H_0 and J_0

H_2SO_4, %	10	20	30	40	50	60	70	80	90	95	98	100
$-H_0$	0.31	1.01	1.72	2.41	3.38	4.46	5.65	6.97	8.27	8.86	9.36	11.10
$-J_0$	0.72	1.92	3.22	4.80	6.60	8.92	11.52	14.12	16.72	18.08	19.64	—

c. VALUES OF ACIDITY FUNCTION H_0 OF ANHYDROUS ACIDS

The review of Paul and Long [83] gives an exhaustive list of work on the determination of acidity functions and summarizes the values of acidity functions of aqueous and nonaqueous solutions of acids determined up to 1957. Table 22 gives the values of H_0 for some anhydrous acids, which are of interest in connection with the use of these acids in the study of hydrogen exchange.

The acidity functions of solutions of acids and bases in liquid hydrogen fluoride and in anhydrous trifluoroacetic acid have also been determined. The following data were obtained for a solution of NaF in liquid HF [87-89]:

NaF, %	NaF, mole/1000 g of HF	$-H_0$
0.45	0.103	9.62
4.32	1.0	8.6

Table 23 gives the results of determinations of the acidity function of solutions of H_2SO_4 in CF_3COOH [77]. Analogous data are given in two more papers [87, 88a]:

H_2SO_4, mole/1000 g of CF_3CO_2H	1.2	1.7	3.7	6.2	9.3
$-H_0$	6.7	7.0	7.7	8.3	9.7

H_2SO_4, M	0.00	0.019	0.075	0.19	0.94	1.87	3.74
$-H_0$	3.33	5.20	5.72	6.01	7.33	8.07	8.86

H_2SO_4, M	5.61	9.35	14.96	16.83	18.70
$-H_0$	9.25	9.33	9.61	9.85	11.10

The following acidity functions were obtained for solutions of HF in CF_3CO_2H [88a]:

HF, %	0	0.9	12.5	22.5	35.1	43.7
$-H_0$	3.03	6.04	6.63	8.40	8.42	8.43

HF, %	49.8	64.1	74.1	76.5	83.8	100.0
$-H_0$	8.39	8.71	8.40	8.73	8.94	9.97

Solutions of an acid at a given concentration in different solvents may differ strongly in the value of H_0. For example, for a molar solution of sulfuric acid in

TABLE 22. Acidity Functions H_0 of Anhydrous Acids

Acid	$-H_0$	Literature reference
HSO_3Cl	12.8	[86]
H_2SO	11.1; 11.0; 10.6	[83, 87, 89]
HF *	10.2	[87, 88, 89, 88a]
H_3PO_4	5.2	[90]
CF_3COOH	3.1; 4.4; 5; 3.03	[91, 87, 88a]
C_3F_7COOH	3.3	[91]
HCOOH	2.2; 1.9; 2.19	[92,87, 92a]
$CHCl_2COOH$	0.7; 1.6 **	[93]

*Hexamethylbenzene and mesitylene were used as indicators. H_0 depends very strongly on the presence of H_2O:

C_{H_2O} (mole/liter)	$-H_0$
0.001	10.2
0.01	9.9
0.12	9.5

** These differences were caused by the use of two different indicators, namely, 4-chloro-2-nitroaniline and 4-nitrodiphenylamine.

TABLE 23. Acidity Functions H_0 of Solutions of H_2SO_4 in CF_3COOH (C in moles of H_2SO_4 per 1000 g of CF_3COOH)

C	$-H_0$	C	$-H_0$
0.01	4.65	0.5	7.10
0.05	5.55	1.0	7.80
0.10	5.90	2.0	8.50
0.20	6.30	3.0	8.87
0.30	6.65	4.0	9.16
0.40	6.90	—	—

water [83] $H_0 = - 0.26$, in glacial acetic acid [83] $H_0 = -3.01$, and in trifluoroacetic acid [87] $H_0 = -6$. This is not surprising as the following onium ions are formed:

$$H_3O^+; \quad CH_3COOH_2^+ ; \quad CF_3COOH_2^+ .$$

and their acidity increases in the given sequence.

d. A CIDITY FUNCTION IN SOLVENTS WITH A LOW DIELECTRIC
CONSTANT

With a change from solvents with a high dielectric constant to solvents with a low
dielectric constant, the acidity function H_0 loses its universal character as the ratio
of the activity coefficients of the ionized and unionized forms of the indicator f_I / f_{IH^+}
does not remain constant and does not become unity at high dilutions (see [94] and the
literature there). Consequently, Hammett's basic postulate does not hold. In addition,
measurement of the acidity is complicated by strong association of ions.*

The literature contains a detailed discussion of the behavior of indicators in solvents with a low dielectric constant, namely, acetic acid (DC = 6.1) and dichloroacetic acid (DC = 8.2). Kolthoff [95] emphasized the inapplicability of Hammett's method to this case and pointed out that in such solvents the relative strength of an acid
varies, depending on the base, as is normal for acids and bases in the wider sense (p. 9).
By spectrophotometric methods it has been demonstrated that an indicator dissolved in
glacial acetic acid participates in ionization and dissociation equilibria. The ionization of a basic indicator is increased in the presence of small amounts of water, though
water is also a base in glacial acetic acid. This anomaly is explained by the presence
of aggregates of three and four ions in the solution. Water influences the degree of
ionic association. With an increase in the water concentration in the solution, the reduction in acidity as a result of dilution of the acid begins to predominate. Thus, if
the color of the indicator is the criterion, with the gradual addition of water there appears to be an acidity maximum, contrary to the normal idea of a monotonic change
in the acidity (see [95a] for more details).

As was shown in the thorough work of Rocek [96], in contrast to solutions of mineral acids in water or other polar solvents, where the differences in the values of H_0
determined with the aid of several indicators rarely amount to 0.1 H_0 unit, when the
same acids are dissolved in acetic acid, the discrepancies reach 0.5 H_0 unit. They are
even greater [93] for solutions in dichloroacetic acid (containing 0.05 mole of water).
Thus, when H_0 is determined with 4-chloro-2-nitroaniline, $H_0 = -0.74$ and with p-ni-
trodiphenylamine, $H_0 = -1.57$, i.e., $\Delta H_0 = 0.84$ (see also p. 53).

Rocek [96] proposed that the acidity of solutions in solvents with low dielectric
constants is expressed by an acidity referred to a definite indicator and denoted it by
$(H_0)_I$.

B. Reaction Rate and Acidity Function

The acidity function has found important application in the description of the
kinetics of reactions catalyzed by concentrated solutions of strong acids [97, 98]. On
several examples, Hammett established [82, 99] (see also [15], p. 100) that there is a
linear relation between the logarithm of the rate constant k and the acidity function
of the solution H_0:

$$\log k + H_0 = \text{const}. \tag{16}$$

---

* The reasons why the method of determining acidity in accordance with Hammett's
concept is approximate are examined in N. A. Izmailov's book "The Electrochemistry
of Solutions" (pp. 784-790). See also Doklady Akad. Nauk SSSR, 127, 104 (1959).

This equation is known as Hammett's relation. Its accuracy was confirmed for many reactions, including hydrogen exchange with acids (p. 59). Long and Paul [97] gave an excellent, exhaustive review of work published up to and including 1957 on the use of the acidity function H_0 in the examination of the kinetics and mechanism of acid catalysis.

Hammett envisaged two mechanisms of acid catalysis, which he arbitrarily denoted as A-1 and A-2. The former corresponds to a mono- and the latter to a bimolecular reaction (S-1 and S-2 according to Ingold).

a. A-1 MECHANISM

The first stage of the reaction consists of the establishment of equilibrium between the electrically neutral substrate molecule S and the conjugate acid SH^+, obtained by the addition of a proton:

$$S + H^+ \rightleftarrows SH^+.$$ (17)

This equilibrium is established rapidly.

Then follows the stage of intramolecular rearrangement:

$$SH^+ \rightarrow X^+,$$ (18)

which also determines the reaction rate. It is completed by the formation of the final product, which is accompanied by the rapid elimination of a proton with the transference of the latter to a molecule Y. For example, in the inversion of sugar, Y = H_2O.

It is assumed that the activated complex X^+ is similar in structure to SH^+ and occupies the same position relative to S as the ionized form of an indicator BH^+ relative to the starting molecule B.

Let us apply the equation proposed by Brønsted to allow for the effect of the medium on the reaction rate (primary salt effect) (see [15, p. 194]) to process (18), on which the reaction rate depends:

$$v = -\frac{dC_S}{dt} = k'C_{SH+}\frac{f_{SH^+}}{f_{X^+}},$$ (19)

where k' is the rate constant of the slow stage of the reaction in an infinitely dilute solution of the acid; f_{SH^+} and f_{X^+} are the activity coefficients of the protonized substrate and transition state for the given solvent.

By using the expression for h_0 (7) as applied to the base S, let us replace $C_{SH^+} \cdot f_{SH^+}$ by the corresponding value:

$$C_{SH^+}f_{SH^+} = \frac{C_S}{K_{SH^+}}a_{H^+}f_S$$ (20)

and substitute it in Equation (19):

$$v = k'\frac{C_S a_{H^+}f_S}{K_{SH^+}f_{X^+}}.$$ (21)

However, for an electrically neutral indicator B,

$$h_0 = a_{H^+} \frac{f_{BH^+}}{f_B} , \tag{22}$$

whence

$$a_{H^+} = \frac{h_0 f_B}{f_{BH^+}} . \tag{23}$$

By substituting the value of a_{H^+} in Equation (21), we obtain:

$$v = \frac{k'}{K_{SH^+}} \frac{C_S h_0 f_S f_{BH^+}}{f_{X^+} f_B} , \tag{24}$$

where K_{SH^+} is the thermodynamic ionization constant of the acid SH^+, conjugate with the substrate S. For the particular case when S is a very weak base, which is ionized insignificantly in the solution in which the reaction occurs (i.e., under conditions where $C_{SH^+} \ll C_S{}^*$), the concentration of the un-ionized form of the substrate practically equals its total concentration C_S^{total}. By definition, the rate constant of a monomolecular reaction is given by the equation:

$$k = -\frac{1}{C} \frac{dC}{dt} , \tag{25}$$

and by applying this to our case, we obtain:

$$k = -\frac{1}{C_S^{total}} \frac{dC_S^{total}}{dt} = h_0 \frac{k'}{K_{SH^+}} \frac{f_S f_{BH^+}}{f_{X^+} f_B} . \tag{26}$$

By taking logarithms of Equation (26):

$$\log k = \log h_0 + \log \frac{k'}{K_{SH^+}} + \log \frac{f_S f_{BH^+}}{f_{X^+} f_B} \tag{27}$$

$$\log h_0 = -H_0; \quad \log \frac{k'}{K_{SH^+}} = \text{const.}$$

Consequently:

$$\log k = -H_0 + \text{const} + \log \frac{f_S f_{BH^+}}{f_{X^+} f_B} . \tag{28}$$

According to Hammett's postulate, $\dfrac{f_S}{f_{X^+}} = \dfrac{f_B}{f_{BH^+}}$. Therefore, the last term equals zero

and we obtain the Hammett relation:

---

* The result for the case when the concentrations C_S and C_{SH^+} are commensurate was also given in Long and Paul's article [97].

$$\log k + H_0 = \text{const.} \tag{16}$$

It is presumed that if a straight line with a slope of 45° ($\tan 45° = 1$) is obtained on a graph of $\log k$ against H_0, a reaction catalyzed by an acid proceeds most probably by an A-1 mechanism.

Examples of reactions for which Hammett's relation is satisfied are the inversion of sucrose, hydrolysis of methylal, $CH_2(OCH_3)_2$, and acetic anhydride, depolymerization of paraldehyde and trioxane (a polymer of formaldehyde) (see Fig. 6), and hydrogen exchange with acids (p. 59).

Deviations from a slope of 45° are regarded as the result of differences in the structure of the substrate and indicator with the consequence that the electrolytes have different effects on the ratio of the activity coefficients of the two forms of the substrate and indicator (see [97, p. 967] and for more details [101]).

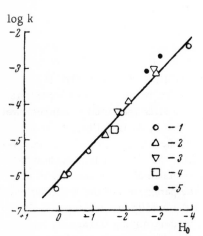

Fig. 6. Check on the Hammett relation on the depolymerization of trioxane in aqueous solutions of acids at 40°. The black points correspond to experiments with solutions of sulfuric acid in glacial acetic acid: 1) H_2SO_4; 2) HCl; 3) $HClO_4$; 4) H_3PO_4; 5) H_2SO_4 in CH_3COOH.

b. A-2 MECHANISM

The first stage of an A-2 catalytic reaction is the same as that in an A-1 reaction, i.e., the rapid establishment of the equilibrium of substrate protonization:

$$S + H^+ \rightleftarrows SH^+. \tag{17}$$

In contrast to an A-1 reaction, where the reaction rate depends wholly on the rate of the intramolecular rearrangement, in this case it is determined by the rate of the bimolecular reaction:

$$SH^+ + Y \rightarrow \text{product} + H^+. \tag{29}$$

The transition state corresponds to a form with a structure differing from SH^+, as a result of which, the Hammett relation does not hold. As the derivation of the kinetic equation shows (see [97, p. 940]), the rate constant is proportional not to the acidity function but to the acid concentration.

An example of a reaction proceeding by an A-2 mechanism is the enolization of acetophenone in an aqueous solution of acid [100].

Depending on whether the rate constant is proportional to the acidity function of the solution or the acid concentration, it is possible to put forward a hypothesis on the probable reaction mechanism (A-1 or A-2).

While recognizing the value of these criteria in the determination of the mechanism of acid-catalyzed reactions, Long [101, 102] still does not consider them unequivocal. It is extremely probable that there are other mechanisms apart from A-1 with which there is proportionality between the reaction rate constant and the acidity function. The same consideration also applies to reactions whose rate constant is proportional to the acid concentration. There is also the possibility of mechanisms other than A-2 for them. In addition there may be transitional forms of mechanism lying between A-1 and A-2. More and more authors are now abandoning the Hammett-Zucker hypothesis [102a-f] (see also Melander's views, which are given on p. 63).

It should also be noted that some reactions involving concentrated acids are described by the acidity function J_0 better than by H_0. These are reactions satisfying scheme (12):

$$ROH + H^+ \rightleftarrows R^+ + H_2O$$

instead of scheme (13)

$$ROH + H^+ \rightleftarrows ROH_2^+ .$$

Nitration is an example of this.

c. TEMPERATURE DEPENDENCE OF ACIDITY FUNCTION H_0

Almost all measurements of the acidity function are made at room temperature. However, the kinetics of acid-catalyzed reactions are often measured at temperatures very different from room temperature. As was reported by A. I. Gel'bshtein and M. I. Temkin [90, 103], with a rise in temperature of 20°, the acidity h_0 of sulfuric and phosphoric acids decreases by a factor of almost two, i.e., the effect of temperature on acidity is commensurate with the temperature dependence of the probability of activation. Therefore, it is impossible to determine the parameters of the Arrhenius equation accurately if the temperature dependences of the acidity are neglected. These authors [90, 103] filled this gap in literature data by determining the acidity function H_0 of concentrated aqueous solutions of hydrochloric, phosphoric, and sulfuric acids over a wide range of temperatures. For example, the following values were obtained for H_0 for 95% H_3PO_4: -5.05 at 4°, -4.77 at 20°, -4.44 at 40°, -4.08 at 60°, and -3.75 at 80°.

By measuring the kinetics of formic acid decomposition in concentrated sulfuric and phosphoric acids and knowing H_0 for the latter acids as a function of temperature, A. I. Gel'bshtein and M. I. Temkin [104] were able to differentiate between the effects of the acidity of the medium on the pre-exponential factor and on the activation energy of the catalytic reaction.

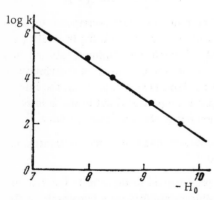

Fig. 7. Rate constant of formic acid decomposition as a function of the acidity of the system $H_2SO_4 - H_2O$.

The graphs in Figs. 7 and 8 show the relation of the logarithm of the rate constant of formic acid decomposition to $-H_0$ at 120° for the systems $H_2SO_4 - H_2O$ and $P_2O_5 - H_2O$. The rate constant k at 20° was obtained by linear extrapolation of experimental data for other temperatures. The slope of the line for decomposition of HCOOH in sulfuric acid equaled 1.6 ($\log k = \text{const} + 1.6\,h_0$). For the second system, linearity was not observed at all: there were two different rate constants for the same value of h_0 (see dashed line).

By analyzing the results obtained and considering the effect of temperature on H_0, the authors came to the conclusion that a variation in the acidity of a definite system affects only the activation energy of the reaction, while the pre-exponent remains constant.

System	Activation energy kcal/mole	log of pre-exponent
$H_2SO_4 - H_2O$ (from 80.7 to 98.2% H_2SO_4)	from 26.6 to 24.5	6.7
$P_2O_5 - H_2O$ (from 72.4 to 83.3% P_2O_5)	from 28.0 to 24.8	9.3

With a change from the system $H_2SO_4 - H_2O$ to $P_2O_5 - H_2O$, the pre-exponent changes very strongly. As a result of this, at the same solution acidities, the rate of HCOOH decomposition in concentrated phosphoric acid solutions is much greater than in sulfuric acid. Consequently, the reaction rate constant not only depends on the acidity of the solution, but the specific nature of the solvent has an effect (Menshutkin effect).

The understanding of the mechanism of catalysis by concentrated acids is deepened by a knowledge of the temperature dependence of the acidity function. It makes it possible to find reasons for apparent deviations from Equation (16) and to determine accurately the order of reaction with respect to acidity [105]. The work of A. I. Gel'bshtein and M. I. Temkin therefore added substantially to our knowledge of the relation between reaction rate and acidity function.

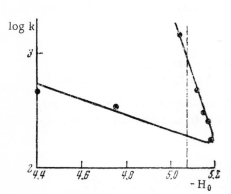

Fig. 8. Rate constant of formic acid decomposition as a function of the acidity of the system $P_2O_5 - H_2O$.

C. Hammett's Relation in Hydrogen Exchange

The dependence of the rate of hydrogen exchange in aromatic hydrocarbons and their derivatives on the acid concentration was first observed qualitatively by Ingold and his co-workers [106]. Gold together with Long and Satchell [107-114], Mackor and his co-workers [74, 77], and Beeck and Stevenson [115] confirmed the applicability of Hammett's relation to catalysis of hydrogen exchange by concentrated acids.

a. DATA ON HYDROGEN EXCHANGE

The first work of Gold on the exchange of deuterium between 9-deuteroanthracene and sulfuric acid (84-90%) was carried out with a heterogeneous system, but the other measurements were made with homogeneous solutions.

In the catalysis of deuterium exchange by strong acids there is normally a very good linear relation between H_0 and $\log k$. However, in addition to the theoretical slope of unity, both smaller and larger slopes have been observed (see pp. 53 and 57 for discussion of a possible reason for these deviations). A summary of results is given in Table 24.

TABLE 24. Slopes of Lines of log k Against H_0

No.	Substance	Acid	Concentration	tan α	Ref.
1	α-D-Naphthalene	H_2SO_4 in CF_3CO_2H + CCl_4	0.1-0.4 mole/1000 g	0.86	[74]
2	o-D-p-Chlorophenol	H_2SO_4 in H_2O	40.5-63.5%	0.90	[108]
3	o-D-p-Nitrophenol	H_2SO_4 in H_2O	64.3-97.6%	0.94	[108]
4	m-D-Anisole	$HClO_4$ in H_2O	10.0-11.7 mole/liter	0.98	[112]
5	o-D-p-Nitrophenol	H_3PO_4 in H_2O	63.6-76.4%	1.00	[110]
6	D-Benzene	H_2SO_4 in CF_3CO_2H	–	1.00	[77]
7	o-D-Toluene	H_2SO_4 in CF_3CO_2H	0.5-4 mole/1000 g	1.00	[77]
8	o-D-p-Cresol	H_2SO_4 in H_2O	9.8-45.3%	1.08	[108]
9	p-D-Anisole	H_2SO_4 in H_2O	3.51-7.59 mole/liter	1.18	[112]
10	o-D-Anisole	$HClO_4$ in H_2O	5.18-7.88%	1.18	[112]
11	D-Benzene	H_2SO_4 in H_2O	66.5-83.2%	1.36	[108]
12	p-D-Toluene	H_2SO_4 in H_2O	53.0-73.4%	1.40	[111]
13	o-D-p-Cresol	HCl in H_2O	7-21.5%	1.42	[110]
14	p-T-Toluene	H_2SO_4 in H_2O	68-75%	1.60	[106a]

The measurements were made with acids whose acidity functions varied from $H_0 = -0.15$ to $H_0 = -9.0$.

Table 24 shows that the slope of lines of log k against H_0 for deuterium exchange between the same substance and different acids may vary quite considerably (cf., for example, No. 8 and No. 13, and No. 6 and No. 11 in Table 24 and Figs. 9 and 10).

As pointed out by Gold and Satchell [110], these discrepancies may be partly ascribed to the inaccuracy of the determination of H_0 of acid solutions. They are

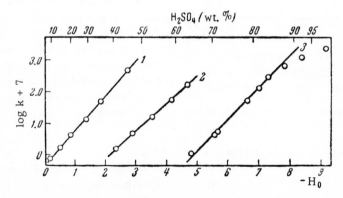

Fig. 9. Relation of deuterium exchange rate to solution acidity. 1) p-Cresol; 2) p-chlorophenol; 3) p-nitrophenol.

mainly explained by different changes in the activity coefficients of the ionized and molecular forms of the substrate and indicator, which differ, moreover, in comparable solutions of mineral acids. Similar deviations in the slopes of the line have also been observed in chemical reactions [101, 116].

The points often lie well on a straight line (H_0 against log k + a) even if the rate constants change by a factor of several thousand as, for example, in deuterium exchange between p-D-toluene and sulfuric acid (Table 25 and Fig. 11).

However, the experimental points sometimes deviate from a straight line. For example, in the deuterium exchange of o-D-p-nitrophenol with concentrated sulfuric

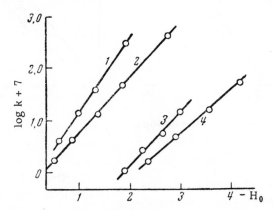

Fig. 10. Relation of deuterium exchange rate to solution acidity. 1) p-Cresol in HCl; 2) p-cresol in H_2SO_4; 3) p-chlorophenol in H_3PO_4; 4) p-chlorophenol in H_2SO_4.

acid this is explained by deactivation of the aromatic ring as a result of the addition of a proton to the hydroxyl group, i.e., a phenomenon analogous to that discussed above, namely, the effect of the dual reactivity of a substance on hydrogen exchange (see p. 36).

A comparison of the rates of deuterium exchange in isomeric deuteroanisoles with solutions of sulfuric acid in water and in glacial acetic acid shows that for solutions with the same value of H_0, the exchange rate is lower in acetic acid (Fig. 12). This is caused by deviations from Hammett's postulate on the constancy of the ratio of the activity coefficients of the ionized and un-ionized forms of an indicator in a solution with a low dielectric constant (p. 54).

Tritium in 1,2,3-trimethoxy-4,6-ditritiobenzene exchanges at different rates with an aqueous solution of potassium bisulfate and with mixtures of trifluoroacetic acid and water with the same value of H_0 [114].

b. INTERPRETATION OF DATA ON HYDROGEN EXCHANGE

Hammett interpreted a linear relation of the reaction rate constant to h_0 as an indication that the slowest stage of the reaction is the intramolecular rearrangement of the protonized substrate (A-1 mechanism, p. 55). Gold and Satchell [108, 109] interpreted the applicability of Hammett's relation to deuterium exchange catalyzed by strong acids in exactly the same way.

On the basis of a kinetic analysis, the authors rejected an intermolecular mechanism in which the deuterium in the molecule of the phenol derivative exchanges   for

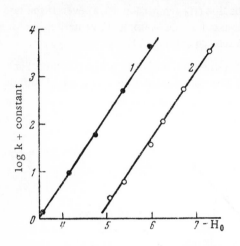

Fig. 11. Relation of deuterium exchange
rate to solution acidity. 1) Toluene; 2)
benzene.

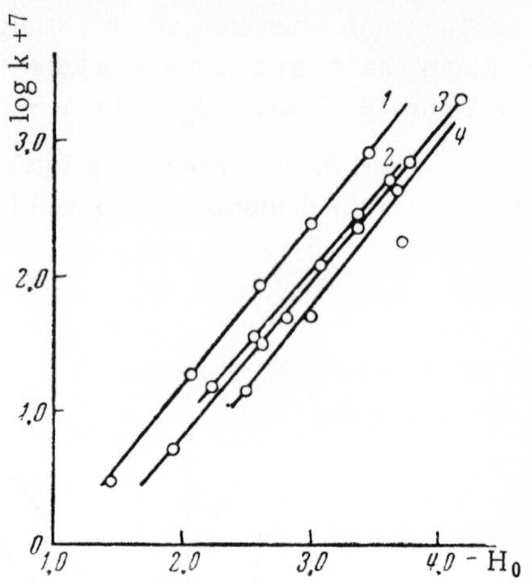

Fig. 12. Relation of deuterium exchange
rate to solution acidity. 1) p-D-Anisole
in $H_2SO_4 - H_2O$; 2) o-D-anisole in H_2SO_4-
H_2O; 3) p-D-anisole in $H_2SO_4 - H_2O$; 4)
o-D-anisole in $CH_3COOH - H_2SO_4$.

protium by reaction with another similar
molecule. They adopted the intramolecular mechanism illustrated by the follow-
ing scheme (for simplicity, the substituents in the aromatic ring are omitted and it is
assumed that one hydrogen atom in benzene is replaced by deuterium):

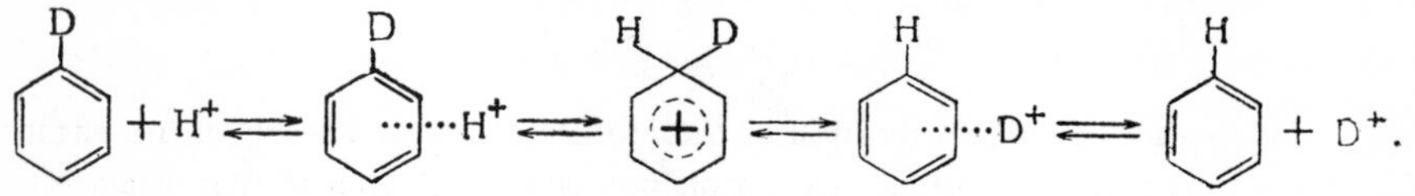

A proton is transferred to the aromatic ring to form the complex (II). In the latter, the
proton is bound by the whole π-electron system of the aromatic ring ("outer" complex
according to Mulliken or π-complex according to Brown and Brady, see p. 15). The
intramolecular rearrangement consists of the transfer of the proton to a definite carbon
atom with the simultaneous migration of the deuteron to the position previously occupied by the proton (IV). This process requires the expenditure of activation energy and
proceeds slowly. It is assumed that there exists an intermediate complex (III) with the
structure of a carbonium ion and hydrogen and deuterium atoms attached to the same
carbon atom. The assumption of its existence is not essential to the explanation of the
reaction kinetics, but in the opinion of the authors, it is quite probable in view of the
similarity between acid hydrogen exchange in the aromatic ring and electrophilic replacements of hydrogen (for example, halogenation and nitration), where such intermediate complexes are generally accepted (p. 6). The carbonium ion (III) has the
structure of an "inner" complex according to Mulliken or σ-complex according to Brown
(p. 15). It is assumed that the last stage of the reaction, namely, the elimination of a
deuteron, proceeds very rapidly.

A similar mechanism (rearrangement of a π-complex into a carbonium ion) was proposed by Taft and his co-workers [117] for the hydration of olefins, whose rate also obeys Hammett's relation.

Melander [116a] considers that merely the existence of proportionality between the rate constant and acidity function is insufficient evidence for assuming that there is a pre-equilibrium with the formation of an intermediate product of the π-complex type.

An investigation based on the theory of absolute reaction rates led this author to the conclusion that the formation of such a complex does not follow from kinetic data. Kinetic analysis shows only that the transition state, which leads to the formation of the carbonium ion, cannot contain a firmly bound water molecule as in this case there would be proportionality between the reaction rate constant k and the concentration of H_3O^+ ions instead of the experimentally observed proportionality between k and h_0.

On the basis of the same kinetic analysis, Malender rejected the mechanisms for hydration of an olefin postulated by Taft.

TABLE 25. Deuterium Exchange Between p-D-Toluene and H_2SO_4

H_2SO_4 concentration, %	$-H_{\bullet}$	$k \cdot 10^7$, sec^{-1}
53.0	3.57	1.36
58.5	4.16	9.40
63.9	4.76	56.6
68.9	5.39	498
73.4	5.96	4250

In Melander's opinion, in explaining the mechanism of a reaction it is generally necessary to avoid assumptions on the formation of intermediate products if they are not required kinetically. For acid hydrogen exchange in aromatic compounds it is natural to adopt a mechanism analogous to that proposed for electrophilic replacement of hydrogen in these compounds, i.e., the slow stage of the reaction is the addition of a proton to the substrate with the direct formation of a carbonium ion:

A similar view is held by Eaborn and Taylor [106a].

There is not always a simple relation between the acidity function H_0 and the rate constant of an exchange reaction, even if the medium has a high dielectric constant. Such a case was found by Mackor and his co-workers [77]. NaF is a strong base in liquid HF. By using anthracene as the indicator, these authors established that $-H_0$ for a molar solution of NaF in liquid HF is a unit lower than in 0.055 M solution, but the rate constant of deuterium exchange in monodeuterobenzene was the same with pure HF and with solutions of NaF at concentrations of 0.055, 1.0, and 2.0 moles per 1000 g of HF. On the other hand, even a small amount of BF_3 strongly accelerated the deuterium exchange.

The authors explained the independence of the reaction rate of H_0 by the fact that in this case ionization does not occur, but an unstable "outer" complex is formed. Consequently, even in hydrogen exchange in such a strong ionizing solvent as liquid hydrogen fluoride, associative processes which Mackor represented schematically by the equation

$$\text{ArD} + \text{HF} \rightleftarrows (\text{ArD·HF}) \rightleftarrows (\text{ArH·DF}) \rightleftarrows \text{ArH} + \text{DF}.$$

may be of considerable importance. Mackor [74] considers that, in general, in acid hydrogen exchange it is necessary to attach more importance to the addition of a molecule of the acid AH to a molecule of the aromatic compound than to the addition of a proton to it with the formation of the conjugate acid, BH^+, which was assumed by Gold and Satchell. This is described by the following equation:

First an unsymmetrical undissociated "salt" $\text{ArHD}^+ \cdot \text{A}^-$ is formed rapidly and then there slowly forms a complex of similar structure in which the deuterium and protium have exchanged places. According to Mackor, the ratio of the rate constants of the forward and reverse reactions k_1/k_{-1} corresponds to the "reduced" basicity constant of the hydrocarbon, which will be discussed later (p. 147). Preliminary ionization is not essential for the change to the activated state and deuterium atoms may occupy positions equivalent to that of the hydrogen atom even in the undissociated "salt" (for example, in $\text{ArHD}^+ \cdot \text{SO}_4\text{D}^-$). In this case, the structure of the acid may be of importance in addition to the acidity of the medium (see also [117a]).

One must agree with Mackor that in hydrogen exchange (basic as well as acid), an important part is played by associative processes, which, in reactions between weak protolytes and in solvents with a low dielectric constant, are not merely to the fore, but become the only possible processes. Support for this view will be given below when we have demonstrated that electrically neutral protolytes are by no means always ionized in an acid–base interaction (see Section IV and Section V, chapter 3).

The brief discussion of the mechanism of hydrogen exchange shows that at the moment it is possible to put forward only more or less probable hypotheses on the details of it. It is evident that the more the different points of view put forward and discussed, the closer we shall approach to understanding the mechanism of one of the simplest, but still quite complex reactions, namely, isotopic exchange of hydrogen.

13. SALT EFFECT IN HYDROGEN EXCHANGE

The presence of electrolytes changes the rate of reactions catalyzed by acids and bases. A quantitative theory of the salt effect for dilute aqueous solutions was put forward by Brønsted and it has withstood thorough experimental checking [15, 60]. The data presented above show conclusively that the isotopic exchange of hydrogen has much in common with other reactions catalyzed by acids and bases. This is also true of the kinetic salt effect in hydrogen exchange, though it has been studied very little as yet. However, we should mention the proportionality between the rate constant of exchange of the hydrogen of phosphine with water and the sodium chloride concentration demonstrated by Weston and Bigeleisen's measurements [79].

A very convenient solvent for studying the salt effect is liquid ammonia. Organic substances and many salts are readily soluble in it. In addition, as a result of the low dielectric constant of liquid ammonia, differences in the intensity of the electrostatic field of ions are much more clearly marked in it than in water.

A detailed study of deuterium exchange in the methyl group of β-naphthyl methyl ketone with liquid ammonia was made in the Isotopic Reactions Laboratory [118]. The reaction rate is six times greater in a 3.9 N solution of $Ca(NO_3)_2$.

The effect of the field of force of ions is very clear and the rate of the exchange reaction increases with an increase in the charge and a decrease in the radius of ions of the salt. If the rate constant (at 25°) without a salt is taken as 100, the following values are obtained for the rate constants in 2.5 N salt solutions:

Nitrates

Ca^{++}	Sr^{++}	Ba^{++}	Li^+	Na^+	NH_4^+
420	350	350	350	230	190

Sodium salts

Br^-	I^-	NO_3^-	CNS^-	ClO_4^-
400	240	230	170	115

The rate constants of the exchange reaction decrease in the sequence: $Ca^{++} > Sr^{++} \cong Ba^{++}$; $Li^+ > Na^+$; $Br^- > I^- > NO_3^- > ClO_4^-$.

The ion series found for deuterium exchange are practically the same as those found previously in studies of the salt effect in ammonolysis of lactones, esters, and halogen compounds in liquid ammonia [119-121]. It was found that another rule is also common to these reactions: there is an increase in the activation energy of deuterium exchange and ammonolysis in the presence of salts. The reaction rate is increased by a simultaneous increase in the energy and entropy of activation (pre-exponential factor in the Arrhenius equation). In 3.9 N solutions of calcium nitrate, the activation energy of hydrogen exchange in β-naphthyl methyl ketone is increased by 3300 kcal, while the pre-exponential factor is increased by a factor of 1500 (the corresponding activation entropy is increased by 15 entropy units).

Brønsted's classical theory of the salt effect cannot be used to interpret the salt effect in deuterium exchange (and ammonolysis) in liquid ammonia. It is based on the Debye-Hückel theory of strong electrolytes. As V. A. Pleskov [122] showed, the latter applies to mono-monovalent electrolytes at concentrations a factor of 10^6 less than the salt concentration in experiments on deuterium exchange. These solutions differ little in composition from ammoniates: there are eight moles of ammonia per mole of salt. The work of Biltz [123] and other authors [124] has shown that the heat of formation of crystalline ammoniates and there stability increases with an increase in the charge and a decrease in the radius of the cation and with the same cation, they increased with an increase in the radius of the anion. There is a parallel increase in the number of molecules present in a higher ammoniate. The same undoubtedly occurs in ammonia solutions of salts. In the latter, the ammonia molecules are bound by the ions and are in an ordered state with the strength of the bond depending on the charge and radius of the ion. With an increase in the salt concentration, there is naturally an increase in the activation energy of a reaction involving the solvent (deuterium exchange and ammon-

olysis): the more strongly the solvent molecules are bound by ions, the higher the activation energy must be. This is actually observed on measuring the kinetics of ammonolysis and deuterium exchange.

Reactions with a solvent are accompanied by disordering of the solution structure.
Therefore, there is a simultaneous increase in the reaction entropy, which likewise depends regularly on the radius and charge of the ions. As a rule, the entropy effect predominates and the reaction is accelerated by the addition of a salt through the pre-exponential factor. If both the energy and the entropy of activation are changed for the
same reasons, it is quite probable that there is a linear relation between them. This

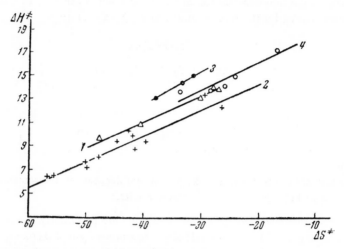

Fig. 13. Relation between activation parameters of reactions.
1) Deuterium exchange between methyl β-naphthyl ketone and
liquid ammonia; 2) ammonolysis of pilocarpine; 3) ammonolysis of 2-chloro-benzothiazole; 4) ammonolysis of 9-phenyl -
9-chlorofluorene in liquid ammonia.

actually exists and is shown graphically in Fig. 13. The enthalpy of activation $H^{\neq}$
(kcal/mole) is plotted along the ordinate axis and the entropy of activation $\Delta S^{\neq}$ along
the abscissa axis.

The facts described are generally explained from the point of view of solvation
effects. However, the phenomena accompanying deuterium exchange in concentrated
ammonia solutions of salts must be more diverse and complex.

14. ACID-BASE CATALYSIS OF HYDROGEN EXCHANGE IN H–H, B–H, N–H, As–H, P–H, P–H, AND S–H BONDS

It is shown below that the rate of hydrogen exchange in bonds of hydrogen with
other elements besides carbon depends on the acidity and basicity of the reagents and
medium.

A. Exchange in H–H Bond

Wirtz and Bonhöffer [125] reported that molecular hydrogen exchanges with a solution of alkali in heavy water during prolonged heating. Abe [126] mistakenly ascribed
the exchange to the presence of colloidal iron. This was disproved by Wilmarth and his

co-workers [127], who made a detailed study of the kinetics of the reaction. He reported a relation between the strength of the base and its catalytic activity and considered that in the first stage of the reaction, the hydrogen molecule fulfills the function of an acid. In accordance with the higher protophilicity of NH_2^- ions in liquid ammonia in comparison with OH^- ions in water, the exchange reaction in an ammonia solution of potassamide at $-53°$ proceeds with a rate constant which is 10^4 greater than that found for the exchange with an aqueous solution of potassium hydroxide at $100°$.

It is considered that a deuterium molecule reacts simultaneously with an OH^- (or NH_2^-) ion and a water (or ammonia) molecule:

$$OH^- + D:D + H:OH \rightarrow HOD + DH + OH^-$$
$$NH_2^- + D:D + H:NH_2 \rightarrow H_2ND + DH + NH_2^-.$$

No isotopic exchange was detected between gaseous deuterium and sulfuric acid or between deuterium and a solution of BF_3 in HF [128].

B. Exchange in B—H Bond

Information recently appeared [128a, 128b] of isotopic exchange of hydrogen in the B—H bond of decaborane, $B_{14}H_{10}$. At room temperature there was rapid exchange between heavy water dissolved in dioxane and a bridging hydrogen atom in the molecule, whose position was demonstrated [128b] by nuclear magnetic resonance measurements. The high rate of the exchange reaction is explained by the acid ionization of decaborane [128c].

$$B_{10}H_{14} + D_2O \rightarrow B_{10}H_{13}^- + HD_2O^+.$$

Deuterium takes the place of hydrogen. Then under the effect of the solvent, acting as a base, there is an intramolecular migration of deuterium with the result that when equilibrium is reached (after 72 hr), eight hydrogen atoms in the decaborane molecule have been replaced by deuterium. It was also found [128d] that six hydrogen atoms in the decaborane molecule may be exchanged for deuterium by the action of deuterium chloride in the presence of aluminum chloride, i.e., under the conditions of electrophilic attack by a very strong acid (p. 42).

C. Exchange in N—H Bond

The hydrogen in N—H bonds normally exchanges very rapidly when it is attached to a trivalent nitrogen atom, whose electron envelope contains a free pair of electrons. Saturation of the latter, for example, by addition of a proton or coordination by a metal ion, retards the exchange in accordance with A. I. Brodskii's rule [129].

By means of nuclear magnetic resonance, Ogg [130] measured the rate of ionization of the N—H bond in an ammonia molecule. It was found that in the liquid phase, the reaction $NH_3 + NH_3 \rightleftharpoons NH_4^+ + NH_2^-$ proceeds very slowly (the rate constant was not reported). It might be considered that the exchange between isotopic forms of ammonia also proceeds at a low rate as was found in the case of water (p. 71). On the other hand, the transfer of a proton from an ammonia molecule to an amide ion in an ammonia solution of potassamide, $NH_3 + NH_2^- \rightleftharpoons NH_2^- + NH_3$ proceeds very rapidly ($k = 4.6 \cdot 10^6$ liter-mole$^{-1} \cdot$ sec^{-1}) in accordance with the fact that the affinity of an NH_2^- ion for a proton is 200 kcal/mole greater than the affinity of an ammonia molecule for a proton.

Retarded hydrogen exchange in ammonium ions has been studied very thoroughly and quantitatively. Swain and his co-workers [131] determined the rate of hydrogen exchange between ammonium ions and the hydroxyl group of alcohols and showed that it is determined by the hydrogen ion concentration. The first series of experiments was carried out with ammonium bromide and methanol in dimethylformamide. The con-clusions drawn were confirmed with other systems. The reaction is first-order with re-spect to each of the reagents. The exchange rate is inversely proportional to the con-centration of hydrogen ions (in the form of protonized solvent molecules). The product of the exchange rate constant and the acid concentration remains constant even with a change in the acid concentration by a factor of 100. The activation energy of the ex-change reaction with triethylammonium chloride in methanol solution varies from 22 kcal at a hydrogen chloride concentration of 0.69 M to 15 kcal in a 0.016 M solution of acid. Thus, inhibition of the reaction by acid has been demonstrated strictly. Ac-cording to Swain, the kinetic data correspond to a termolecular mechanism for the ex-change. A proton (from an ammonium ion or an alcohol molecule) adds to a solvent molecule to form a complex in which the ammonia (or amine) and alcohol are con-nected by a hydrogen bond:

$$\text{solvent} + CH_3OH + NH_4^+ \rightleftharpoons \text{solvent} . H^+ + \text{complex}$$
where the complex has the composition:

$$CH_3OH \ldots NH_3 \text{ or } CH_3O^- \ldots \overset{+}{H}NH_3 .$$

The complex then decomposes:

$$\text{complex} \rightleftharpoons CH_3OH + NH_3 .$$

Grunwald, Löwenstein, and Meiboom [132] used proton magnetic resonance to meas-ure the rate of protolysis of alkylammonium salts in aqueous solution (over the pH range from 1.0 to 8.6). In this work, in which the experimental and theoretical level was high a complete kinetic analysis was given of the elementary reactions determining the over-all protolytic reaction. The work is also of great interest for the examination of hydro-gen exchange in N−H bonds. Within the limits of experimental error, the rate constant of the over-all protolysis reaction at a constant amine salt concentration was found to be inversely proportional to the hydrogen ion concentration, like the rate constant of the exchange according to Swain. Measurements with methylammonium chloride were reported in the first communication. The half-time of the protolytic reaction varied over the range of 0.002-0.2 sec. There was no exchange of protons at pH = 0.98. The rate and mechanism of protolysis of di- and trimethylammonium ions were examined in the second communication.

By determining the change in the over-all rate constant of the protolytic reaction k in relation to the concentration of the alkylammonium salt and the concentration of hydrogen ions, the authors calculated the rate constants of the individual reactions from the equation:

$$k = k_1 + \frac{k_2 K_W}{[H_3O^+]} + (k_3 + k_4) K_A \frac{[R_3NH^+]}{[H_3O^+]} .$$

These reactions were as follows:

1. $R_3NH^+ + H_2O \xrightarrow{k_1} R_3N + H_3O^+$

2. $R_3NH^+ + OH^- \xrightarrow{k_2} R_3N + H_2O$

3. $R_3NH^+ + NR_3 \xrightarrow{k_3} R_3N + R_3NH^+$

4. $R_3NH^+ + O-H + NR_3 \xrightarrow{k_4} R_3N + H-O + HNR_3^+$
   $\quad\quad\quad\quad |\quad\quad\quad\quad\quad\quad\quad\quad\quad |$
   $\quad\quad\quad\quad H\quad\quad\quad\quad\quad\quad\quad\quad\quad H$

$$K_W = [H_3O^+][OH^-]; \quad K_A = \frac{[H_3O^+][R_3N]}{[R_3NH^+]},$$

where R_3 may be CH_3H_2, $(CH_3)_2H$, and $(CH_3)_3$.

The following values were obtained for the rate constants of the individual reactions:

	k_1	k_2	k_3	k_4
$CH_3NH_3^+$	$< 0.4 \cdot 10^{-2}$	$< 10^{11}$	$2.5 \cdot 10^8$	$3.4 \cdot 10^8$
$(CH_3)_2NH_2^+$	$< 0.4 \cdot 10^{-2}$	$< 10^{11}$	$0.4 \cdot 10^8$	$5.6 \cdot 10^8$
$(CH_3)_3NH^+$	$< 5.5 \cdot 10^{-2}$	$< 10^{11}$	$< 2 \cdot 10^8$	$3.1 \cdot 10^8$

When the proton from an alkylammonium ion is transferred to an amine molecule (3) or a hydroxyl ion (2) instead of a water molecule (1), the rate constant is increased sharply. The affinity of the reacting particle for a proton obviously affects the rate of transfer of a proton. The rate constants of the analogous elementary reactions of mono-, di-, and trialkylammonium ions differ appreciably.

In acid solutions (at pH 3-5), the over-all protolysis process is dominated by the termolecular reaction (4) in which an amine molecule and a water molecule participate in addition to an alkylammonium ion. This mechanism is in agreement with the scheme proposed by Swain for hydrogen exchange between an amine salt and an alcohol, which is also assumed to involve a solvent molecule.

New data on the kinetics of hydrogen exchange in ammonium salts are given in [132a, b].

We should mention the work of Briscoe, Anderson, et al. [133], who measured the kinetics of hydrogen exchange between cobalt, platinum, and palladium ammoniates and heavy water. The rate of this pseudomonomolecular reaction is inversely proportional to the activity of the hydrogen ions in solution. Over the pH range from 3.55 to 5.59, the half-time of hydrogen exchange in cobalt hexaammoniate fell from 79 to 8.5 min. The activation energy of the reaction E = 28.1 kcal/mole.

The authors explained the strong change in the exchange rate with a change in the acid concentration by the fact that the exchange reaction is caused by acid ionization of the ammoniate:

$$Co(NH_3)_6^{+++} \rightleftarrows Co(NH_3)_5NH_2^{++} + H^+,$$

which is suppressed by hydrogen ions. Acid dissociation of an ammoniate in water was demonstrated by A. A. Grinberg [134].

A detailed paper on the kinetics of deuterium exchange between water and the cobalt ammoniates $[CO(NH_3)_6]Cl_3$ and $[Co(NH_3)_4C_2O_4]Cl$ appeared recently. The rate of the reaction in a series of buffer solutions is proportional to the concentration of hydroxyl ions. Block and Gold [134a] consider that the acidity of the $Co(NH_3)_6^{+++}$ ion is caused by the fact that it adds a hydroxyl ion rather than its capacity to lose a proton, as was assumed previously.

Nilsen [134b] measured the rate of the exchange reactions:

$$CH_3CONHCH_3 + D_2O \rightarrow CH_3CONDCH_3 + HDO$$
$$\text{and } CH_3CONDCH_3 + H_2O \rightarrow CH_3CONHCH_3 + HDO.$$

The minimum rate of hydrogen exchange corresponded to pD = 5.4 and pH = 5.1. The reactions are catalyzed by acids.

D. Exchange in P–H Bond

Weston and Bigeleisen [79] found that hydrogen exchange between phosphine and heavy water proceeds extremely slowly in the absence of a catalyst. The reaction is catalyzed by acids and bases. In the pH range from 3.4 to 5.0, the rate constant is a linear function of the H_3O^+ concentration. Over this pH range, the constant varies from $12.9 \cdot 10^{-4}$ to $0.4 \cdot 10^{-4}$ sec^{-1} at 27°. At a pH of 10.0-11.9, the rate constant of the exchange is proportional to the concentration of hydroxyl ions ($k \cdot 10^4$ sec^{-1} varies from 0.6 to 22.6). The activation energy of the reaction catalyzed by an acid or an alkali equals 17.6 kcal/mole.

The following mechanism is proposed:

a) in acid solutions:

$$PH_3 + DA \rightleftarrows PH_3D^+ + A^-$$

$$A^- + PH_3D^+ \rightleftarrows PH_2D + HA$$
$$HA + HDO \rightleftarrows H_2O + DA$$

b) in alkaline solutions:

$$PH_3 + B^- \rightleftarrows PH_2^- + BH$$
$$BH + HDO \rightleftarrows BD + H_2O$$
$$PH_2^- + BD \rightleftarrows PH_2D + B^-.$$

Phosphine participates in the exchange reaction as either an acid or a base, whose dissociation constants have been roughly estimated as 10^{-29} and 10^{-28}.

E. Exchange in As–H Bond

Zeltmann and Gerold [134c] studied the equilibrium of hydrogen exchange between heavy water and arsine (AsH_3) at 25.4°. The time required for the equilibrium to be reached could be shortened by the addition of a base or an acid, which consequently catalyze hydrogen exchange in the As–H bond.

F. Exchange in O−H Bond

The rate of hydrogen exchange in O−H bonds could be measured by means of proton magnetic resonance. Swain and Halmann [135] established that the equilibrium of exchange between tritiated triphenylcarbinol and normal methanol in toluene is reached in 0.01 sec. During this time, the reaction of triphenylcarbinol with tert-bu-tanol was only approximately 10% complete. Consequently, the exchange rate depends on the structure of the alcohol molecule. Hydrogen exchange between hydroxyl groups of alcohols is catalyzed by acids and bases.

The rate of hydrogen exchange between hydrogen peroxide and water was meas-ured in the thorough work of Anbar, Löwenstein, and Meiboom [136]. The reaction pro-ceeds very rapidly and the half-time of exchange at a solution pH of 4.5 equals only $5 \cdot 10^{-4}$ sec. Measurements were made at various pH values and several hydrogen per-oxide concentrations. Acids and alkalis catalyze the exchange. At pH < 4, its rate is proportional and at pH > 5, inversely proportional to the hydrogen ion concentration. The reaction is first-order with respect to hydrogen peroxide in an acid medium and second-order in an alkaline medium.

In examining the reaction mechanism, the authors considered that the following molecules and ions were present in the reaction system: H_2O, H_2O_2, H_3O^+, $H_3O_2^+$, OH^-, HO_2^-. They consider that a bimolecular reaction with H_2O_2 and H_3O^+ is most probable in the presence of acid and a termolecular reaction with H_2O_2, H_2O, and HO_2^- in an alkaline medium. The rate constant is of the order of 10^7 sec^{-1}. The activated com-plex formed in the case of the termolecular reaction is assigned the structure of a six-membered ring in which the proton is transferred by a Grotthus mechanism.

Valuable information is available on the kinetics of very fast reactions involving rupture of the O−H bond and this is important in the examination of hydrogen exchange. Modern experimental techniques made it possible for Eigen and his co-workers [137-139] to measure the rate of reactions that are complete in times of the order of 10^{-11} sec. Normal solutions of strong acids and bases are neutralized at this rate. The rate of neutralization is commensurate with the frequency of collisions between ions in an aqueous solution. The rate constant was determined by two independent methods and the same value was obtained, namely, $k = 1.3 \cdot 10^{11}$ sec^{-1}. This reaction is usually represented by the equation: $H_3O^+ + OH^- \rightleftharpoons H_2O + H_2O$. Eigen considers that its ex-ceptionally high rate is explained by a particular mechanism, differing from the mech-anism of recombination of other oppositely charged ions, which requires a considerable activation energy for displacement of the water envelope of the hydrated ion. At least three particles participate in the reaction between hydroxonium and hydroxyl ions:

Knowing the rate of neutralization and the equilibrium constant of ionization of water, they calculated the rate of self-ionization of water. The rate constant of the reaction $H_2O + H_2O \rightleftharpoons H_3O^+ + OH^-$ at 25° equals $2.6 \cdot 10^{-5}$ sec^{-1}. This rate corresponds to a mean life of a water molecule of approximately 14 hr. It is possible that such a low rate of self-ionization of water is the reason (as already pointed out by A. I. Brodskii

[140]) for the very slow deuterium exchange between deuterium oxide and normal water discovered by R. E. Mardaleishvili, G. K. Lavrovskaya, and V. V. Voevodskii [141]). The rate constant of the transfer of a proton of a water molecule to a more protophilic ammonia molecule ($H_2O + NH_3 \rightleftharpoons NH_4^+ + OH^-$) is ten orders higher ($k = 5 \cdot 10^5$). The rate constant of the transfer of a proton from a water molecule to a hydroxyl ion is even five orders greater ($k = 1.3 \cdot 10^{10}$ sec^{-1}). The time required for its completion is commensurate with the time in which a hydrogen bond is formed or broken (10^{-12} - 10^{-13} sec). Eigen illustrated this reaction schematically in the following way:

A review of Eigen's work is given in [141a].

G. Exchange in S—H Bond

Deuterium exchange between methanol and liquid hydrogen sulfide proceeds at a readily measurable rate at a low temperature. Thus, at −115°, the half-time of exchange in a mixture of 1 mole of CH_3OD and 0.7 mole of H_2S reaches 100 min ($k = 10^{-4}$ sec^{-1} and the activation energy equals 7.5-8 kcal/mole). These data were given in a paper by Geib [142] almost twenty years ago and are often quoted. However, hardly anyone notes the important fact that the reaction is catalyzed by acid. Thus, at −115° (with a component ratio of 1 of CH_3OH to 0.5 of D_2S) the half-time of exchange equals 30 min. This is reduced to 5 min in the presence of 0.011 mole of HCl.

In measuring the kinetics of exchange between liquid water and gaseous hydrogen sulfide, Garand and Amanrich [142a] observed acceleration of the reaction by ammonia and amines.

15. HETEROGENEOUS ACID—BASE HYDROGEN EXCHANGE

According to S. Z. Roginskii and I. I. Ioffe [143], reactions catalyzed by acids and bases in solution are also catalyzed under heterogeneous conditions by solids with acid—base properties. New experimental confirmation of this relation between homogeneous and heterogeneous acid—base catalysis is given in [144, 145].

The relation between homogeneous and heterogeneous acid—base catalysis also applies to hydrogen exchange. It should be regarded as additional confirmation of the acid—base nature of this exchange.

In parallel with deuterium exchange between saturated hydrocarbons and concentrated sulfuric acid, which was discovered by Ingold and studied in detail by Burwell, D. N. Kursanov, and V. N. Setkina, Beeck, and Stevenson, and others (for literature see [129, p. 300] [146, 147]), we should mention deuterium exchange between hydrocarbons with a tertiary carbon atom and deuterated solid catalysts (aluminosilicate, etc. [148, 149]). Heterogeneous deuterium exchange with aromatic hydrocarbons proceeds quite readily [150].

To estimate the acidity of solid catalysts, Benesi [151] used indicators which are normally used to determine the acidity function H_0 of liquid acids. It was found that the acidity of an aluminosilicate catalyst ($H_0 = -8.2$) is similar to that of 95% sulfuric acid; the similarity of their behavior in hydrogen exchange is not surprising.

The reaction between hydrocarbons (mainly aromatic) and solid acid catalysts was studied by an adsorption method (A. V. Kiselev, et al. [152, 153]) and by infrared spectroscopy (A. N. Terenin, et al. [154]). It was regarded as an acid−base reaction in this work.

The first reports on heterogeneous deuterium exchange catalyzed by solid bases has also appeared. Hydrogen exchange in hydrocarbons (2-methyl-1-butene and benzene) occurs at a considerable rate even at 70° on deuterated potassium and calcium amides [155] and is analogous to the reactions in ammonia solutions of potassamide which were studied in detail previously (see Section III). Hart [156] observed the transfer of deuterium from the alkyl group of ethylbenzene to another alkylbenzene molecule when a mixture of them was heated with sodium hydride.

In connection with heterogeneous deuterium exchange with solid bases, we should mention Morton's observation [157] that the latter (in particular, organosodium compounds) change the color of azo compounds. This is in accordance with measurements of the absorption spectra of solutions of azo indicators in liquid ammonia with potassamide added [158] which remained unknown to Morton.

16. SUMMARY

1. Hydrogen exchange in bonds of hydrogen with any element is catalyzed by acids or bases. In accordance with this, hydrogen exchange should be divided into acid, basic and amphoteric (between like particles of different isotopic composition) exchange.

2. Heterolytic hydrogen exchange may occur in solutions and on solid catalysts with the properties of acids or bases.

3. The rate of hydrogen exchange varies over a very wide range, depending on the structure of the substance and the properties of the solvent and the catalyst. By the use of very strong catalysts, namely, bases and acids in protophilic and protogenic solvents, respectively, (for example, KND_2 in liquid ND_3, $AlBr_3$ in liquid DBr, and BF_3 in liquid DF) it is possible to increase the rate of reactions by up to approximately twenty orders relative to their rate in amphoteric solvents (D_2O and C_2H_5OD) without a catalyst.

4. In addition to the protolytic properties of the solvent and catalyst, the rate of hydrogen exchange depends to a large extent on the dielectric constant of the solvent and the degree of polarity of its molecules, on the charge of the substrate, and catalyst, and on steric hindrance.

5. When a substance shows dual reactivity, which may appear, for example, when its molecules contains a heteroatom with a free pair of electrons in conjugation with the π-electron system of an aromatic ring, the normal ratios between exchange rates in amphoteric and protophilic solvents may be disturbed. Dual reactivity is the reason for acceleration of acid hydrogen exchange by a base and basic hydrogen exchange by an acid and also the inhibition of acid exchange by an acid.

6. As a rule, Brønsted's and Hammett's relations apply to hydrogen exchange. The reasons for deviations from these relations are discussed and some considerations on the mechanism of acid hydrogen exchange in aromatic compounds are presented.

7. Acidlike aprotic substances accelerate hydrogen exchange because of the formation of complexes with hydrogen acids, which are stronger acids than the original hydrogen acids.

8. As in other heterolytic reactions in solution, a salt effect is observed in hydrogen exchange.

LITERATURE CITED

1. A. I. Shatenshtein, Uspekhi Khimii 28, 3 (1959).
2. A. I. Brodskii, Chemistry of Isotopes [in Russian] (Izd. AN SSSR, Moscow, 1957).
3. O. Reitz, Isotopenkatalyse in Lösung, Handbuch der Katalyse, Vol. II (Springer, 1940) pp. 272-348.
4. C. K. Ingold and C. L. Wilson, Z. Elektrochem. 44, 62 (1938).
5. M. S. Kharash, W. G. Brown, and J. McNab, J. Org. Chem. 2, 36(1937); 4, 442(1939).
6. A. I. Shatenshtein, Ukr. Khim. J. 22, 3 (1956).
7. A. I. Shatenshtein, Uspekhi Khimii 16, 548 (1947).
8. A. I. Shatenshtein and E. N. Zvyagintseva, Doklady Akad. Nauk SSSR 117, 852 (1957).
9. A. I. Shatenshtein and E. A. Izrailevich, Zhur. Fiz. Khim. 28, 3 (1954).
10. W. K. Wilmarth, J. C. Dayton, and J. M. Flournoy, J. Am. Chem. Soc. 75, 4549, 4553 (1953).
11. A. I. Shatenshtein, L. N. Vasil'eva, N. M. Dykhno, and E. A. Izrailevich, Doklady Akad. Nauk SSSR 85, 381 (1952).
12. A. I. Shatenshtein, E. N. Zvyagintseva, E. A. Yakovleva, E. A. Izrailevich, Ya. M. Varshavskii, M. G. Lozhkina, and A. V. Vedeneev, Collection: "Isotopes in Catalysis" [in Russian] (Izd. AN SSSR, Moscow, 1957) p. 218.
13. S. R. Palit, M. N. Das, and G. R. Somayajula, Nonaqueous Titration [Russian translation] (Goskhimizdat, Moscow, 1958).
14. L. Audrieth and J. Kleinberg, Nonaqueous Solvents [Russian translation] (IL, Moscow, 1955).
15. A. I. Shatenshtein, Theories of Acids and Bases [in Russian] (Goskhimizdat, Moscow, 1949).
16. N. A. Izmailov, Electrochemistry of Solutions [in Russian] (Izd. KhGU, Khar'kov, 1959).
17. P. A. Small and J. H. Wolfden, J. Chem. Soc. 1811 (1936).
18. M. Koizumi, Bull. Chem. Soc. Japan 14, 353 (1939).
19. A. I. Shatenshtein, A. V. Vedeneev, and P. P. Alikhanov, Zhur. Obshchei Khim. 28, 2638 (1958).
20. W. G. Brown, K. E. Wilzbach, and W. H. Urry, Cand. J. Res. 27, 398 (1949).
20a. P. F. Tryon, W. G. Brown, and M. S. Kharasch, J. Am. Chem. Soc. 70, 2003 (1948).
21. Ya. M. Varshavskii, M. G. Lozhkina, and A. I. Shatenshtein, Zhur. Fiz. Khim. 31, 1377 (1957).
22. A. V. Vedeneev, Hydrogen Exchange in Some Oxygen-Containing Organic Compounds [in Russian] (Dissertation, Karpov Physicochemical Institute, Moscow, 1955).
23. A. I. Shatenshtein, K. I. Zhdanova, V. R. Kalinichenko, and L. N. Vinogradov, Doklady Akad. Nauk SSSR 102, 779 (1955).
24. Ya. M. Varshavskii, V. R. Kalinichenko, S. E. Vaisberg, and A. I. Shatenshtein, Zhur. Fiz. Khim. 30, 1647 (1956).
25. K. I. Zhdanova, V. M. Basmanova, and A. I. Shatenshtein, Zhur. Obshchei Khim. 31, 250 (1961).
26. A. I. Shatenshtein, V. R. Kalinichenko, and Ya. M. Varshavskii, Zhur. Fiz. Khim. 30, 2093, 2097 (1956).
27. A. I. Shatenshtein and Ya. M. Varshavskii, Doklady Akad. Nauk SSSR 85, 157(1952).

28. Ya. M. Varshavskii, S. E. Vaisberg, and M. G. Lozhkina, Zhur. Fiz. Khim. 29, 750 (1955).

29. A. I. Shatenshtein, K. I. Zhdanova, and V. M. Basmanova, Doklady Akad. Nauk SSSR 133, 1117 (1960).

30. A. I. Shatenshtein, Uspekhi Khimii 21, 914 (1952).

31. A. I. Shatenshtein and E. A. Yakovleva, Zhur. Obshchei Khim. 28, 1713 (1958).

32. A. I. Shatenshtein and E. A. Yakovleva, Doklady Akad. Nauk SSSR 105, 1024 (1955).

33. A. I. Shatenshtein and E. N. Zvyagintseva, Zhur. Obshchei Khim. 29, 1751 (1959); 31, 1432 (1961).

34. J. Brønsted and E. Vance, Z. phys. Chem. 163, 240 (1933).

35. J. Brønsted, A. L. Nickolson, and A. Delbanco, Z. phys. Chem. 169, 379 (1934).

36. A. I. Shatenshtein and A. V. Vedeneev, Zhur. Obshchei Khimii 28, 2644 (1958).

37. W. Klatt, Z. anorg. Chem. 222, 225 (1935).

38. V. Gold and F. A. Long, J. Am. Chem. Soc. 75, 4543 (1953).

39. M. A. Reid, J. Am. Chem. Soc. 76, 3264 (1954).

40. M. E. Vol'pin, K. I. Zhdanova, D. N. Kursanov, V. N. Setkina, and A. I. Shatenshtein, Izvest. Akad. Nauk SSSR, Otdel. Khim. Nauk (1959) p. 754.

41. A. I. Shatenshtein, N. M. Dykhno, E. A. Izrailevich, L. N. Vasil'eva, and M. Faivush, Doklady Akad. Nauk SSSR 89, 479 (1951).

42. A. N. Nesmeyanov, Uch. Zap. MGU 132, 5 (1950).

43. A. N. Nesmeyanov and M. I. Kabachnik, Collection: "Problems in Chemical Kinetics, Catalysis, and Reactivity" [in Russian] (Izd. AN SSSR, Moscow, 1955) p. 644.

44. C. K. Ingold, C. G. Raisin, and C. L. Wilson, J. Chem. Soc. 1637 (1936).

45. M. Koizumi and T. Titani, Bull. Chem. Soc. Japan 13, 681 (1938).

46. R. Elderfield, Heterocyclic Compounds [Russian translation] (IL, Moscow, 1955) Vol. 4, pp. 62 and 64.

47. T. I. Abramovich, I. P. Gragerov, and V. V. Perekalin, Doklady Akad. Nauk SSSR 121, 295 (1958).

48. W. G. Brown and N. J. Letang, J. Am. Chem. Soc. 61, 2547 (1939); 63, 358 (1941).

48a. S. Bruckenstein and I. M. Kolthoff, J. Am. Chem. Soc. 78, 10 (1956).

49. K. F. Bonhöffer, K. H. Geib, and O. Reitz, J. Chem. Phys. 7, 664 (1939).

50. H. C. Brown and F. R. Jensen, J. Am. Chem. Soc. 80, 2291 (1958).

51. R. J. Gillespie and J. A. Leisten, Quart. Rev. 8, 40 (1954).

52. L. A. Flexser, L. P. Hammett, and A. Dingwall, J. Am. Chem. Soc. 57, 2103 (1935).

53. E. F. Caldin, Nature 172, 583 (1953).

53a. M. Avramoff and J. Sprinzak, J. Amer. Chem. Soc. 82, 4953 (1960).

53b. A. I. Shatenshtein, E. A. Yakovleva, and É. S. Petrov, Doklady Akad. Nauk SSSR 186, 882 (1961).

53c. A. I. Shatenshtein, E. A. Yakovleva, and É. S. Petrov, Zhur. Obshchei Khim. 31, 12 (1961).

54. J. Kenner, M. Polanyi, and P. Szego, Nature 135, 267 (1935).

55. A. Klit and A. Langseth, Nature 135, 956 (1935); Z. phys. Chem. 176, 65 (1936).

56. K. I. Zhdanova, V. M. Basmanova, and A. I. Shatenshtein, Zhur. Fiz. Khim. 33, 1438 (1959).

57. A. I. Shatenshtein and P. P. Alikhanov, Zhur. Obshchei Khim. 30, 982 (1960).

58. D. P. N. Satchell, J. Chem. Soc. (1958) p. 3910.

59. D. Bethell, V. Gold, and D. P. N. Satchell, J. Chem. Soc. (1958) pp. 1918, 1927.

60. R. P. Bell, Acid−Base Catalysis (Oxford, 1941).

61. G. Burhart, W. G. Ford, and E. Singleton, Nature 136, 688 (1935); J. Chem. Soc. (1936) p. 20.

62. R. P. Bell, J. Phys. Chem. 55, 885 (1951).

62a. R. P. Bell, Proton in Chemistry, Cornell University Press (New York, 1959) p. 163.

63. L. P. Hammett, Chem. Rev. 17, 125 (1935).

64. J. E. Leffler, The Reactive Intermediates of Organic Chemistry (New York, 1956).

65. R. G. Pearson and R. L. Dillon, J. Am. Chem. Soc. 75, 2439 (1953).

66. R. G. Pearson, Discuss. Farad. Soc. 17, 187 (1954).

67. A. A. Frost and R. G. Pearson, Kinetics and Mechanism (New York, 1953).

68. K. F. Bonhöffer, K. H. Geib, and O. Reitz, J. Chem. Phys. 7, 664 (1939).

69. J. Hochberg and K. F. Bonhöffer, Z. phys. Chem. 184, 420 (1939).

69a. A. J. Kresge and Y. Chiang, J. Am. Chem. Soc. 83, 2877 (1961).

69b. J. Colapietro and F. A. Long, Chem. Industry (1960) p. 1056.

70. A. I. Shatenshtein and Yu. P. Vyrskii, Doklady Akad. Nauk SSSR 70, 1029 (1950); Zhur. Fiz. Khim. 25, 1206 (1951).

71. J. B. Conant and G. W. Wheland, J. Am. Chem. Soc. 45, 1212 (1932).

72. W. K. McEwen, J. Am. Chem. Soc. 58, 1125 (1936).

73. G. Dallinga, A. A. V. Stuart, P. J. Smit, and E. L. Mackor, Z. Elektrochem. 61, 1019 (1957).

74. E. L. Mackor, A. Hofstra, and J. H. van der Waals, Trans. Farad. Soc. 54, 66 (1954).

75. E. L. Mackor, A. Hofstra, and J. H. van der Waals, Trans. Farad. Soc. 54, 186 (1954).

76. M. Kilpatrick and F. E. Luborsky, J. Am. Chem. Soc. 75, 577 (1953).

77. E. L. Mackor, P. J. Smit, and J. H. van der Waals, Trans. Farad. Soc. 53, 1309 (1957).

78. J. Hine, R. C. Peek, and B. D. Oakes, J. Am. Chem. Soc. 76, 827 (1954).

79. R. E. Weston and J. Bigeleisen, J. Am. Chem. Soc. 75, 2439 (1953).

80. A. I. Shatenshtein, Collection: "Problems in Physical Chemistry" [in Russian] (Goskhimizdat, Moscow, 1958), No. 1, p. 202.

81. L. P. Hammett and A. J. Deyrup, J. Am. Chem. Soc. 54, 2721 (1932).

82. L. P. Hammett, Physical Organic Chemistry (New York, 1940).

83. M. A. Paul and F. A. Long, Chem. Rev. 57, 1 (1957).

84. F. H. Westheimer and M. S. Kharasch, J. Am. Chem. Soc. 68, 1871 (1946).

85. N. C. Deno, J. J. Juruzelski, and A. Schriesheim, J. Am. Chem. Soc. 77, 3044 (1955).

86. V. A. Pal'm, Doklady Akad. Nauk SSSR 108, 270 (1956).

87. M. Kilpatrick and H. H. Hyman, J. Am. Chem. Soc. 80, 77 (1958).

88. H. H. Hyman, M. Kilpatrick, and J. J. Katz, J. Am. Chem. Soc. 79, 3668 (1957).

88a. H. H. Hyman and R. A. Garber, J. Am. Chem. Soc. 81, 1847 (1959).

89. E. L. Mackor, A. Hofstra, and J. H. van der Waals, Trans. Farad. Soc. 54, 66 (1958).

90. A. I. Gel'bshtein, G. G. Shcheglova, and M. I. Temkin, Zhur. Neorg. Khim. 1, 282 (1958).

91. G. V. Tiers, J. Am. Chem. Soc. 78, 4165 (1956).

92. L. P. Hammett and A. J. Deyrup, J. Am. Chem. Soc. 54, 4239 (1932).

92a. R. Stewart and T. Mathews, Canad. J. Chem. 38, 602 (1960).

93. D. P. N. Satchell, J. Chem. Soc. (1958) p. 3904.

94. D. S. Noyce and W. A. Pryor, J. Am. Chem. Soc. 77, 1397 (1955).

95. J. M. Kolthoff and S. Bruckenstein, J. Am. Chem. Soc. 78, 1, 10 (1956).

95a. S. Bruckenstein, J. Am. Chem. Soc. 82, 307 (1960).

96. J. Rocek, Collection of Czechoslovakian Chemical Communications 22, 1 (1957).

97. F. A. Long and M. A. Paul, Chem. Rev. 57, 934 (1957).

98. M. A. Paul and F. A. Long, Symposium on pH Measurement, American Society for Testing Materials, Special Technical Publication (1957) No. 190, p. 80.

99. L. P. Hammett and M. A. Paul, J. Am. Chem. Soc. 56, 830 (1934).

100. L. Zucker and L. P. Hammett, J. Am. Chem. Soc. 61, 2791 (1939).

101. F. A. Long and D. McIntyre, J. Am. Chem. Soc. 76, 3240 (1954).

102. F. A. Long, Proc. Chem. Soc. (1957) p. 220.

102a. J. Koskikailo and E. Whally, Trans. Farad. Soc. 55, 815 (1959).

102b. G. Archer and R. P. Bell, J. Chem. Soc. 3228 (1959).

102c. R. H. Boyd, R. W. Taft, A. P. Wolf, and D. R. Christman, J. Am. Chem. Soc. 82, 4729 (1960).

102d. J. F. Bunnett, J. Am. Chem. Soc. 82, 499 (1960).

102e. N. S. Deno, P. T. Groves, and G. Saines, J. Am. Chem. Soc. 81, 5790 (1959).

102f. A. J. Kresge and Y. Chiang, Proc. Chem. Soc. (London) (1961) p. 81.

103. A. I. Gel'bshtein, G. G. Shcheglova, and M. I. Temkin, Doklady Akad. Nauk SSSR 107, 109 (1956); Zhur. Neorg. Khim. 1, 506 (1956).

104. A. I. Gel'bshtein, G. G. Shcheglova, and M. I. Temkin, Zhur. Fiz. Khim. 30, 2267 (1956).

105. A. I. Gel'bshtein and M. I. Temkin, Doklady Akad. Nauk SSSR 118, 740 (1958).

106. C. K. Ingold, C. G. Raisin, and C. L. Wilson, Nature 134, 734 (1934); J. Chem. Soc. 915, 1637 (1936).

106a. C. Eaborn and R. Taylor, J. Chem. Soc. (1960) p. 3301.

107. V. Gold and F. A. Long, J. Am. Chem. Soc. 75, 4543 (1953).

108. V. Gold and D. P. N. Satchell, J. Chem. Soc. (1955) p. 3609.

109. V. Gold and D. P. N. Satchell, J. Chem. Soc. (1955) p. 3619.

110. V. Gold and D. P. N. Satchell, J. Chem. Soc. (1955) p. 3622.

111. V. Gold and D. P. N. Satchell, J. Chem. Soc. (1956) p. 2743.

112. D. P. N. Satchell, J. Chem. Soc. (1956) p. 3911.

113. D. P. N. Satchell, J. Chem. Soc. (1957) p. 2878.

114. D. P. N. Satchell, J. Chem. Soc. (1958) p. 3904.

115. O. Beeck, J. W. Otvos, D. P. Stevenson, and C. D. Wagner, J. Chem. Phys. 17, 418 (1949).

116. F. A. Long and D. McIntyre, J. Am. Chem. Soc. 76, 3243 (1954).

116a. L. Melander and P. C. Myre, Ark. för Kemi 13, 507 (1959).

117. R. W. Taft, E. L. Purlee, P. Riesz, and C. A. DaFazio, J. Am. Chem. Soc. 77, 1584 (1955).

117a. A. J. Kresge and Y. Chiang, J. Am. Chem. Soc. 81, 5509 (1959).

118. A. I. Shatenshtein, Yu. P. Vyrskii, and E. A. Rabinovich, Doklady Akad. Nauk SSSR 124, 146 (1959).

119. A. I. Shatenshtein, G. S. Markova, and E. A. Izrailevich, Zhur. Fiz. Khim. 7, 26 (1936); 8, 696, 13, 1175 (1939); 17, 24 (1943); Doklady Akad. Nauk SSSR 35, 73 (1942).

120. L. F. Audrieth et al., J. Am. Chem. Soc. 60, 579 (1938); 61, 2387 (1939); 64, 2498 (1942).

121. R. C. Anderson and G. W. Watt et al., J. Am. Chem. Soc. 63, 1953 (1941); 64, 467 (1942); 65, 49 (1943); 66, 376 (1944).

122. V. A. Pleskov, Zhur. Fiz. Khim. 10, 601 (1937).

123. W. Biltz, Z. Anorg. Chem. 130, 93 (1923).

124. van Arkel, The Chemical Bond [Russian translation] (Gostekhimizdat, Leningrad, 1934) p. 147.

125. K. Wirtz and K. F. Bonhöffer, Z. Phys. Chem. 177, 1 (1936).

126. S. Abe, Pap. Inst. Phys.-Chem. Res. (Tokyo) 38, 287 (1941); Chem. Abstr. 35, 6177 (1941).

127. W. K. Wilmarth, J. C. Dayton, and J. M. Flournoy, J. Am. Chem. Soc. 75, 4549, 4553 (1953).

128. S. W. Weller and G. A. Mills, Advances in Catalysis, Vol. VIII (1956).

128a. J. J. Miller and M. F. Hawthorne, J. Am. Chem. Soc. 80, 754 (1958); 81, 4501 (1959).

128b. J. Shapiro, M. Lustig, and R. B. Williams, J. Am. Chem. Soc. 81, 838 (1959).

128c. G. A. Guter and G. W. Schoeffer, J. Am. Chem. Soc. 78, 3546 (1956).

128d. J. A. Dupont and M. F. Hawthorne, J. Am. Chem. Soc. 81, 4998 (1959).

129. A. I. Brodskii, Izvest. Akad. Nauk SSSR, Otdel. Khim. Nauk (1949) p. 3.

130. R. A. Ogg, Discuss. Farad. Soc. 17, 215 (1954).

131. C. G. Swain, M. M. Labes, J. T. McKnight, and V. P. Kreiter, J. Am. Chem. Soc. 79, 1084, 1088 (1957).

132. E. Grunwald, A. Löwenstein, and S. Meiboom, J. Chem. Phys. 27, 630, 1067 (1957).

132a. S. Meiboom, Z. Elektrochem. 64, 50 (1960); M. T. Emerson, E. Grunwald, and R. A. Kromhout, J. Chem. Phys. 33, 547 (1960).

132b. M. T. Emerson, E. Grunwald, M. L. Kaplan, and R. A. Kromhout, J. Am. Chem. Soc. 82, 6307 (1960).

133. J. S. Anderson, H. V. A. Briscoe, N. L. Spoor, and L. H. Cobb, J. Chem. Soc. (1943) pp. 361, 367.

134. A. A. Grinberg, Introduction to the Chemistry of Complex Compounds [in Russian] (Goskhimizdat, Moscow, 1951).

134a. H. Block and V. Gold, J. Chem. Soc. (1959) p. 956.

134b. S. O. Nilsen, Biochem. et Biophys. Acta 37, 146 (1960); C. A. 54, 11658 (1960).

134c. A. N. Zeltmann and G. Gerold, J. Chem. Phys. 31, 889 (1959).

135. C. G. Swain and M. Halmann, Bull. Res. Council Israel 5A, 184 (1956).

136. M. Anbar, A. Löwenstein, and S. Meiboom, J. Am. Chem. Soc. 80, 2630 (1958).

137. M. Eigen, Discuss. Farad. Soc. 17, 194 (1954).

138. M. Eigen and J. Schoen, Z. Phys. Chem. (N.F.) 3, 126 (1955); Z. Elektrochem. 59, 483 (1955).

139. M. Eigen and L. De Mayer, Z. Elektrochem. 59, 986 (1955).

140. A. I. Brodskii, J. Chim. Phys. (1958) p. 26.

141. R. E. Mardaleishvili, G. K. Lavrovskaya, and V. V. Voevodskii, Zhur. Fiz. Khim. 28, 2195 (1954).

141a. M. Eigen, Z. Elektrochem. 64, 115 (1960).

142. K. H. Geib, Z. Elektrochem. 45, 648 (1939).

142a. J. Garand and R. Amanrich, J. Chim. Phys. 56, 532 (1959).

143. I. I. Ioffe and S. Z. Roginskii, Zhur. Fiz. Khim. 31, 612 (1957).

144. G. I. Andrianova and S. Z. Roginskii, Transactions of the All-Union Conference on Catalysis [in Russian] (Izd. AN SSSR, Moscow, 1957).

145. O. V. Krylov and E. A. Fokina, Transaction of the All-Union Conference on Catalysis [in Russian] (Izd. AN SSSR, Moscow, 1957).

146. D. N. Kursanov and V. V. Voevodskii, Uspekhi Khimii 23, 641 (1954).

147. D. N. Kursanov, Ukr. Khim. Zhur. 22, 34 (1956).

148. S. G. Hindin, G. A. Milles, and A. G. Ollad, J. Am. Chem. Soc. 73, 278 (1951); 77, 538 (1955).

149. R. C. Hansford, P. G. Waldo, J. C. Drake, and R. E. Honig, Ind. Eng. Chem. 44, 1108 (1952).

150. G. M. Panchenkov, Z. V. Gryaznova, V. M. Emel'yanova, and L. Ganichenko, Doklady Akad. Nauk SSSR 109, 325, 546 (1956).

151. H. A. Benesi, J. Am. Chem. Soc. 78, 5490 (1956); J. Phys. Chem. 61, 970 (1957).

152. A. V. Kiselev, Uspekhi Khimii 25, 705 (1956).

153. Yu. A. El'tekov, Collection: "Surface Chemical Compounds and Their Role in Adsorption Phenomena" [in Russian] (Izd. MGU, Moscow, 1957) p. 260.

154. A. I. Terenin, Collection: "Surface Chemical Compounds and Their Role in Adsorption Phenomena" [in Russian] (Izd. MGU, Moscow, 1957) pp. 206 and 270.

155. A. I. Shatenshtein, Yu. G. Dubinskii, E. A. Yakovleva, I. V. Gostunskaya, and B. A. Kazanskii, Izvest. Akad. Nauk SSSR, Otdel, Khim. Nauk 104 (1958).

156. H. Hart, J. Am. Chem. Soc. 78, 2619 (1956).

157. A. A. Morton and F. H. Bolton, J. Am. Chem. Soc. 75, 1146 (1953).

158. A. I. Shatenshtein and E. A. Izrailevich, Zhur. Fiz. Khim. 26, 377 (1952).

HYDROCARBONS AS ACIDS AND BASES

INTRODUCTION

Even recently hydrocarbons were regarded as an example of chemical "neutrality." The combination of the words "hydrocarbon" with "acid" or "base" would have sounded very incongruous to chemists. It is true that individual examples of acid–base reactions of hydrocarbons had already been known for several decades. For example, Kraus and his co-workers prepared metal salts of hydrocarbons (triphenylmethane, etc.), which conducted an electric current well in liquid ammonia. They were prepared by the action of a solution of an alkali metal or metal amide in ammonia on the hydrocarbon. Some chemists (Conant, Wheland, and Morton) regarded Shorygin's reaction, which consists of the metallation of hydrocarbons by organo-alkali compounds, as the displacement of a weak acid from its salt by a stronger acid. In carrying out physicochemical investigations of solutions of organic substances in liquid hydrogen fluoride in Fredenhagen's laboratory, Klatt observed the high electrical conductivity of an anthracene solution, which was difficult to explain by anything other than the ionization of this hydrocarbon like a base dissolved in an acid. Nonetheless, such observations were isolated for a long time because reagents such as a solution of potassamide in liquid ammonia, liquid hydrogen fluoride, or, even more so, a solution of boron trifluoride in it, whose reversible reactions with some hydrocarbons have a clearly expressed acid–base character, were too exotic for chemists. Methods of detecting weaker protolytic reactions either did not exist or were not readily available.

Facts indicating the acidic and basic properties of hydrocarbons have multiplied continuously in recent years and the idea of hydrocarbons as acids and bases has spread increasingly. It has been developed as a general concept in the Isotopic Reactions Laboratory of the L. Ya. Karpov Physicochemical Institute in connection with systematic investigations of hydrogen exchange in nonaqueous solutions. The combination of two factors, namely, the high protophilicity or protogenicity of the solvents and catalysts discussed in Section II and the sensitivity of the isotope method, has been extremely favorable to the elucidation of the acidity and basicity of hydrocarbons.

In the next two chapters we will consider mainly the reactions of hydrocarbons which have much in common with the reactions of normal acids and bases, for example, the replacement of hydrogen by a metal and the conversion of a hydrocarbon molecule to an onium salt by the addition of a proton to it. On many examples it will be shown that there are no fundamental differences between hydrocarbons and other acids and bases.

In parallel we will summarize facts on basic and acidic exchange of hydrogen in hydrocarbons, which are now quite numerous. This makes it possible to follow once again the continuous connection between hydrogen exchange in solutions and the protolytic properties of the substrate and reagent.

Hydrocarbons are among the weakest acids and bases and therefore the inadequacy of the Brønsted-Lowry scheme for describing acid—base interaction appears particularly clearly in the discussion of the many protolytic reactions of hydrocarbons. It is not by chance that the need for a reexamination of the definitions of acids and bases arose. In particular, from a detailed knowledge of exchange reactions of hydrocarbons in protophilic and protogenic media.

A comparison of the reactions of the same substances with acids and bases is also essential to complete our knowledge of the reactivity of hydrocarbons and the intereffect of the atoms in their molecules.

The problem of the intereffect of atoms in molecules of organic substances arose at the same time as the theory of chemical structure. In the first article on this theory, published in 1862, its originator A. M. Butlerov, in defining the idea of chemical structure, emphasized that atoms connected in a molecule "affect each other either indirectly or directly" [1]. The same idea was expressed in a more developed form in a course in organic chemistry: "The chemical structure of each elementary part (or as we now say, atom — A. Sh) in a complex body is determined by its nature and the mode of its chemical arrangement in the particle, on the one hand, and the nature, number, and chemical disposition of the other parts present in the particle, on the other" [2].

The outstanding student of A. M. Butlerov, V. V. Markovnikov [3], who successfully developed precisely this aspect of the theory of chemical structure, wrote in his celebrated dissertation "Data on the intereffect of atoms in chemical compounds"(1869): "The theory of structure is the outer mechanism, whose action is directed and regulated by the theory of the intereffect of atoms." According to V. V. Markovnikov, the character of a complex substance depends not only on the elements constituting it, but also on their intereffect. This is demonstrated by the whole field of hydrocarbon chemistry.

In discussing the intereffect of atoms in hydrocarbon molecules it is necessary to consider first of all the nature of the bonds between carbon atoms and the sequence of the bonds. The conception of conjugation of bonds, which confers special properties on molecules in which multiple and single bonds alternate (dienes, polyenes, and aromatic molecules), arose in chemistry long ago. The heats of formation of such molecules are greater than the sums of the bond energies. The experimental refraction differs from that calculated from the atomic and structural increments (exaltation of refraction). The light absorption maximum is displaced toward longer wavelengths, while the absorption intensity is increased in comparison with the light absorption intensity of saturated hydrocarbon molecules. In molecules with conjugated multiple bonds, the interatomic distances change and tend to become equal. Such molecules have a high capacity for redistribution of electron density in the force field of a reagent, i.e., have a high polarizability.

In the last two decades the idea of the conjugation of bonds has become even more general. In 1936, Baker and Nathan [4-6] proposed the existence of a peculiar type of conjugation of single and multiple bonds (hyperconjugation). This type includes one of the first examples of the intereffect of atoms, which was discovered by V. V. Markovnikov and expressed in a rule for the addition of a hydrogen halide to an ethylenic hydrocarbon (Markovnikov's rule). A broad generalization of the concept of the conjugation of bonds was used by A. N. Nesmeyanov [7] to explain the dual reactivity of many organic compounds. A. N. Nesmeyanov proposed a classification of conjugation systems. Conjugation between bonds involving π-electrons of aromatic π-electron systems is denoted as π,π-conjugation. Conjugation between a multiple and a single bond (i.e., formed by σ-electrons alone) is called π,σ-conjugation. This concept is equivalent to the concept of hyperconjugation. Conjugation between a multiple bond and the p-electrons of a heteroatom is called p,π-conjugation, etc.

Together with conjugation effects, it is also necessary to consider the effects of inductive displacement of electrons of σ-bonds when the molecule contains atoms or groups of atoms with a high electron-donor or electron-acceptor capacity.

A detailed discussion of the structure and reactivity of organic compounds is a special subject which is beyond the scope of this book. In subsequent discussion we will only touch upon certain points connected with data obtained on the acidic and basic properties of hydrocarbons.

LITERATURE CITED

1. A. M. Butlerov, Selected Works on Organic Chemistry [in Russian] (Moscow, Izd. AN SSSR, 1951) p. 72.
2. A. M. Butlerov, Introduction to a Complete Study of Organic Chemistry [in Russian] (Kazan', 1864) p. 672.
3. V. V. Markovnikov, Data on the Intereffect of Atoms in Chemical Compounds [in Russian] (Kazan', 1869).
4. J. W. Baker, Hyperconjugation (Oxford, 1952).
5. F. Becker, Hyperconjugation, Fortschr. Chem. Forsch. 3, 187 (1955).
6. Hyperconjugation Conference, Tetrahedron 5, Nos. 2-3 (1959).
7. A. N. Nesmeyanov, Uch. Zap. MGU 132, 5 (1950).

CHAPTER 1. HYDROCARBONS AS ACIDS

In many hydrocarbons it is possible to replace hydrogen by a metal, as is characteristic of acids. The replacement is sometimes produced by the direct action of a metal on the hydrocarbon and is accompanied by the liberation of hydrogen; more often it occurs during reactions with bases. Organometallic compounds are formed. In this chapter we will first consider reactions forming organoalkali compounds [1-9], which are closest to the reactions of normal acids with metals and with bases. The properties of some organoalkali compounds are similar to those of alkali metal salts. Data are available which make it possible to estimate the relative strengths of hydrocarbons as acids by means of metallation reactions.

The second part of this chapter is a brief summary of results characterizing hydrocarbons as acids from hydrogen isotope exchange reactions with bases, largely with liquid ammonia and a solution of potassamide in it.

At the end of the chapter is a brief description of work on isomerization of olefins and alkylations, in which the acidity of hydrocarbons appears.

1. METALLATION OF HYDROCARBONS

A. Reactions of Hydrocarbons with Alkali Metals

The possibility of metallating a hydrocarbon depends on the lability of the hydrogen in it, the metal used in the reaction and the reaction conditions. Metallation is effected by means of a molten or powdered metal. Amalgams of alkali metals and a liquid eutectic mixture of sodium and potassium are also used. Some hydrocarbons dissolved in ether, ligroin, or benzene react with a metal suspended in the solution. As As alkali dissolve in liquid ammonia, in which hydrocarbons, apart from saturated ones, are readily soluble, there is a method of metallation in ammonia solution [10, 11]. At the same time and to a greater or lesser extent, there may also be reduction (hydrogenation)* of unsaturated hydrocarbons, addition of the metal to ethylenic and aromatic compounds, and also ammonolysis [13]. Some side processes can be avoided if the hydrocarbons are metallated in liquid ammonia by means of potassamide (see p. 85).

Acetylene and finely dispersed sodium suspended in xylene, dioxane, etc., react to form the acetylide [14]. Cyclopentadiene reacts very vigorously with potassium in benzene [15, 16] and phenylacetylene reacts with potassium or sodium in ether. Indene does not react under these conditions [15]. The reaction of metallic sodium with indene occurs above 140° and that with fluorene at 190-200° [17]. If acetylene is passed over sodium at a temperature above 100°, sodium acetylide is formed. Above 210°, both hydrogen atoms are replaced and the carbide C_2Na_2 is obtained [18]. At approximately the same temperature, triphenylmethane and potassium form triphenylmethylpotassium [19], while dipehnylmethane is converted into the potassium derivative at 230°. Toluene is metallated to a very small extent (3%) by boiling with potassium for 3 hr [20]. Toluene reacts with metallic cesium at the melting point of cesium (30°) with the liberation of hydrogen and the formation of benzylcesium [21].

When sodium oxide is added, metallic potassium metallates alkylbenzenes comparatively rapidly [22]. It is considered that sodium oxide acts as a hydrogen acceptor:

$$2C_6H_5CH_3 + 2K + Na_2O \rightarrow 2C_2H_5CH_2K + NaH + NaOH .$$

With treatment at 90° for 3 hr, the hydrogen atoms in the α-position are replaced in the following yields: toluene − 87%, ethylbenzene − 31%, cumene − 11%, m-xylene− 71%, and p-xylene − 31%. The replacement of the hydrogen in the methyl group of toluene proceeds with more difficulty than the corresponding reaction with α- and β-methylnaphthalenes.

A solution of an alkali metal in liquid ammonia readily metallates many hydrocarbons even at a low temperature (−33°) (see reviews [10, 11]). The reactions of a solution of sodium with acetylene are described in [18, 23, 24]. Cyclopentadiene reacts with a solution of lithium [25]. Fluorene in liquid ammonia reacts with magnesium [26], while triphenylmethane gives traces of a barium salt with barium [27].

In contrast to triphenylmethane, diphenylmethane reacts slowly and incompletely with sodium in liquid ammonia at room temperature; its sodium derivatives undergoes ammonolysis. Benzyl- and phenylsodiums and potassiums cannot be obtained in liquid

---

* See [12] on the relation between the acidity of hydrocarbons and their capacity to be reduced by solutions of metals in liquid ammonia.

ammonia. It was noted [28] that for the replacement of hydrogen in a CH bond by a metal under these conditions, it is necessary for the carbon atom to be connected to two phenyl groups as in $(C_6H_5)_2CHCH_2C_6H_5$ and $(C_6H_5)_2CHCH(C_6H_5)_2$, for example. Consequently, aromatic hydrocarbons containing a benzhydryl group $(C_6H_5)_2CH-$ react (benzhydryl rule).

Thus, hydrocarbons in which the carbon of the aliphatic CH bond is attached to an aromatic ring or adjacent to a triple bond are able to react with alkali metals with the liberation of hydrogen.

B. Reactions of Hydrocarbons with Bases

Literature data show that hydrocarbons are metallated by caustic alkalis, amides and hydrides of alkali metals, and also organoalkali compounds which are derivatives of hydrocarbons with weaker acidic properties. All these reactions may be reduced to the reaction between an acid and a strong base, whose part is played by OH^-, NH_2^-, and H^- ions and carbanions.

The reactions between hydrocarbons and these bases may be represented schematically by the equations:

$$RH + OH^- \rightleftarrows R^- + H_2O$$
$$RH + NH_2^- \rightleftarrows R^- + NH_3$$
$$RH + H^- \rightleftarrows R^- + H_2 *$$
$$RH + R_1^- \rightleftarrows R^- + R_1H$$

a. CAUSTIC ALKALIS

Acetylene reacts even with an aqueous solution of an alkali. Fluoradene (indenofluorene), which is one of the most acidic hydrocarbons, reacts readily with dilute aqueous alkali [29]. Its ionization constant in aqueous solution pK = 11 ± 0.5. An alkaline solution of fluoradene is red. The hydrogen at the position marked with an asterisk is replaced by a metal.

Caustic alkalis are able to metallate cyclopentadiene, indene, and fluorene (see [1], p. 314). For example, when fused with potassium hydroxide at 250-280°, fluorene is converted into fluorenylpotassium with the simultaneous elimination of water [15]. It was found that potassium hydroxide will metallate fluorene under much milder conditions: with vigorous stirring in ether solution at room temperature, the yield of the metallation product reaches 50% after 3 hr. This is explained by the fact that excess

---

* See p. 66 on the acidity of molecular hydrogen. It is a stronger acid than benzene as it reacts with phenylpotassium at room temperature according to the equation C_6H_5K + $+ H_2 \rightarrow C_6H_5 + KH$ [30].

potassium hydroxide binds the water formed in the reaction. Sodium hydroxide is much less efficient than potassium hydroxide as a metallating agent [31].

A series of organic compounds have been metallated with potassium hydroxide in liquid ammonia. Kraus [32] mentioned the reaction with triphenylmethane. It was also observed [33] that if solid alkali is present in an ammonia solution of quinaldine, as a result of metallation, the solution acquires a pink color, which becomes an intense cherry red in time. Indene, fluorene, diphenylmethane, and even α-methylnaphthalene were metallated by this method. Metallation by caustic alkali in liquid ammonia is apparently promoted by the fact that ammonia binds water firmly. This has been used to dehydrate salts [34]. The fact that ammonia binds water firmly is indicated by the absence of reaction between metallic sodium and small amounts of water dissolved in liquid ammonia [35].

Alcohols dissolved in liquid ammonia react slowly with alkali metals with the reaction rate depending on the degree of acidity of the alcohol: $CH_3OH > C_2H_5OH > tert-C_4H_9OH$ [35a] (see p. 228).

A. A. Vainshtedt [35b] observed the formation of colored organopotassium compounds when a solution of potassium alcoholate in alcohol reacted with many derivatives of fluorene in acetone and especially in pyridine. For example, 9-phenylfluorene (I), di-α-naphthofluorene (II), bifluorenyl (III), etc., react in this way. As was established later [35c], there is an analogous reaction with perinaphthalene (IV), which has a labile hydrogen atom in the position marked with an asterisk. This substance condenses with benzaldehyde and is more acid than fluorene, but less acid then cyclopentadiene.

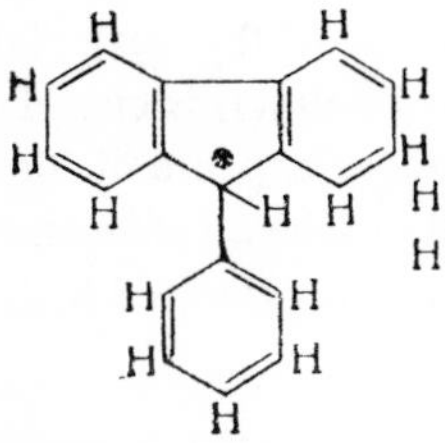

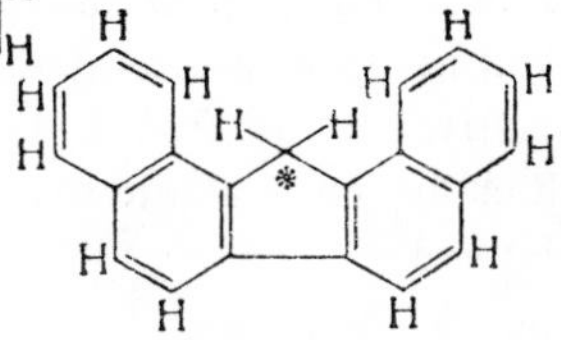

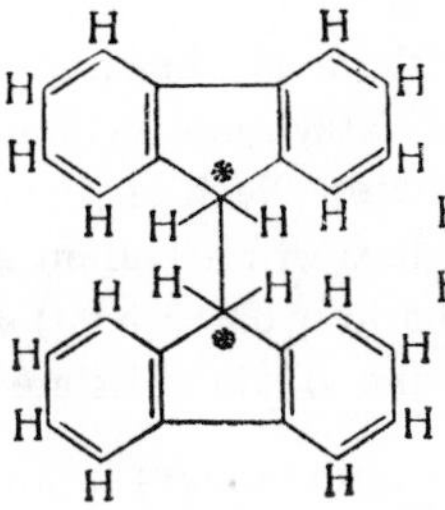

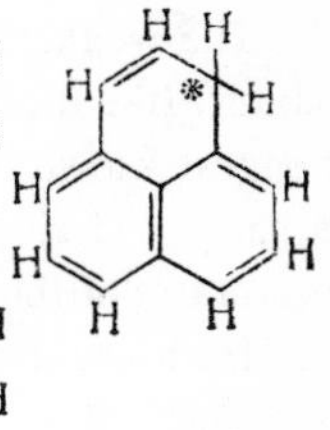

b. AMIDES OF ALKALI METALS

The hydrocarbons which are metallated by a solution of potassium in liquid ammonia are metallated even at low temperatures (−33°) by a solution of potassamide in ammonia [36].

c. HYDRIDES OF ALKALI METALS

Hydrides of alkali metals metallate acetylene at 40° [18, p. 79]. Sodium hydride metallates alkylbenzenes [37]

$$C_6H_5 \overset{|}{\underset{|}{C}}H + NaH \rightarrow C_6H_5 \overset{|}{\underset{|}{C}}Na + H_2.$$

d. ORGANOALKALI COMPOUNDS

The metallation of hydrocarbons discovered by P. P. Shorygin in 1907 [38] is of fundamental and of exceptionally great practical importance. P. P. Shorygin established that the reaction between ethylsodium and benzene forms phenylsodium and liberates ethane:

$$C_6H_6 + C_2H_5Na \rightarrow C_6H_5Na + C_2H_6 .$$

In the reaction of ethylsodium with alkylbenzenes (toluene, xylene isomers, ethylbenzene, mesitylene, and p-cymene) and diphenylmethane, the replacement of one hydrogen atom in the methyl or α-methylene group was observed.

The metallation of hydrocarbons by organoalkali compounds, which is known as Shorygin's reaction, is analogous to the normal displacement by a strong acid of a weaker acid from its salt, for example:

$$HCl + NaCN \rightleftarrows NaCl + HCN$$
$$CH_3COOH + C_6H_5ONa \rightleftarrows CH_3COONa + C_6H_5OH .$$

These reactions are examples of an acid–base process:

$$acid_1 + base_2 \rightleftharpoons base_1 + acid_2$$

$$HCl + CN^- \rightleftarrows Cl^- + HCN$$
$$CH_3COOH + C_6H_5O^- \rightleftarrows CH_3COO^- + C_6H_5OH.$$

The weaker the acid, the stronger is its conjugate base (anion of the salt). Ethylsodium, like other alkylalkalis (their saltlike nature is discussed below), is one of the strongest known bases. Such organoalkali compounds are therefore capable of forming carbanions (i.e., ions with trivalent negative carbon) from hydrocarbons (aromatic and unsaturated aliphatic hydrocarbons) which are more acidic than the saturated hydrocarbons from which the alkylalkalis are derived.

Shorygin's reaction makes it possible to synthesize many organoalkali compounds. A detailed summary of the latter has been given by K. A. Kocheshkov and T. V. Talalaeva [1, pp. 56-79, 172-204, 313-323]. Below (p. 124), we will discuss the use of Shorygin's reaction for comparing the acidity of hydrocarbons.

2. SALTLIKE NATURE OF ORGANOALKALI COMPOUNDS

Many organoalkali compounds are similar to normal salts in their physicochemical properties (see [1, pp. 3 and 4], [3, pp. 445-447], [5, p. 219], [6, pp. 46-58], and [39]). For example, alkyl- and arylpotassiums and sodiums are colorless crystalline substances which have comparatively high melting points or decomposed without melting; they are practically insoluble in hydrocarbons. Organoalkali compounds in which the metal is attached to a carbon atom which, in its turn, is attached to an aromatic ring, are intensely colored (normally orange or red). Examples of these are benzyllithium, triphenylmethylsodium, fluorenylpotassium, etc. The color of these compounds is an inherent property of the carbanions. The compounds are soluble in ether, pyridine, and liquid ammonia.

An approximate calculation by means of the Born-Haber cycle showed [40] that
the lattice energy of methylsodium (170 kcal/mole) is close to the lattice energy of
typical ionic crystals, namely, sodium bromide (175 kcal) and sodium iodide (164 kcal).
Thus, the C—Na bond in this organometallic compound is mainly of an ionic nature. *

The saltlike nature of many organoalkali compounds is confirmed by the electrical
conductivity of their solutions.

A. Electrical Conductivity of Solutions

The slight electrical conductivity of a solution of triphenylmethylsodium in ether
was first reported by Schlenk [41]. According to the latter measurements of Ziegler
[42], the equivalent electrical conductivity of this solution λ varies over the range from
0.07 to 0.006 ohm^{-1} cm^2 at 0° and a dilution V varying from 14 to 380 liter/g-equiv.
Under the same conditions, the electrical conductivity of the typical salt, LiI, rose from
0.01 to 0.04 ohm^{-1} cm^2. The dissociation constant of triphenylmethylsodium in ether
at 25° is 10^{-12} [43, 44, 44a]. Because of the low dielectric constant of ether (DC = 4.5),
the measurements were repeated [42] with solutions in pyridine (DC = 12.6). Over approximately the same range of concentrations, the electrical conductivity of pyridine
solutions was 100 times greater (λ = 10-16). The electrical conductivities of solutions
of potassium salts of diphenylmethane, fluorene, and indene in pyridine were measured
in the same work. According to the experiments of Kraus [45], the electrical conductivity of a solution of triphenylmethylsodium in liquid ammonia (DC = 22) is still greater by a factor of ten (λ = 157 at V = 370 liter/g-equiv. at −33°). At the same dilution,
the equivalent electrical conductivity of a solution of sodium nitrate in the same solvent λ = 194 ohm^{-1} cm^2 [46].

The electrical conductivities of solutions of sodium and lithium acetylides in liquid
ammonia were measured [47]. These compounds are weaker electrolytes like AgCN.
At −33.5°, the equivalent electrical conductivity of a solution of sodium acetylide λ =
= 13 (V = 170 liter/g-equiv.). The metal was liberated at the cathode during electrolysis.

The exceptionally high reactivity of alkylalkalis and their low solubility in readily
accessible solvents made it difficult to examine their electrolyte properties. Hein [48,
49] was successful in this and used diethylzinc as the solvent. The measurements were
made at 50°. The increase in the equivalent electrical conductivity with an increase
in the radius of the alkali metal ion was caused by the decrease in the interionic interaction. The data obtained are given below.

Substance	C, liter/g-equiv.	λ, ohm^{-1} cm^2
LiC$_2$H$_5$	3.13	0.13
NaC$_2$H$_5$	4.44	4.01
KC$_2$H$_5$	5.41	0.49
RbC$_2$H$_5$	5.52	9.38

The electrical conductivity of the compounds given is practically the same as that of
a solution of the typical salt tetraethylammonium bromide in diethylzinc (λ = 8.45 at
C = 1.06 N and 20°). Solutions of phenyllithium and benzyllithium in diethylzinc also
conduct a current.

---

* The C—Li bond in alkyllithium is much closer to a covalent bond.

Faraday's law is obeyed well in the electrolysis of solutions of alkylalkalis. It was established that $C_2H_5^-$ ions are discharged on the anode. Tetraethyllead was formed in almost quantitative yield on the current on a lead electrode. A hydrocarbon formed from the two radicals (R−R) or by their disproportionation was liberated on an anode with which the ions discharged did not react.

Similar measurements were later made by Jander [50]. In his experiments, the equivalent electrical conductivity of solutions of LiC_2H_5 in $Zn(C_2H_5)_2$ with concentrations of from 1.3 to 59.9 mole/liter at 20° varied from $4 \cdot 10^{-5}$ to $8.8 \cdot 10^{-3}$ $ohm^{-1} cm^2$. It was approximately two orders below the electrical conductivity of tetra-n-butyl-ammonium iodide in the same solvent, which changed from $2 \cdot 10^{-3}$ to $2 \cdot 10^{-1}$ ohm^{-1} cm^2 with a rise in concentration from 1.5 to 57.3 mole/liter. It was considered that there was ionization of a complex with the solvent, for example:

$$LiC_2H_5 + Zn(C_2H_5)_2 \rightarrow Li C_2H_5 \cdot Zn (C_2H_5)_2 \rightleftarrows Li^+ + Zn (C_2H_5)_3^-$$

or complexes of more complicated composition. Consequently, in the case of alkyl-metals, the electrical conductivity of a solution cannot be used as a direct measure of the extent to which the R−Me bond is ionic.

B. Ionic Reactions

When solutions of NaC_2H_5 and $(C_5H_{11})_4NI$ in diethylzinc are mixed, NaI precipitates instantaneously [51]. This ionic reaction may be followed conductimetrically. A conductimetric titration has also been carried out [50] between solutions of LiC_2H_5 and C_2H_5ZnI in diethylzinc:

$$LiC_2H_5 + C_2H_5ZnI = LiI + Zn(C_2H_5)_2 .$$

3. CARBANIONS

A. Formation of Carbanions

The electrical conductivity of solutions of many organoalkali compounds and the ionic reactions which they give indicate that these compounds consist of ions with the metal as the cation and the hydrocarbon in the state of an anion. As already mentioned, the latter is called a carbanion.

In the metallation reactions described above, carbanions are formed by the replacement of hydrogen by a metal (Me or $Me^+ + e^-$) and by acid−base reactions with strong bases (: B), which are organoalkali compounds or amides of alkali metals. The schemes and examples of the corresponding reactions are given below:

$$-\overset{|}{\underset{|}{C}} : H + Me \rightarrow -\overset{|}{\underset{|}{C}} :^{\ominus} Me^+ + H \tag{I}$$
$$(C_6H_5)_3 CH + K \rightarrow (C_6H_5)_3 C^-K^+ + H$$

$$-\overset{|}{\underset{|}{C}} : H + :B \rightarrow -\overset{|}{\underset{|}{C}} :^{\ominus} + B : H^+ \tag{II}$$
$$(C_6H_5)_3 CH + NH_2^- \rightarrow (C_6H_5)_3 C^- + NH_3 .$$

Carbanions are also obtained by the rupture of other bonds besides C−H bonds, for example, by the rupture of C−Hal, C−O, and C−C bonds [1-5]:

$$-\overset{|}{\underset{|}{C}}:Hal + 2Me \rightarrow -\overset{|}{\underset{|}{C}}:^{\ominus} Me^+ + Me^+ + Hal^-$$

$$(C_6H_5)_3CCl + 2K \rightarrow (C_6H_5)_3 C^-K^+ + K^+ + Cl^- \tag{III}$$

$$-\overset{|}{\underset{|}{C}}:O:\overset{|}{\underset{|}{C}}- + 2Me \rightarrow -\overset{|}{\underset{|}{C}}:^{\ominus} Me^+ + -\overset{|}{\underset{|}{C}}-O^- Me^+$$

$$C_6H_5OC_6H_5 + 2K \rightarrow C_6H_5^-K^+ + C_6H_5O^-K^+ \tag{IV}$$

$$-\overset{|}{\underset{|}{C}}:\overset{|}{\underset{|}{C}}- + 2Me \rightarrow 2 -\overset{|}{\underset{|}{C}}:^{\ominus} Me^+$$

$$(C_6H_5)_2CH-CH(C_6H_5)_2 + 2K \rightarrow 2(C_6H_5)_2 CH^-K^+. \tag{V}$$

Carbanions can also be formed by the addition of a metal or an anion at a double bond. Anion-radicals are formed in the first case (aromatic compounds may be used instead of compounds with a double bond):

$$\overset{|}{\underset{|}{C}}::\overset{|}{\underset{|}{C}} + Me \rightarrow \cdot\overset{|}{\underset{|}{C}}:\overset{|}{\underset{|}{C}}:^{\ominus} Me^{\oplus} \tag{VI}$$

An ion is formed in the second case:

$$\overset{|}{\underset{|}{C}}::\overset{|}{\underset{|}{C}} + B \rightarrow :^{\ominus}\overset{|}{\underset{|}{C}}:\overset{|}{\underset{|}{C}}:B \tag{VII}$$

$$C_6H_5CH = CH_2 + NH^-{}_2 \rightarrow C_6H_5\overset{\ominus}{C}H - CH_2NH_2.$$

In this reaction, the unsaturated compound fulfills the function of an acidlike substance. Carbanions are formed according to scheme (VII) during anionic polymerization of vinyl and diene compounds. Anion-radicals are formed during the reduction of unsaturated compounds with alkali metals and are able to initiate polymerization.

The formation of a carbanion by the reaction of a metal with a free radical is also possible.

$$-\overset{|}{\underset{|}{C}}\cdot + Me \rightarrow \overset{\ominus}{\underset{|}{\overset{|}{C}}}:Me^{\oplus} \tag{VIII}$$

$$(C_6H_5)_3C\cdot + K \rightarrow (C_6H_5)_3C^-K^+$$

B. Spectra of Carbanions

The ionization of a hydrocarbon molecule is very often accompanied by a strong displacement of the light absorption maximum into the visible region of the spectrum.

By observing the change in color of solutions of phenylated alkanes in liquid ammonia on treatment with alkali metals (or metal amides), Wooster and Mitchell [28, 52] determined which of the hydrocarbons were ionized by this treatment. On the basis of the results obtained, they formulated the so-called benzhydryl rule (p. 84).

Some qualitative data on the color of carbanions are given in Leffler's summary [53, p. 177] and also in the articles [22, 54-57]. The carbanions of the following substances have the following colors: acetylene and phenylacetylene — colorless; indene — yellow; fluorene — orange; phenylfluorene and diphenylmethane — orange-red; triphenylmethane and xanthene — red; naphthyldibiphenylmethane — violet; tribiphenylmethane — blue; dinaphthylmethane and diphenylnaphthylmethane — green, etc.

Anderson [58] measured the absorption spectrum of the triphenylmethyl anion and compared it with the spectra of triphenylmethyl as a cation and a radical. A subsequent check [56, 57] revealed the inaccuracy of the measurements. The first detailed investigation of the absorption spectra of carbanions, which were obtained by the action of potassamide or potassium hydroxide on the hydrocarbon in liquid ammonia, was carried out in the Isotopic Reactions Laboratory [56, 57]. A solution of potassamide itself absorbs light in the near ultraviolet (λ_{max} = 345 mμ) and therefore measurements were made only in the visible region of the spectrum with excess of this base. The completeness of the conversion of the substance into a carbanion was checked by determining whether the absorption obeyed Beer's law. For measurements in the ultraviolet, the hydrocarbons were ionized by means of potassium hydroxide (p. 84).

TABLE 26. Light Absorption by Carbanions $ArCH_2^-$

Anion	λ_{max} $m\mu$	$\log \varepsilon_{max}$
$C_6H_5CH_2^-$	408, 450, 462	—
α - $C_{10}H_7CH_2^-$	560	2.9
α - $(C_5H_4N)CH_2^-$	445	3.0
α - $(C_9H_6N)CH_2^-$	495	3.3
γ - $(C_9H_6N)CH_2^-$	465	3.7
9 - $(C_{13}H_8N)CH_2^-$	530	3.8

Table 26 gives data on the light absorption of carbanions of the type $ArCH_2^-$. The values of λ_{max} for the benzyl anions were taken from the literature, where they were obtained by calculation [59-61].

The ionization of a hydrocarbon produces a displacement of the light absorption maximum toward longer wavelengths by up to 350 mμ (see Fig. 14). Lengthening of the conjugation chain (with a change from benzyl to naphthyl and from the anion of methylpyridine to methylquinoline and methylacridine) is accompanied by displacement of the maximum toward longer wavelengths and an increase in the absorption intensity. Table 27 gives λ_{max} and $\log \varepsilon_{max}$ for the absorption curves of carbanions of the type Ar_2CH^-. A comparison of the positions of the maxima on the absorption curves of the carbanions of diphenylmethane and its para-derivatives shows that carbanions of the p-phenyl and p-benzyl derivatives absorb light at longer wavelengths.

There is also a higher absorption intensity, which may be related to the electrophilicity of these substituents. The carbanions of other para-derivatives of diphenylmethane give

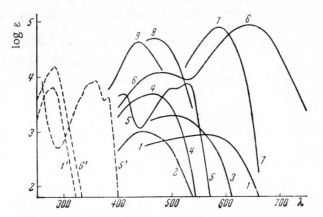

Fig. 14. Absorption spectra of carbanions in liquid ammonia.
1) $\alpha\text{-}C_{10}H_7CH_2^-$; 2) $\alpha\text{-}(C_5H_4N)CH_2^-$; 3) $\alpha\text{-}(C_9H_6N)CH_2^-$; 4) $\gamma\text{-}$
$(C_9H_6N)CH_2^-$; 5) $9\text{-}(C_{13}H_8N)CH_2^-$; 6) $\alpha,\alpha'\text{-}(C_{10}H_7)_2CH^-$; 7)
$(C_6H_5C_6H_4)_2CH^-$; 8) $(C_6H_5CH_2C_6H_4)_2CH^-$; 9) $(C_6H_5)_2CH^-$.

spectra which differ from that of the diphenylmethane carbanion mainly in intensity. Carbanions in which the substituents have the most clearly expressed nucleophilicity, such as $N(CH_3)_2$ and $N(C_2H_5)_2$, have the lowest light absorption intensity.

The absorption intensity of the fluorenyl ion is much lower than that of the diphenylmethyl ion. This is because of the bond between the two benzene rings in the former.

Carbanion	λ_{max}, mμ	$\log \varepsilon_{max}$
Diphenylmethyl..	440	4.6
Fluorenyl ..... {	480	3.1
	367	3.9

As a comparison of the following values shows, the spectrum of the $(C_6H_5)_3C^-$ ion is quite close to that of the $(C_6H_5)_2CCH_3$ ion ($\lambda_{max} = 495$ mμ; $\log \varepsilon_{max} = 4.4$ and $\lambda_{max} = 480$ mμ; $\log \varepsilon_{max} = 4.3$). The first work on the infrared absorption spectra of carbanions was carried out in the same laboratory [55]. The ionization of the hydrogen in the methyne group of triphenylmethane leads to the disappearance of the aliphatic CH bond frequency of 3022 cm^{-1} from its spectrum. The ionization of the hydrogen of the methylene group in indene and diphenylmethane is accompanied by weakening of the bands in the region of 2900 cm^{-1}, which are characteristic of the valence vibration in the CH_2 group, and the appearance of a band with a frequency of about 3000 cm^{-1}, which may be ascribed to the valence vibration of C$-$H in the carbanion. The infrared spectra of toluene and bibenzyl in ammonia change insignificantly under the influence of potassamide. It was difficult to describe them reliably as the absorption intensities were estimated visually in the work cited.

4. COMPARISON OF THE STRENGTHS OF CARBO ACIDS
BY METALLATION REACTIONS
A. General Comparison

A knowledge of the conditions under which different hydrocarbons react with alkali metals and with bases of different strengths makes it possible to draw some conclusions

TABLE 27. Light Absorption of Carbanions Ar_2CH^-

Carbanion	λ_{max}	$\log \varepsilon_{max}$
$\alpha,\alpha'- (C_{10}H_7)_2CH^-$	645	4.9
$(C_6H_5C_6H_4)_2CH^-$	590	4.9
$(C_6H_5CH_2C_6H_4)_2CH$	470	4.7
$(C_6H_5)_2CH^-$	440	4.6
$(C_6H_5OC_6H_4)_2CH^-$	450	4.5
$[(H_3C)_3CC_6H_4]_2CH^-$	440	4.5
$(H_3CC_6H_4)_2CH^-$	445	4.4
$(H_3COC_6H_4)_2CH^-$	430	4.1
$[(H_3C)_2NC_6H_4]_2CH^-$	445	3.5
$[(H_5C_2)_2NC_6H_4]_2CH^-$	450	3.3
$C_6H_5\bar{C}HCH{=}CH_2$	420	4.2

on the relative acidity of hydrocarbons. From the data presented above (p. 83, 85, and 86) it follows that hydrocarbons may be arranged in the following sequence in order of acidity: fluoradene > cyclopentadiene > indene > perinaphthalene > fluorene > triphenylmethane > diphenylmethane > toluene > benzene > ethane.

Particularly valuable information on the relative acidity of hydrocarbons is given by successive applications of Shorygin's reaction. Thus, triphenylmethylsodium metallates fluorene smoothly [62]:

In its turn, triphenylmethylsodium is formed by metallation of triphenylmethane with diphenylmethylsodium [3, p. 448].

$$(C_6H_5)_2CHNa + (C_6H_5)_3CH \rightarrow (C_6H_5)_2CH_2 + (C_6H_5)_3CNa.$$

Diphenylmethane is metallated by benzylsodium [6, p. 16]:

$$C_6H_5CH_2Na + (C_6H_5)_2CH_2 \rightarrow C_6H_5CH_3 + (C_6H_5)_2CHNa.$$

Phenylsodium is obtained by the reaction between ethylsodium and benzene [38, p. 383]. This confirms the above sequence of acids with respect to decreasing strength.

As reported by K. A. Kocheshkov and T. V. Talalaeva [1, p. 175], conclusions on the relative acidity of hydrocarbons may also be drawn from the relative rates of metallation of different hydrocarbons by the same reagent by comparing the yields of organometallic compounds and the yields of products formed during the metallation of two hydrocarbons by an inadequate amount of metallating reagent [competing metallation]. Many such studies were carried out by Morton, Gilman, et al.

In the literature there are qualitative series of relative acidity (see, for example, the table in K. A. Kocheshkov and T. V. Talalaeva's book [1], p. 179, and also [2, 3, 5]). Conant and Wheland attempted to use Shorygin's reaction for the quantitative estimation of the strengths of hydrocarbons as acids in the form of conventional dissociation constants [63-66] by determining the concentrations of components in the reaction mixture (for example, by using a colorimetric method if the carbanion were colored). Knowing the ratio of the concentrations of ionized and un-ionized forms and assuming the dissociation constant of one of the acids, they calculated the dissociation constant of the second acid. The negative logarithms of these dissociation constants (i.e., the values pK = − log K) are given below in brackets. As the beginning of this scale it was assumed that pK = 20 [64] for acetophenone (first number) or pK = 16 for methanol (second number): phenylacetylene, indene, and phenylfluorene (22; 21), fluorene (24; 25), triphenylmethane (28.5; 33), diphenyl-α-naphthylmethane (−; 34), diphenylmethane (29.5; 36), diphenylmethylethylene (−; 36), and isopropylbenzene (−; 37).* It was also found that the strength of carbo acids falls in the series: diphenylmethane, phenylcyclopentane, isopropylbenzene, phenylcyclohexane [65].

Here and above the lability of the hydrogen at the α-carbon atom attached to aromatic rings was considered mainly. As was shown by the work of P. P. Shorygin (p. 86), it is precisely these hydrogen atoms which are replaced first. Later work made it possible to differentiate between the acidities of different alkylbenzenes.

B. Classes of Hydrocarbons

a. ALKYLBENZENES

It was found that the tendency toward metallation in the α-position falls from toluene to ethylbenzene and then to isopropylbenzene [67]. Thus, in the reaction with ethylpotassium, the respective degrees of metallation in the α-position are 100, 50, and 13% of the total replacement of hydrogen by the metal in these alkylbenzenes. Amylsodium [68], in contrast to ethylpotassium [69, 70], does not touch the hydrogen in the isopropyl group at all. On the other hand, amyl- and phenylpotassiums metallate isopropylbenzene almost exclusively in the α-position [71]. When methyl and isopropyl groups are present simultaneously in the molecule (p-cymene), only the former is metallated ([70], see also [72]). Thus, the replacement of the hydrogen atoms in the alkyl group of toluene by electron-donor methyl groups progressively reduces the lability of the hydrogen atoms remaining in it.

Experiments on competitive metallation of benzene and isopropylbenzene [67] show that the aromatic ring of the latter is deactivated and that the effect is greatest in the ortho-, less in the meta-, and least in the para-position. The relative rates of metallation in these positions with respect to the rate of metallation of benzene are 0.10, 0.37, and 0.48, while the corresponding value for the methyne group is 0.32.

---

* The pK of fluoradene equals 11 ± 0.5 [29]. For methane there is the very rough value of pK = 40 and according to other data, pK = 50.

The aromatic ring of tert-butyl-benzene is deactivated somewhat more than the ring of isopropylbenzene (see also [70]). The ratios of the yields of individual isomers (o, m, p, and α) in the metallation of isopropylbenzene with ethylpotassium at 20° are given by the figures 9.0, 48.5, 29.5, and 13.0 [67].

Xylenes are metallated with greater difficulty and at a higher temperature than toluene [73]. In metallation by amylpotassium, the hydrogen in the methyl group of m- and p-xylenes is replaced by metal more readily than in o-xylene; the latter; in contrast to the other isomers, is partly metallated in the nucleus. The relative yields in the metallation of p-, m-, and o-xylene are 54, 32, and 22%, respectively. Dimet-allation occurs in the two methyl groups of p- and m-xylene and with much more dif-ficulty in o-xylene (yields: 36, 37, and 19%).

Though one hydrogen atom of the methyl group is generally replaced in the met-allation of toluene, K. A. Kocheshkov and T. V. Talalaeva found conditions under which there is not only dimetallation [73], but even polymetallation of toluene [74]. Heating toluene with ethyllithium at 90-100° leads to the formation of crystalline tetra- and pentalithium compounds of toluene. Tetralithiumdiphenylmethane has also been isolated.

Thus, although the hydrogen atoms at the α-carbon atom of an aliphatic substit-uent are much more labile than the hydrogen atoms of the aromatic nucleus, it is pos-sible to replace even several hydrogen atoms in the latter by a metal.

b. AROMATIC HYDROCARBONS

Naphthalene is metallated more readily than benzene [75-77]. Together with α- and β-monosubstituted derivatives, the metallation of naphthalene also forms 1,3-, 1,8-, and 2,6-disubstituted derivatives and also a trisubstituted derivatives of undeter-mined composition.

c. ALKYNES

It has been established that the acidity of alkynes falls in the series: $C_6H_5C \equiv$ $\equiv CH > C_5H_{11}C \equiv CH > HC \equiv CH$ [76]. By metallation in liquid ammonia, it was shown that fluorene is a more acid hydrocarbon than acetylene [54].

d. ALKENES

Ethylene is a weaker acid than acetylene; it is metallated only by strong bases such as amylsodium [68]. Morton and his co-workers made a detailed study of the metallation of alkenes from ethylene to octene inclusive [78-81]. A vinyl group ac-tivates a hydrogen atom at an adjacent carbon atom. Propylene is therefore metal-lated much more readily than ethylene and the metal replaces hydrogen in the methyl group [81] (see also 72, 79]). By dimetallation it was possible to obtain $NaCH = CHCH_2Na$ [82].

The hydrogen atoms in isobutylene are even more labile than in propylene and the former is metallated readily not only by amylsodium, but even by benzylsodium, i.e., is a stronger acid than toluene, to form the mono- and disubstituted products $CH_2 = C(CH_3)CH_2Na$ and $CH_2 = C(CH_2Na)_2$ [81] (see also [72, 78, 79, 82]).

The acidity of an ethylenic hydrocarbon is lower if methyl (or, in general, alkyl) groups are present on both sides of the double bond as in 2-butene [72], in which only the hydrogen in the alkyl position is replaced. The weakening of the acidic properties is even greater in 1-butene, 1-pentene, and other ethylenic hydrocarbons [72]. However, first of all the hydrogen in the allyl position is replaced preferentially (60-70%) by the metal and then metallation occurs with much more difficulty in the vinyl group [72, 81]. The same was observed [78] in the metallation of 1-hexene. In alkenes with a long chain, the lability of the corresponding hydrogen atom is considerably lower than in the methyl group of propylene or, in particular, of isobutylene, but it is still more active than a hydrogen atom in ethylene as 1-hexene is metallated by vinylsodium comparatively readily [78]. Trimethylethylene and tetramethylethylene have moderate reactivity [72]. Thus, the relative acidity of alkenes decreases in the sequence:

$$(CH_3)_2C = CH_2 > CH_3CH = CH_2 > CH_3CH = CHCH_3 >$$
$$> CH_3CH_2CH = CH_2 > (CH_3)_2C = CHCH_3 > (CH_3)_2C = C(CH_3)_2 .$$

There are reports on the metallation of alkenes with a branched chain [78].

Diallyl is very reactive [78, 79]; it is metallated readily by benzylsodium. The double bond migrates during metallation. This is also observed in the metallation of a series of alkenes (see p. 116 on the isomerization of alkenes). The following products are obtained during the metallation of diallyl:

$$CH_2 = CHCHNaCH_2CH = CH_2, \quad NaCH = CHCH_2CH_2CH = CH_2$$
$$NaCH = CHCH_2CH = CHCH_3, \quad NaCH = CHCH_2CH = CHNa .$$

In comparing ethylenic and aromatic hydrocarbons, we have already mentioned that diallyl and isobutylene are stronger acids than toluene. Vinyl- and allylsodium metallate benzene, i.e., the latter is more acid than ethylene and even propylene [79, 81].

The metallation of alkenes was studied in order to determine the mechanism of diene polymerization by sodium, which is used to prepare synthetic rubber. Increasing attention has been paid in recent years to the anionic polymerization of unsaturated hydrocarbons; it is induced by carbanions in ether and in liquid ammonia and by amide ions (NH_2^-) in liquid ammonia [54, 83]. The addition of potassamide to ammonia solutions of diphenylethylene, α-methylstyrene, and styrene forms intensely colored solutions. It was found that there is a direct relation between the rate of anionic polymerization and the strength of the base catalyzing the reaction. The anions of very weak acids (indene, fluorene, acetylene, triphenylmethane, etc., [54]) act as catalysts.

e. SATURATED CYCLIC HYDROCARBONS

In the case of saturated hydrocarbons, it should be noted that decalin is metallated very readily by amylsodium [75]. A series of products of hydrogen substitution, including tetrasubstituted derivatives, are formed. This is the first case of the metallation of a hydrocarbon which does not contain an aromatic ring or a multiple bond.

f. ALKANES

It is possible to draw indirect conclusions on the relative acidity of saturated aliphatic hydrocarbons by comparing the efficiency with which the same substance is metallated by alkylsodiums and lithiums with straight and branched chains of carbon atoms and with different numbers of carbon atoms.

In the metallation of benzene and toluene, the reactivity falls in the series amyl-, butyl-, and propylsodium [84]. In the replacement of hydrogen by a metal in toluene, the following sequence of reactivities was observed: tert-butyl- > n-butyllithium and in the metallation of dibenzofuran: tert-butyl- > isopropyl- > butyl- > methyllithium [85]. Consequently, elongation and branching of the chain decrease the acidity of saturated hydrocarbons, which is generally expressed extremely weakly (see also [86, p. 52] and [8, 87]).

5. ISOTOPIC EXCHANGE OF HYDROGEN

A. Amphoteric Solvents

The isotopic exchange of hydrogen in organic compounds has already been studied for more than 25 years, but until recently, our knowledge of the exchange of hydrogen in CH bonds was very limited as scientists have used almost exclusively amphoteric deuterated solvents, namely, heavy water and deuteroalcohol, with which the hydrogen in CH bonds of only a few substances exchanges and then largely when catalyzed by bases [88, 89]. Such substances generally contain electronegative atoms or groups (Cl, NO_2, $COOH$, and SO_3H), which confer on the substance acidic properties that are clearly marked even in water or alcohol solution (see Table 14). Thus, there is base-catalyzed exchange of hydrogen in nitromethane [90], in sym-trinitrobenzene [91], and in the methyl group of p-nitrotoluene [92].

Substances have also been synthesized in which CH bonds are so activated that the hydrogen exchanges quite rapidly wihout the addition of an acid or a base (i.e., the solvent participates in an acid−base interaction). For example, in thiazolonium salts (I), the hydrogen atom in the position marked with an asterisk exchanges with a half-exchange period of 20 min at 28° [93]. Such ready exchange results from the fact that there are two adjacent electronegative atoms (N^+ and S), which very strongly increase the lability of the hydrogen in the CH bond marked with an asterisk.

Two hydrogen atoms in pyrone (II) exchange with heavy water in the absence of a catalyst [94].

At present there is only one known case of the ready exchange of hydrogen of an unsubstituted hydrocarbon with an amphoteric solvent without a catalyst and this is the case of fluoradene, one of the most acidic hydrocarbons (p. 84). Hydrocarbons of lower acidity (indene, phenylacetylene, and acetylene) exchange their most labile hydrogen atoms with heavy water or alcohol only in the presence of an alkaline catalysts. Acids do not catalyze the exchange.

A 0.5 N solution of sodium hydroxide induces exchange of the hydrogen in acetylene even at 0° [95]. When indene is shaken with heavy water at 100° for 5 hr, there is practically no exchange. The two hydrogen atoms of the methylene group exchange with a 1 N solution of alkali in water. When the alkali concentration is raised to 3 N, the time required for the exchange of the two hydrogen atoms is reduced to an hour [96]. There was no exchange when fluorene was treated with deuteroalcohol for 672 hr at 110°, but about two hydrogen atoms exchanged in the presence of 0.02 N alkali[91] and more than four hydrogen atoms exchanged after 45 hr in the presence of a 0.1 N solution of alkali in heavy water at 110° [96a, p. 286].

Quantitative measurements were recently made on the rate of exchange of hydrogen in acetylene and in the methyne group of phenylacetylene at different pH values of the aqueous solution. At pH < 10, the exchange rate was independent of the rate at which the solution was stirred. Alkali is a catalyst as is indicated by the increase in the rate constant of the exchange of hydrogen in phenylacetylene with an increase in the pH:

pH	$k \cdot 10^6$, sec^{-1}	pH	$k \cdot 10^6$, sec^{-1}
2.0	no exchange	8.6	1300
5.0	2	9.0	4600
7.0	71	9.5	12000
7.9	510		

The authors considered that the rate-determining stage of the exchange is the abstraction of a proton by the base [95a, 95b].

This essentially exhausts the information on hydrogen exchange between hydrocarbons and amphoteric solvents. To it we can add the extremely interesting results obtained by D. N. Kursanov and Z. N. Parnes on the exchange between cyclopentadiene and heavy water dissolved in dioxane.

On the recommendation of Ya. L. Gol'dfarb, D. N. Kursanov, and Z. N. Parnes[97] used as the catalyst the strong organic base N,N'-dimethyl-α-pyridonimine:

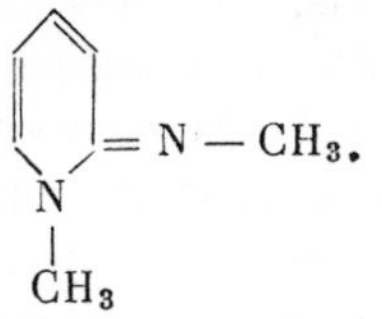

The exchange reached equilibrium in 20 hr at 20° and not only the hydrogen atoms of the methylene group, but also all the other atoms were replaced by deuterium. The case of total exchange is explained by the fact that the base converts the hydrocarbon into the cyclopentadienyl anion:

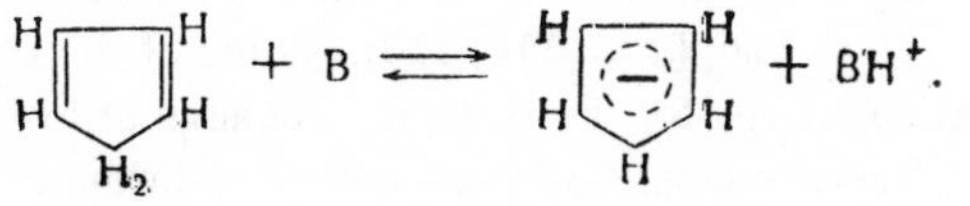

$$\text{(I)}$$

The cyclopentadienyl anion is stabilized because of its aromaticity. All the hydrogen atoms in it are completely alike and the exchange is effected by numerous repetitions of the acid−base reactions (I) and (II):

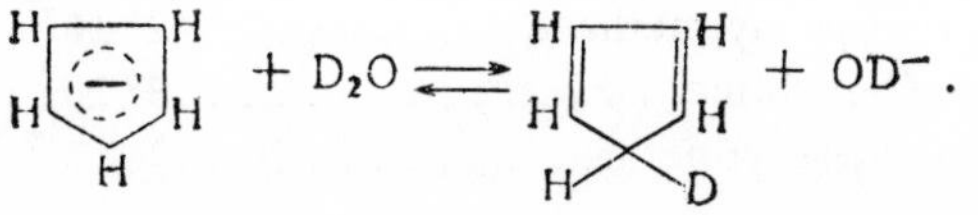

$$\text{(II)}$$

D. N. Kursanov and Z. N. Parnes successfully used the above catalyst to study hydrogen exchange between heavy water and unsaturated and aliphatic-aromatic ketones and aldehydes [98-103].

B. Liquid Ammonia

a. GENERAL COMPARISON OF THE STRENGTHS OF ORGANIC SUBSTANCES AS ACIDS

The investigation of hydrogen exchange in ammonia solutions for determining the strengths of hydrocarbons as acids which was begun in the Isotopic Reactions Laboratory in 1947, was based on the increase in the acidity of substances produced by protophilic solvents. In the first work [104, 105], it was shown that there is a parallelism between the exchange rate constants and the conditional dissociation constants of hydrocarbons of Conant and Wheland, which were measured by means of metallation reactions (p. 93). The conception of the strong analogy between metallations by organoalkali compounds and hydrogen exchanges with bases was developed in subsequent work. At the same time, it was discovered that potassamide is a powerful catalyst of hydrogen exchange in liquid ammonia. This offered completely new possibilities of determining the acidic properties of hydrocarbons, as in the presence of potassamide, deuterium replaces hydrogen atoms of not only aliphatic-aromatic, aromatic, acetylenic, and ethylenic hydrocarbons, but even a series of saturated hydrocarbons and their derivatives [106-126] (see also the reviews [127-133]).

The question arose as to what scale to use for the strengths of acids which could be measured by means of hydrogen exchange reactions in ammonia solutions. It is difficult to give a complete answer to this problem, but some considerations can be put forward and individual data presented.

Ammonia can exist in the liquid state over the temperature range from $-78°$ (freezing point) to $132°$ (critical point). Consequently, by varying only the temperature it is possible in principle to study reactions whose rates vary over approximately six orders if it is assumed that a rate constant doubles with a change in temperature of ten degrees. A greater range may be covered by the use of alkaline catalysis by amide ions, a normal solution of which accelerates a reaction in ammonia by a minimum of ten orders (p. 24). At temperatures above $0°$, by varying the duration of experiments over a practicable range, it is possible to determine rate constants differing by up to six orders (from 10^{-3} to 10^{-9} sec^{-1}). Hence, the total range of rate constants and consequently, the parallel range of acid strengths are very considerable.

The exchange rate constants (sec^{-1}) given below were obtained by measuring the kinetics of hydrogen exchange in ammonia solutions without the addition of catalyst at temperatures from 25 to $120°$ [105, 116, 117]. Some of the constants were corrected to $120°$. The activation energy for the isotopic exchange of hydrogen in the methylene groups of indene and fluorene equals 12 kcal/mole and in the methyl group of acetophenone, 11 kcal/mole.

Substance	k
Indene	$4 \cdot 10^{1}$
Fluorene	$2 \cdot 10^{-2}$
Acetophenone.	$4 \cdot 10^{-3}$
Triphenylmethane.	$2 \cdot 10^{-7}$
Quinaldine.	$8 \cdot 10^{-8}$
Diphenylmethane	$7 \cdot 10^{-9}$

When potassamide is added to ammonia solutions of all the substances listed in Table 26, they are readily converted into salts, whose anions are intensely colored (p. 90). Toluene is a considerably weaker acid than the substances listed and is not ionized appreciably under these conditions. This is explained by the fact that its mole-cule contains three aliphatic CH bonds and only one electronegative phenyl group, which attracts the electrons, while in triphenylmethane, on the other hand, there are three phenyl groups and only one aliphatic CH bond with the result that the hydrogen in it is much more positive than in the methyl group of toluene and is readily proton-ized. The formation of a bond between the rings of diphenylmethane, converting the latter into a fluorene molecule, energetically stabilizes the anionic form: fluorene is much more acidic than diphenylmethane and the rate constant of hydrogen exchange in the C_2H_2 group of fluorene is six or more orders higher than in diphenylmethane.

To induce hydrogen exchange between the methyl group of toluene and liquid am-monia, it is necessary to introduce potassamide into the solution. The rate constants of the exchange in a 0.02 N solution of potassamide ($k_0° = 5 \cdot 10^{-4}$ sec^{-1}) [112, 122] is close to the rate constant of hydrogen exchange in the methylene group of indene with-out a catalyst ($k_0° = 4 \cdot 10^{-4}$ sec^{-1}). This is a further demonstration of the exceptionally high catalytic activity of amide ions.[*] The lability of the hydrogen atoms of the aro-matic CH bonds of toluene is much lower than in the methyl group (the rate constants of deuterium exchange differ by a factor of 250 [122]) and at the same time it is less than in benzene (the rate constant is a factor of 4 less).

For further comparison of the acidity of CH bonds, the rate constant of hydrogen exchange in benzene may be taken arbitrarily as unity. The ratio of the rate constants of exchange in the ortho-CH bonds of benzene derivatives to the rate constants of hy-drogen exchange in benzene shows the extent to which these bonds become more acidic when various substituents are introduced into the benzene ring.

Substituent	$k_{ortho}/k_{C_6H_6}$
H	1
C_6H_5	5
CH_3O	$5 \cdot 10^2$
C_6H_5O	$2 \cdot 10^4$
CF_3	$> 10^5$
F	$> 10^6$

It is much more difficult to make a comparison in the case of an acidity lower than that of benzene. For this purpose, let us consider hydrocarbon molecules wtih the same number of carbon atoms as benzene, but a different structure (Table 28) [108,111, 131].

The saturated hydrocarbon exchanges hydrogen much more slowly. Consequently, its acidic properties are expressed incomparably more weakly than those of aromatic and ethylenic hydrocarbons.

b. ALIPHATIC-AROMATIC HYDROCARBONS

In the metallation of alkylbenzenes, the hydrogen atom at the α-carbon atoms of the alkyl group is generally replaced by an alkali metal. The same hydrogen atoms

---

[*] The conditional ionization constants of toluene and indene, determined by metalla-tion, differ by not less than 15 orders (p. 92).

also participate in hydrogen exchange. This is usually explained by the displacement of the electrons of the aliphatic CH bonds toward the aromatic ring.

TABLE 28. Comparison of the Rates of Hydrogen Exchange in Hydrocarbons with Six Carbon Atoms

Hydrocarbon	Temp., °C	Time, hr	$C_{KND_2, N}$	n	k, sec^{-1}
Benzene	25	1	0.06	3	$2 \cdot 10^{-4}$
Hexene	25	4	0.06	2.5	$2 \cdot 10^{-5}$
Cyclohexene	25	4	0.06	1.7	$1 \cdot 10^{-5}$
Hexane	120	1300	3.0	2.7	—

The rate constants of deuterium exchange in alkylbenzenes, catalyzed by a 0.02 N solution of potassamide, were measured (Table 29).

TABLE 29. Rate Constants of Deuterium Exchange in Benzene and Mono-alkylbenzenes

Temp., °C	Benzene	Toluene	Ethyl-benzene	Isopropyl-benzene	tert-Butyl-benzene
		Alkyl	group		
10	—	$1.4 \cdot 10^{-3}$	$2 \cdot 10^{-4}$	$4 \cdot 10^{-5}$	
0	—	$5 \cdot 10^{-4}$	$7 \cdot 10^{-5}$	—	
		Aromatic	ring		
25	$8.5 \cdot 10^{-5}$	$2.1 \cdot 10^{-5}$	$2.0 \cdot 10^{-5}$	$1.7 \cdot 10^{-5}$	$1.4 \cdot 10^{-5}$
10	$1.9 \cdot 10^{-5}$	—	—	$4 \cdot 10^{-6}$	—
0	$7.0 \cdot 10^{-6}$	$(3 \cdot 10^{-6})$	$(3 \cdot 10^{-6})$	—	—

By taking the ratio of the rate constants of exchange of different hydrogen atoms (in the aromatic ring and methyl, methylene, and methyne groups) of alkylbenzenes to the rate constants of hydrogen exchange in benzene [108] at corresponding temperatures, we obtained the following values for the relative rate constants:

$$\begin{array}{cccc} \text{Ring} & \equiv \text{CH} & = \text{CH}_2 & -\text{CH}_3 \\ 0.25 & 2 & 10 & 70 \end{array}$$

These numbers characterize the relative reactivity of the hydrogen atoms in the aromatic ring and in the α-CH bonds of the alkyl groups of isopropylbenzene, ethyl-benzene, and toluene. The same sequence was found by metallation (p. 93). Table 29 gives the mean rate constants of the exchange of hydrogen atoms in the aromatic ring of monoalkylbenzenes. There is an appreciable decrease in the mean rate constant of hydrogen exchange in the ring with branching of the alkyl group. The activation energy equals approximately 16-17 kcal/mole.

Analogous data were recently obtained by Streitwieser, who catalyzed the exchange by a solution of cyclohexyllithium in cyclohexylamine. The relative rates of exchange of the α-hydrogen atoms of toluene, ethylbenzene, and isopropylbenzene are given by the figures: 1.00 : 0.11° 0.009 (see footnote 34 in [143b]).

To determine the reaction rate of hydrogen atoms in different positions relative to the methyl group of toluene, the authors used monodeuterotoluene isomers [124] and obtained the results given in Table 30.

TABLE 30. Rate Constants of Deuterium Exchange in Monodeuterotoluene Isomers (C_{KNH_2} = 0.02 N)

Temp., °C	Isomer		
	ortho-	meta-	para-
25	$1.6 \cdot 10^{-5}$	$2.4 \cdot 10^{-5}$	$2.4 \cdot 10^{-5}$
0	$1.0 \cdot 10^{-6}$	$2.3 \cdot 10^{-6}$	$(1.6 \cdot 10^{-6})$

The differences in the constants are very small, but the hydrogen in the ortho-CH bond definitely exchanged more slowly than that in the meta-position.

Thus, the hydrogen atoms of the ring and the aliphatic α-CH bond of monoalkylbenzenes exchange at low potassamide concentration and at temperatures of 0-25°. This rule is also observed with other compounds of similar structure,* as is shown by Table 31, which gives the total number of hydrogen atoms exchanged for deuterium at equilibrium. This number equals the sum of the hydrogen atoms in the aromatic ring (H_{Ar}) and in the α-position in the alkyl group (H_α). If the α-carbon atom in the alkyl group is primary, the total number of hydrogen atoms exchanging (n) equals seven, when a secondary carbon atom is present, n = 6, and in a hydrocarbon a tertiary carbon atom n = 5. This rule may be used to determine the nature of the branching of the alkyl group. It also applies to phenylated alkanes, in which the total number of hydrogen atoms exchanging for deuterium equals $H_{Ar} + H_\alpha$, and to partly hydrogenated hydrocarbons containing an aromatic ring, such as hydrindene (I), tetralin (II), and phenylcyclohexane (III):

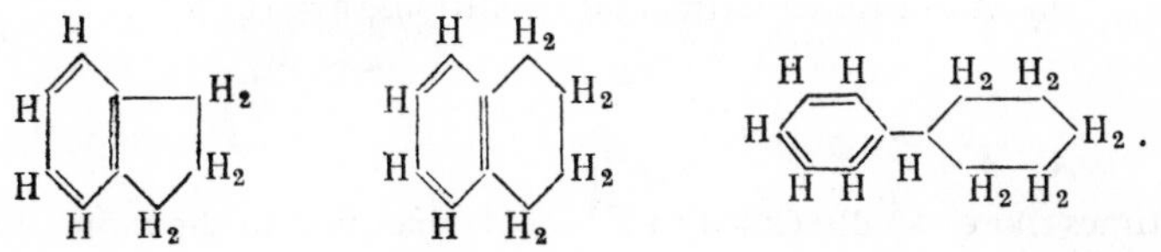

The tetralin molecule may be represented as o-diethylbenzene in which there is ring closure between the ethyl groups. There is much in common between hydrogen

* According to qualitative observations, the hydrogen in the ring of pentamethylbenzene exchanges under more drastic conditions.

exchange in tetralin and monoalkylbenzenes. According to approximate measurements [110] (20°, C_{KND_2} = 0.05 N), the hydrogen in the α-methylene groups exchanges a

TABLE 31. Number of Hydrogen Atoms (n) Exchanging for Dueterium in Alkylbenzenes and Phenylated Alkanes

Hydrocarbon	n	Literature reference
$C_6H_5CH_3$	8	[105, 109]
$C_6H_4(CH_3)_2$	10	[105, 135]
$C_6H_3(CH_3)_3$	12	[105, 135]
$C_6H_2(CH_3)_4$	14	[133]
$C_6H(CH_3)_5$	16	[133]
$C_6(CH_3)_6$	18	[133]
α-$C_{10}H_7CH_3$	10	[105]
β-$C_{10}H_7CH_3$	10	[105]
$C_6H_5CH_2CH_3$	7	[121]
$C_6H_5CH_2CH_2CH_3$	7	[121]
$C_6H_5CH_2(CH_2)_2CH_3$	7	[121]
$C_6H_5CH(CH_3)_2$	6	[121]
$C_6H_5C(CH_3)_3$	5	[121]
$C_6H_5CH_2C_6H_5$	12	—
$C_6H_5CH_2CH_2C_6H_5$	14	[112,121]
$C_6H_5CH_2CH_2CH_2C_6H_5$ *	14	—
$C_6H_5CH_2(CH_2)_2CH_2C_6H_5$	14	[121]
Hydrindene	8	[105]
Tetralin	8	[110]
Phenylcyclohexane	6	[115]

*Under drastic experimental conditions, n = 16.

hundred times more rapidly (k = $1 \cdot 10^{-3}$ sec^{-1}) than that in the aromatic ring of tetralin (k = $1 \cdot 10^{-5}$ sec^{-1}), while the rate of exchange of the latter is a factor of five less than in benzene. The hydrogen of the β-CH bonds is not touched even during prolonged heating (100 hr, 120°, C_{KND_2} = 0.6 N). The exchange of hydrogen in bibenzyl, $C_6H_5CH_2CH_2C_6H_5$, proceeds similarly and the rate of exchange in the aliphatic CH bonds is approximately two orders greater than in the aromatic bonds, while exchange in the aromatic ring of bibenzyl is slower by a factor of two than in benzene [112].

Table 32 gives a summary of the kinetics of hydrogen exchange in methylene and phenyl groups of aliphatic-aromatic hydrocarbons (temperature 25°, C_{KND_2} = 0.02 N).

As it is electronegative, the phenyl ring produces a displacement of electrons in the adjacent aliphatic CH bonds with the result that the hydrogen atoms in them are protonized more readily (see discussion of hydrogen exchange in di- and triphenylmethane above, p. 98).

TABLE 32. Rate Constants of Hydrogen Exchange in α-CH_2 and C_6H_5 Groups of Alkylbenzenes

Hydrocarbon	α-CH_2-	C_6H_5-	Literature reference
Benzene	—	$8.2 \cdot 10^{-5}$	[108]
Ethylbenzene	$8 \cdot 10^{-4}$	$2.0 \cdot 10^{-5}$	[122]
Bibenzyl	$\sim 10^{-3}$	$4.5 \cdot 10^{-5}$	[11]
Tetralin*	$5 \cdot 10^{-4}$	$4 \cdot 10^{-6}$	[110]

*At 20°.

Let us return to alkylbenzenes. In very long experiments, a slight increase in n over the sum of $H_{Ar} + H_\alpha$ was observed [121] with some alkylbenzenes. It was considered that this phenomenon was caused by the beginning of exchange of the β-hydrogen atoms of the alkyl groups. When the temperature was raised to 100-120° and the potassamide concentration to 0.8-2.0 N, it was also possible to exchange the hydrogen atoms in the β-position, as is shown by the total number of hydrogen atoms exchanging in alkylbenzenes.

Hydrocarbon	n
$C_6H_5CH_2CH_2CH_3$	9
$C_6H_5CH_2CH_3$	10
$C_6H_5CH(CH_3)_2$	12
$C_6H_5C(CH_3)_3$	13 *
$(C_6H_5)_3CCH_3$	18

There was also more extensive exchange in n-butyl- and n-hexylbenzene, where up to 12 hydrogen atoms exchanged. The experiments were usually carried out in ampoules of chrome-nickel steel, but the exchange of the β-atoms was not caused by the catalytic action of the ampoule walls, but by potassamide. This was demonstrated by control experiments in glass ampoules.

The question arose as to whether these facts indicate that the aromatic ring has an effect on the lability of hydrogen atoms of the alkyl group remote from the ring. It is difficult to give an unequivocal answer to this question as yet since the exchange of hydrogen in saturated hydrocarbons is observed under similar conditions (p. 114). Nonetheless, it is probable that this effect occurs as the exchange proceeds more readily in aliphatic-aromatic hydrocarbons than in alkanes. For example, complete exchange of the hydrogen in ethylbenzene was reached after 50 hr at 120° and an amide concen-

* At complete exchange n = 14.

tration of 0.7 N. It should also be noted that the lability of the hydrogen atoms of CH bonds in the β-position increases with an increase in the number of phenyl groups in the hydrocarbon molecule. Thus, despite the low solubility of 1,1,1-triphenylethane, $(C_6H_5)_3CCH_3$, in liquid ammonia, its hydrogen exchanges completely in a time which is only sufficient for the β-atoms of tert-butyl-benzene to begin to exchange. All the hydrogen atoms in sym-diphenylpropane could be replaced, though only $H_{Ar} + H_\alpha$ exchanged in parallel experiments with sym-diphenylbutane (the latter substance, however, is less soluble). Clear signs of the effect of aliphatic CH bonds in the β-position on the change in electron density in the ring were found in a study of acid hydrogen exchange with phenylated alkanes [133a].

c. AROMATIC HYDROCARBONS

The rate of exchange of hydrogen in benzene was measured at different temperatures and at several potassamide concentrations [108, 125]. The rate constants (sec^{-1}) are given in Table 33.

TABLE 33. Rate Constants of Hydrogen Exchange in Benzene

Temp., °C	Potassamide concentration, N					
	0.010	0.014	0.021	0.059	0.19	0.43
—30	—	—	—	$4 \cdot 10^{-7}$	—	—
0	$4 \cdot 10^{-6}$	—	$7.3 \cdot 10^{-6}$	—	—	$8.9 \cdot 10^{-5}$
25	$4.4 \cdot 10^{-5}$	$5.7 \cdot 10^{-5}$	$8.6 \cdot 10^{-5}$	$1.8 \cdot 10^{-4}$	$4.2 \cdot 10^{-4}$	—
40	$1.6 \cdot 10^{-4}$	$3.1 \cdot 10^{-4}$	—	—	—	—

In experiments with dilute solutions of potassamide, the rate constant was approximately proportional to the catalyst concentration, but with an increase in the concentration, the increase in the rate constant fell off. Thus, with an increase in the amide concentration by a factor of 43, the rate increased by a factor of only 23. It is evident that the fall in catalytic activity is explained by strong interionic interaction and the formation of ion pairs. It is quite probable that the catalytic activities of different alkali metal amides differ ($CsNH_2 > RbNH_2 > KNH_2*$). This has not been checked experimentally.

The activation energy of the exchange reaction at $C_{KND_2} = 0.01$ and 0.02 N equals $15,800 \pm 500$ kcal/mole. The rate constant of the reaction is expressed by the following equation:

$$k = 10^A e^{-15800/1.986T},$$

where A, the index in the pre-exponential factor, has the following values for different potassamide concentrations:

$$C_{KND_2}, N : 0.01 \ 0.02 \ 0.06 \ 0.19 \ 0.43$$
$$A : 7.24 \ 7.52 \ 7.85 \ 8.22 \ 8.62$$

---

* Sodium and lithium amides are sparingly soluble in liquid ammonia.

They were calculated on the assumption that the activation energy is independent of the potassamide concentration. This assumption is known to be inaccurate as the activation energy changes in the presence of electrolytes (see p. 64 on the salt effect). Therefore, the equation given applies strictly only in the calculation of rate constants at low amide concentrations. At $-30°$ and $C_{KNH_2} = 0.06$ N, measurement of the hydrogen exchange rate gave a value $k = 4 \cdot 10^{-7}$ and the calculated value was $4 \cdot 10^{-7}$ sec^{-1}. Roberts [134] obtained a value of about 10^{-7} sec^{-1} for the rate constants of hydrogen exchange in benzene at $-33°$ and $C_{KND_2} = 0.6$ N, i.e., an order lower than the value obtained in the Isotopic Reactions Laboratory [125]. It is possible that the lower rate of exchange is explained by partial decomposition of the amide during long experiments in the inadequately sealed apparatus used by Roberts.

As in the case of metallation, hydrogen exchange in naphthalene proceeds more rapidly than in benzene. The mean rate of hydrogen exchange in naphthalene was measured first [108] (Table 34). The rate constants of the exchange (in sec^{-1}) at a potassamide concentration of 0.01 N obey the equation $k = 10^{7.0} \cdot e^{-14,200/1.986T}$. The rate

TABLE 34. Rate Constants of Hydrogen Exchange in Naphthalene

Temp., °C	Potassamide concentration, N	
	0.010 N	0.021 N
10.5	$9.9 \cdot 10^{-5}$	—
25	$3.4 \cdot 10^{-4}$	$5.9 \cdot 10^{-4}$
40	$1.3 \cdot 10^{-3}$	—

constants of the exchange of nonequivalent (α and β) hydrogen atoms were determined later [124] by the use of the appropriate isomers of monodeuteronaphthalene ($C_{KND_2} = 0.02$ N; $25°$):

	k, sec^{-1}
Monodeuteronaphthalene	
α-	$8.0 \cdot 10^{-4}$
β-	$3.6 \cdot 10^{-4}$
Benzene	$4.4 \cdot 10^{-5}$

Little is known as yet on hydrogen isotope exchange in other polycyclic hydrocarbons with condensed rings, as anthracene rapidly forms tars even with a low amide concentration, while phenanthrene is sparingly soluble in ammonia. However, even qualitative experiments show that the hydrogen in it exchanges more rapidly than in naphthalene. The rate of acid exchange also increases in the direction benzene < naphthalene < phenanthrene (anthracene) (p. 169). The explanation of this lies in the considerable and different polarizabilities (α) of the molecules of aromatic hydrocarbons (Table 35). Regardless of whether the reagent is an electron donor or an electron acceptor, it polarizes the phenanthrene molecule more than the naphthalene molecule and the latter more than the benzene molecule. This treatment is in agreement with the higher reactivity of α-deuteronaphthalene in comparison with the β-isomer in exchange both with an acid and with a base [124] in accordance with the high polarizability of the naphthalene molecule in the α-position [136].

TABLE 35. Polarizability of the Molecules of Aromatic Hydrocarbons

Hydrocarbon	n_{20}^D	$\alpha \cdot 10^{24}$ cm^3
Benzene	1.501	10
Naphthalene	1.582	18
Phenanthrene	1.657	22

Among polycyclic hydrocarbons in which the aromatic rings are connected linearly it is possible to make measurements only with biphenyl as p- and m-terphenyl are sparingly soluble in liquid ammonia. As in the case of naphthalene, the mean exchange rate constant was determined first [112]: $k = 2.8 \cdot 10^{-4}$ sec^{-1} at 25° and $C_{KND_2} = 0.02$ N. The rate constants of the exchange of the ortho-, meta-, and para-atoms in the biphenyl molecule were then determined [124] (Table 36). They were so similar to each other that the differences were almost inappreciable at 25°, but became more marked when the temperature was reduced to 0°. Measurements at lower temperatures were impossible due to the low solubility of biphenyl in liquid ammonia.

TABLE 36. Rate Constants of Deuterium Exchange with Monodeuterobiphenyl Isomers

Temp., °C	ortho-	meta-	para-
25	$3.1 \cdot 10^{-4}$	$2.5 \cdot 10^{-4}$	$2.5 \cdot 10^{-4}$
10	$8.2 \cdot 10^{-5}$	$7.5 \cdot 10^{-5}$	$5.4 \cdot 10^{-5}$
0	$3.4 \cdot 10^{-5}$	$2.6 \cdot 10^{-5}$	$2.3 \cdot 10^{-5}$

The mean rate of hydrogen exchange in biphenyl is a factor of seven greater than in benzene with the reactivity of the hydrogen atoms decreasing with an increase in the distance of the CH bond from the substituent (phenyl group), i.e., ortho > meta > > para. The exchange reaction with an acid appears quite different (p. 169). Therefore, although the biphenyl molecule is polarized more strongly than the benzene molecule ($\alpha = 21 \cdot 10^{-24}$ cm^3), the explanation proposed for naphthalene is unsuitable. The data on biphenyl fit in well with the general system if allowance is made for the inductive displacement of electrons produced by one phenyl group in the other, which is attacked by the reagent (for more details see p. 278).

Isotopic exchange with deuteroammonia catalyzed by potassamide was used for the first time to prepare a completely deuterated hydrocarbon on the example of biphenyl [135, 137, 138].

d. UNSATURATED ALIPHATIC AND ALICYCLIC HYDROCARBONS

The hydrogen atoms in acetylene are so labile that they are replaced by a metal and exchanged for deuterium in aqueous solutions of alkali. Isotopic exchange of the hydrogen in ethylene in liquid ammonia, catalyzed by potassamide, proceeds more readily than in saturated hydrocarbons, but is still slow [107]. Thus, the lability of the hydrogen atom of a CH bond of an aliphatic hydrocarbon varies very considerably, depending on whether the carbon atom is bound by a triple, double, or single bond, i.e., a very important part is played by the type of valence hybridization of the carbon [139], on which depends the degree of electropositivity of a hydrogen atom attached to it; the hydrogen in acetylene compounds is most electropositive.

With the successive replacement of hydrogen atoms in an ethylene molecule by methyl groups there is a decrease in the heat of hydration of the alkene Q (Table 37).

TABLE 37. Constants of Ethylene and Its Homologs

Hydrocarbon	Q, kcal/mole	I, ev	$\mu \cdot 10^{18}$
$CH_2 = CH_2$	32.8	10.62	0.00
$CH_3CH = CH_2$	30.1	9.84	0.34
$(CH_3)_2C = CH_2$	26.4	9.35	0.54
$(CH_3)_2C = CH(CH_3)$	26.9	8.85	0.47
$(CH_3)_2C = C(CH_3)_2$	26.6	8.30	0.00

The magnitude of ΔQ gives an idea of the energy stabilization of the molecule accompanying the formation of a σ,π-conjugation system. It is also illustrated by the decrease in the ionization potential I, which gives the number of electron volts (ev) of energy required for the removal of a π-electron from the alkene molecule to convert it into a positive ion. Measurement of the dipole moments (μ) of unsymmetrically methylated ethylenes gives a direct indication of the uneven distribution of electron density in their molecules.

The applicability of the Markovnikov-Zaitsev rule to the addition of a hydrogen halide at a double bond is the result of the polarity and polarizability of molecules of unsymmetrical unsaturated hydrocarbons. In a reaction with a hydrogen halide, the halogen adds to the carbon atom with the lower electron density, which is the carbon atom bearing the least hydrogen atoms, while the hydrogen atom adds to the carbon atom of the double bond, which has the greater number of hydrogen atoms, for example, the propylene molecule $CH_3CH = \overset{*}{C}H_2$ is converted into $CH_3CHClCH_3$, by the addition of HCl. If both carbon atoms of the double bond have the same number of hydrogen atoms, but one of them is also attached to a methyl group, it is to precisely this one that the halogen of a hydrohalic acid adds, for example $CH_3CH = \overset{*}{C}HCH_2CH_3 + HI \rightarrow$ $\rightarrow CH_3CHICH_2CH_2CH_3$. The atom marked with an asterisk is charged more negatively than the others and therefore the electropositive hydrogen is directed toward it.

According to work on the metallation of olefins, a metal replaces most readily a hydrogen atom in an alkyl position, which is consequently protonized more readily than the others (p. 94). Experiments on deuterium exchange in liquid ammonia catalyzed by potassamide [113] are in general agreement with the facts presented. Three hydrogen atoms in the propylene molecule exchange at a high rate and these are evidently the atoms of the methyl group, which forms the positive end of the dipolar molecule. The two hydrogen atoms of the methylene group exchange at a lower rate, while the last hydrogen atom of the methyne group is replaced by deuterium very slowly. Analogously, the rate of exchange of six hydrogen atoms (of the two methyl groups) in isobutylene is approximately an order higher than that of the other two hydrogen atoms.

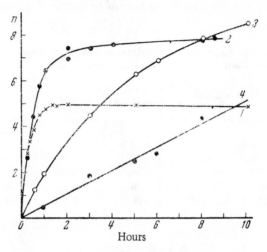

Fig. 15. Rate of hydrogen exchange between ethylenic hydrocarbons and a solution of KND_2 and ND_3. 1) Propylene; 2) isobutylene; 3) trimethylethylene; 4) tetramethylethylene.

Thus, the hydrogen atoms in the allyl position are actually most labile. With a change from isobutylene to tetramethylethylene, whose molecule contains the same grouping of atoms, but symmetrically arranged and therefore without a dipole moment, the rate of the exchange reaction falls by almost two orders. The hydrogen in trimethylethylene, whose dipole moment is close to that of isobutylene, exchanges at a much higher rate than that in tetramethylethylene. Table 38 gives the rate constants (sec^{-1}) of isotopic exchange in the methyl and methylene groups of these hydrocarbons at 25° and $C_{KND_2} =$ = 0.05 N from a very rough estimate on the basis of the kinetic curve (Fig. 15), where n is the number of hydrogen atoms exchanging.

Table 39 gives a summary of results with ethylenic hydrocarbons containing up to 16 carbon atoms in the chain. The mean rate constants of the exchange reaction, i.e., without allowance for the nonequivalence of the hydrogen atoms in the alkene molecule (sec^{-1} at C_{KND_2} = 0.06 N), decrease with an increase in the length of the carbon atom chain (Table 40).

Although all the hydrocarbons listed contain only one double bond, which is at the end of the carbon chain in most cases, all the hydrogen in their molecules under-

TABLE 38. Approximate Rate Constants of Hydrogen Exchange in Methylated Ethylenes

Hydrocarbon	CH_3 group	CH_2 group
Propylene	$3 \cdot 10^{-3}$	$4 \cdot 10^{-4}$
Isobutylene	$7 \cdot 10^{-4}$	$9 \cdot 10^{-5}$
Trimethylethylene	$5 \cdot 10^{-5}$	—
Tetramethylethylene	$1 \cdot 10^{-5}$	—

goes exchange and this is evidently due to the presence of the multiple bond. For example, parallel experiments with cetene and cetane showed [113] that there is no exchange with the saturated hydrocarbon. The difference in the rates of isotopic reactions with hexene and hexane was mentioned above (Table 28).

TABLE 39. Number of Hydrogen Atoms Exchanging for Deuterium in Olefins

Hydrocarbon	n	Literature reference
Propylene, C_3H_6	6	[113]
1-Pentene, C_5H_{10}	10	[107, 111]
2-Pentene, C_5H_{10}	10	[106, 111]
1-Hexene, C_6H_{12}	12	[107, 111]
1-Octene, C_8H_{16}	16	[107, 111]
2-Octene, C_8H_{16}	16	[106, 111]
4-Nonene, C_9H_{18}	18	[115]
4-Tridecene, $C_{13}H_{26}$	26	[115]
6-Tridecene, $C_{13}H_{26}$	(24.5)	[115]
1-Cetene, $C_{16}H_{32}$	(28)	[111, 113]

It hardly seems likely that the effect of a double bond can extend far along a chain of carbon atoms. The isomerization of olefins catalyzed by an amide [111] (see p. 118 for more details) suggests that the exchange occurs as a result of migration of the double bond along the chain of carbon atoms.* This was checked by experiments [111, 113] with hydrocarbons containing quaternary carbon atoms, which form a barrier to isomerization [140].

----

* There is no migration of the double bond in experiments with propylene as otherwise the five hydrogen atoms would exchange at the same rate.

It was actually found that in such hydrocarbons, deuterium replaces only hydrogen atoms in the part of the molecule containing the double bond and separated by the quaternary carbon atom (N is the total number of hydrogen atoms):

Hydrocarbon	n	N

$$CH_2 = C - CH_2 - \underset{\underset{CH_3}{|}}{\overset{\overset{CH_3}{|}}{C}} - CH_3 \qquad\qquad 7 \qquad 16$$

with CH_3 also on the second carbon

$$CH_2 = CH - CH_2 - \underset{\underset{CH_3}{|}}{\overset{\overset{CH_3}{|}}{C}} - \underset{\underset{CH_3}{|}}{\overset{\overset{CH_3}{|}}{C}} - CH_2CH = CH_2 \qquad 10 \qquad 22$$

In 2,4,4-trimethyl-1-pentene, which is isomer of octene, the chain of carbon atoms along which the hydrogen atoms exchange is shortened and the mean exchange rate is an order higher than for 1-octene.

$$k, \text{sec}^{-1}$$

2,4,4-Trimethylpentane . . $3 \cdot 10^{-4}$
1-Octene . . . . $3 \cdot 10^{-5}$

Thus, in catalysis by potassamide in deuterated ammonia there are two phenomena: firstly, ethylenic hydrocarbons isomerize with migration of the double bond to form an equilibrium mixture of isomers with the double bond at different carbon atoms and sec-

TABLE 40. Rate Constants of Hydrogen Exchange in Alkenes

Hydrocarbon	$k_{25°}$	$k_{50°}$
1-Pentene	—	$2 \cdot 10^{-4}$
2-Pentene	$4 \cdot 10^{-5}$	—
1-Hexene	$2 \cdot 10^{-5}$	$1 \cdot 10^{-4}$
1-Octene	—	$5 \cdot 10^{-5}$

ondly, in the molecule of each of the isomers there is hydrogen exchange with the hydrogen in the allyl position replaced most readily. The superposition of these reactions, which are similar to mechanism but proceed at different rates, leads to the exchange of all the hydrogen atoms in the molecule or in that part of it where migration of the double bond is possible.

The isomerization of unsaturated hydrocarbons under heterogeneous catalyses, which was studied in detail by N. D. Zelinskii's school, has much in common with the isomerization catalyzed by potassamide. R. Ya. Levina and her co-workers [140, 141] demonstrated the migration of a double bond from the side chain into cyclohexane and cyclopentane rings. If such an isomerization occurs under alkaline catalysis in liquid ammonia, then if the explanation of the mechanism of deuterium exchange in olefins is correct, the complete exchange of hydrogen in such hydrocarbons is to be expected. This conclusion is confirmed by the results of experiments with the following hydro-carbons:

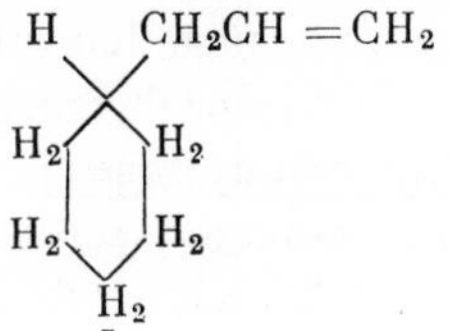

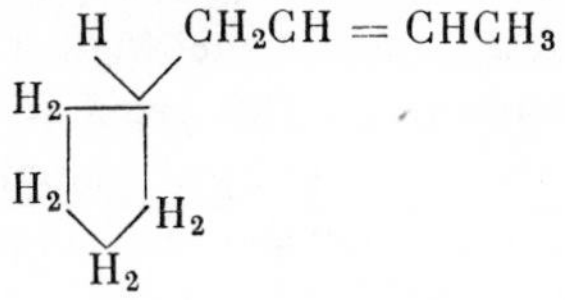

Hydrocarbon	n
I. Allylcyclohexane, C_9H_{16}	16
II. Crotylcyclopentane, C_9H_{16}.	16
III. Propylcyclohexene, C_9H_{16}.	16
IV. Methylcyclopentene, C_6H_{10}	10
V. Cyclohexene, C_6H_{10}.	10

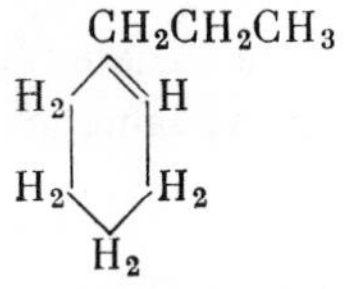

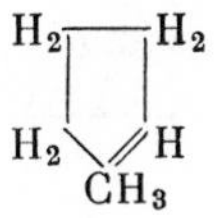

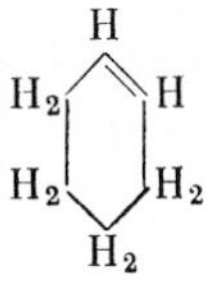

A double bond may migrate from a side chain into a saturated ring and from a ring into a side chain. The lability of a double bond in a ring is demonstrated by the com-plete and comparatively rapid (Table 28) exchange of hydrogen in cyclohexene.

e. CYCLOPROPANE HYDROCARBONS

A characteristic of cyclopropane hydrocarbons is the fact that although their mole-cules contain no multiple bonds, with respect to properties they occupy an intermediate position between ethylenic and saturated hydrocarbons [142]. They add a hydrogen halide in accordance with Markovnikov's rule, i.e., the three-membered ring is opened most readily between the carbon atoms with the greatest and least numbers of hydrogen atoms with the halogen adding to the most alkylated carbon atom, for example:

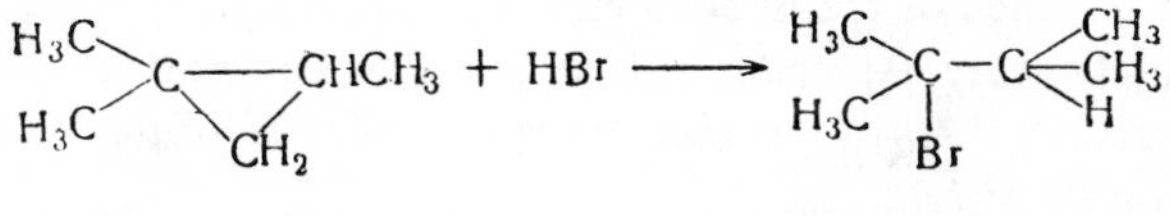

Like ethylenic hydrocarbons, cyclopropanes are hydrogenated under comparatively mild conditions. With an increase in the number of carbon atoms in the ring, there is a rise in the temperature at which the hydrocarbon is catalytically hydrogenated (B. A. Kazan-skii and M. Yu. Lukina):

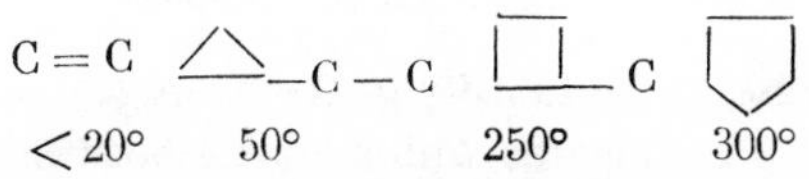

The unsaturation of the three-membered ring is also confirmed by spectral measurements; for example, the frequency of the CH bonds in the vibration spectrum of the
cyclopentane ring is close to the corresponding frequencies in the spectra of aromatic
and ethylenic hydrocarbons. The unsaturation of cyclopropane compounds is also indicated by the observed refractions, dipole moments, etc. [143]. Chemical and physical
data indicate that a cyclopropane ring participates in conjugation with a double bond
and with an aromatic ring. This peculiarity of cyclopropane hydrocarbons made them
an interesting subject for a study of isotopic exchange of hydrogen [123]. Since the hydrogen in unsaturated compounds exchanges with bases more readily than that in saturated compounds (p. 99), it was to be expected that this rule would apply to cyclopropane hydrocarbons.

Ethylcyclopropane (I) and isopentane (II) were used to determine the difference in
the reactivities of a cyclopropane and a paraffin with the same number of carbon atoms.

$$\begin{array}{cc} \underset{H_2C}{\overset{H_2C}{>}}CHCH_2CH_3 & \underset{H_3C}{\overset{H_3C}{>}}CHCH_2CH_3. \\ \\ I & II \end{array}$$

Five hydrogen atoms exchanged in ethylcyclopropane at $120°$ and $C_{KND_2} = 1$ N
and the number of atoms exchanging changed little even after 200 hr. It is evident that
the hydrogen atoms of the cyclopropane ring exchanges first. At the same potassamide
concentration there was a slow exchange ($k = 2 \cdot 10^{-6}$ sec^{-1}) at $25°$, while in experiments with isopentane at this temperature, hydrogen exchange was not detected even
after 240 hr.

The unsaturation of the cyclopropane group was demonstrated no less clearly by a
comparison of hydrogen exchange in phenylcyclopropane (III) and isopropylbenzene (IV).

$$\begin{array}{cc} \underset{H_2C}{\overset{H_2C}{>}}CHC_6H_5 & \underset{H_3C}{\overset{H_3C}{>}}CHC_6H_5. \\ \\ III & IV \end{array}$$

The exchange of the hydrogen in the β-CH bonds of the isopropyl group of (IV)
requires prolonged heating with 0.8 N potassamide solution at $100°$ (p. 103). At the
same time, the presence of the three-membered ring results in complete exchange of
the hydrogen in phenylcyclopropane (III) at a similar potassamide concentration (1 N)
after only 6 hr at $25°$; if C_{KND_2} is reduced to 0.05 N, 80% of the hydrogen is exchanged
at room temperature after a day.

The ready exchange of hydrogen in phenylcyclopropane is the result of not only the
unsaturation of the three-membered ring, but also an increase in its reactivity because
of conjugation with the π-electrons of the phenyl group. In ethylcyclopropane, where
conjugation is absent, the hydrogen exchanges slowly: n = 0.3 after 6 hr at $25°$ with a
potassamide concentration $C_{KND_2} = 1$ N, while all the hydrogen atoms exchanged (n =
= 10) in a parallel experiment with phenylcyclopropane.

An analogous increase in the lability of the hydrogen atoms of a three-membered
ring is observed when it is conjugated with a double bond in vinylcyclopropane (V).

$$\begin{array}{cc} \underset{H_2C}{\overset{H_2C}{\diagdown}}CHCH=CH_2 & \underset{H_2C}{\overset{H_2C}{\diagdown}}CHCH_2CH_3. \\ V & VI \end{array}$$

Experiments with ethylcyclobutane [123] were also carried out for comparison. The mean value of the rate constant of exchange (without allowance for the nonequivalence of the hydrogen atoms) at 120° and C_{KND_2} = 1 N, k $\simeq 3 \cdot 10^{-7}$ sec^{-1}, while k $\simeq$ $\simeq 10^{-4}$ sec^{-1} was obtained for ethylcyclopropane under the same conditions.

f. SATURATED HYDROCARBONS

In contrast to acid hydrogen exchange, hydrogen exchange in saturated hydrocarbons with bases has been studied very little as yet. This is all the more unfortunate as in general, there are very few data on the electrophilic reactivity of saturated hydrocarbons.

Data obtained on hydrogen exchange with a solution of potassamide in liquid ammonia [197, 113, 126, 131] are of a qualitative nature, mainly because of the low solubility of saturated hydrocarbons in liquid ammonia in contrast to unsaturated hydrocarbons. A summary of the results is given in Table 42, where n is the number of atoms exchanging. In contrast to other tables, the concentrations of deuterium (at. %) in the

TABLE 41. Approximate Rate Constants of Hydrogen Exchange in Cyclopropane Hydrocarbons and Isopentane

Hydrocarbon	Temperature 25°		Temp. 120°; C_{KND_2} = 1N
	C_{KND_2}=0.05N	C_{KND_2}=1 N	
$\underset{H_3C}{\overset{H_3C}{\diagdown}}CHCH_2CH_3$	—	No exchange	$(\sim 10^{-7})$
$\underset{H_2C}{\overset{H_2C}{\diagdown}}CHCH_2CH_3$	—	$\sim 10^{-6}$	$\sim 10^{-4}$
$\underset{H_2C}{\overset{H_2C}{\diagdown}}CHCH=CH_2$	$3 \cdot 10^{-6}$—$6 \cdot 10^{-6}$	—	—
$\underset{H_2C}{\overset{H_2C}{\diagdown}}CHC_6H_5$	$2 \cdot 10^{-4}$—$2 \cdot 10^{-5}$	—	—

solvent (C_s) and in the water from combustion of the hydrocarbon after the experiment (C_w) are also given here. The table shows that it was possible to introduce a considerable amount of deuterium (up to 18 at. %) into some substances at 120°. The last experiment, which is marked with an asterisk, was carried out in glass. This shows that the hydrogen exchange was induced by potassamide and not caused by catalysis by chrome-nickel steel, from which the ampoules used in the other experiments were made.

Very interesting experiments [126] were carried out with two samples of isobutane containing deuterium in different positions:

$$
\begin{array}{cc}
\begin{array}{l} H_3C\diagdown \\ H_3C- CH \\ DH_2C\diagup \end{array} &
\begin{array}{l} H_3C\diagdown \\ H_3C- CD \\ H_3C\diagup \end{array} \\
\mathbf{I} & \mathbf{II}
\end{array}
$$

A solution of potassamide in normal ammonia was used for the exchange.

TABLE 42. Hydrogen Exchange in Saturated Hydrocarbons (at 120°)

Hydrocarbon	Hours	C_{KND_2}, N	C_s	C_w	n
Isopentane	330	1	13.8	3.6	3.3
n-Hexane	1300	3	10.5	1.8	2.7
n-Heptane	170	0.8	95	13.1	2.1
"	500	1	95	17.6	3.0
"	1300	3	10.5	2.1	3.5
Decalin	180	0.8	95	3.5	1.0
"	1300	3	10.5	1.6	1.7
Cyclohexane	180	0.8	95	5.8	0.7
Cyclopentane	1300	3	10.5	3.3	3.4
Methylcyclohexane	500*	1	50	7.5	2.0

When the substances were heated (120°) for 100 hr with 1 N potassamide solution, half of the deuterium in substance (I) exchanged, while there was no appreciable exchange in deuterium in substance (II). Consequently, the hydrogen of the methyl group is much more labile than that of the methyne group. This conclusion was also drawn from a number of other examples. [Compare the lability of the hydrogen in methyl, methylene, and methyne groups of monoalkylbenzenes (p. 100) and also the results given below on the stability of primary, secondary, and tertiary carbanions (p. 119 and 121).]

C. Hydrogen Exchange Between Hydrocarbons

The possibility of deuterium exchange between hydrocarbons with the participation of carbanions was suggested by the author of this book [129, p. 391] before such reactions were achieved experimentally.

G. A. Razuvaev and his co-workers [143a] obtained carbanions by the addition of metallic sodium to benzyl- and phenylmercury. Phenylsodium exchanges hydrogen with deuterobenzene and the exchange reaches equilibrium after 2 hr at 80°. Benzylsodium reacts analogously with deuterotoluene, $C_6D_5CH_3$.

In order to provide support for a mechanism for the alkylation of aromatic hydrocarbons, Hart later undertook an investigation of deuterium exchange between alkylbenzenes in the presence of additives which partly converted them into organoalkali compounds (see p. 122). This investigation was continued in order to compare the acidities of alkylbenzenes [143b].

A mixture of ethylbenzene containing deuterium in the methylene group with two other aliphatic-aromatic hydrocarbons was heated in a sealed ampoule at 150° in the presence of metallic potassium and the relative rate at which deuterium was transferred into the α-position of the side chain of each of the hydrocarbons was measured.

It is assumed that potassium reacts with the hydrocarbon at a high temperature to form a carbanion:

$$ArCHR_1R_2 + K \rightarrow Ar\overline{C}R_1R_2.K^+ + H.$$

The carbanion abstracts a proton or a deuteron according to the equation:

$$Ar\overline{C}R_1R_2 + ArCDR_3R_4 \rightarrow ArCDR_1R_2 + Ar\overline{C}R_3R_4$$

the more readily, the more acid the reacting hydrocarbon is.

The authors consider that the metallation is the stage controlling the exchange reaction as a whole, while the transfer of a proton (or deuteron) from a hydrocarbon molecule to a carbanion proceeds rapidly. Therefore, measurement of the deuterium exchange rates is a means of estimating the acidity of hydrocarbons in the same way as in the method using metallation (p. 93).

The relative exchange rates of the hydrogen atom in the methyne group of alkylbenzenes with the general formula $C_6H_5CHRCH_3$ were as follows:

R	$k_{rel.}$
CH_3	1.87
CH_2CH_3	1.00
$CH_2CH_2CH_3$	0.90
$CH(CH_3)_2$	0.24
$C(CH_3)_3$	0.14

i.e., acidity of the methyne hydrogen decreases with successive substitution of CH_3 groups for the hydrogen atoms in the methyl group attached to the same carbon atom as the methyne hydrogen.

A decrease in the exchange rate was also observed when the two CH_3 groups were replaced by C_2H_5 groups:

Hydrocarbon	$k_{rel.}$
$C_6H_5CH(CH_3)_2$	1.87
$C_6H_5CH(C_2H_5)_2$	0.23

The absolute value of the rate constant of hydrogen exchange in sec-butylbenzene, $C_6H_5CH(CH_3)C_2H_5$, $k_{150°} = 2.85 \cdot 10^{-6}$ sec^{-1}.

In some of the latest work in the Isotopic Reactions Laboratory [143c], hydrogen exchange between hydrocarbons was effected under very mild conditions (25° with CH_3OK as the catalyst and $CH_3OCH_2CH_2OCH_3$ as the solvent). As already mentioned on p. 41, it was established previously that there is a sharp increase in the catalytic activity of an alcoholate in hydrogen exchange in triphenylmethane if ethylenediamine is used as a solvent instead of an alcohol. This effect is the result of an increase in the electron-donor capacity of the alcoholate, which is caused by more energetic solvation of the cation by ethylenediamine in comparison with alcohol. The formation of a strong solvate of the cation with solvent molecules disrupts interionic interaction, which involves noncoulombic forces of a donor-acceptor nature, and the anion becomes more strongly basic. The absence of a hydrogen bond between the solvent molecule and the anion has an analogous effect.

Dimethoxyethane (DME) is a solvent which forms strong solvates with alkali-metal cations. This is precisely why alkali metals dissolve in dimethoxyethane in the same way as in ethylenediamine (and liquid ammonia). The high solvation of DME with respect to alkali-metal ions was demonstrated by measurements of the intensity of electron paramagnetic resonance spectra of benzylpotassium [143d], which showed the high stability of anion-radicals in DME. Carbanions are also stabilized in this solvent as is demonstrated, for example, by the results of spectrophotometric measurements which show that an alcoholate is able to metallate fluorene in the presence of DME [143e]. As this solvent promotes the conversion of hydrocarbons into carbanions, hydrogen exchange becomes possible even under the action of alcoholates. The addition of 0.5 mole of CH_3OK to a mixture of equimolecular amounts of indene and 9-deuterofluorene made it possible to reach exchange equilibrium rapidly at 25° with the first pair of substances and at 50° with the second pair of substances.

6. ISOMERIZATION OF UNSATURATED HYDROCARBONS
A. Nature of Reaction

The isomerization of unsaturated hydrocarbons (alkynes, alkenes, and dienes) with migration of the multiple bond (or bonds) in the molecule is catalyzed by bases such as caustic alkalis, alcoholates, amides, and hydrides of alkali and alkaline earth metals, and organoalkali compounds. There is a direct relation between the degree of acidity of a hydrocarbon, determined by means of metallation or deuterium exchange, and its capacity to isomerize. In addition the more acidic the hydrocarbon, the weaker is the base which is capable of isomerizing it. It is natural to assume that the isomerization begins with an acid—base reaction between the hydrocarbon and the catalyst. B. A. Kazanskii and I. V. Gostunskaya [144-147] suggested that the first stage is an isomerization with migration of a multiple bond is protonization of the hydrogen in the CH bond in the allyl position relative to the multiple bond:

$$RCH_2\text{—}CH = CH_2 + B \rightarrow [RCH = CH = CH_2]^- + BH^+.$$

It might be considered that as a result of the high stability of a secondary as compared with a tertiary carbanion, the negative charge is concentrated at the methylene group and therefore the proton adds in this position to form an isomer of the olefin which is more stable thermodynamically than the original olefin:

$$[RCH = CH = CH_2]^- + H^+ \rightarrow RCH = CH \text{—} CH_3.$$

The same scheme was later put forward by American authors [148, 148a, 148b].

This scheme of anionic isomerization of unsaturated hydrocarbons was confirmed experimentally in the Isotopic Reactions Laboratory by spectral measurements on the isomerization allylbenzene ⇌ propenylbenzene [56, 148c]. It is known from the literature that both allylbenzene and propenylbenzene undergo acid ionization to form red solutions of organoalkali compounds when treated with potassamide in liquid ammonia [149-151]. These reactions may be represented by the following equations:

$$C_6H_5CH_2 - CH \rightleftharpoons CH_2 + NH_2^- \rightleftharpoons [C_6H_5CH \cdots CH \cdots CH_2]^- + NH_3$$

$$C_6H_5CH \rightleftharpoons CH - CH_3 + NH_2^- \rightleftharpoons [C_6H_5CH \cdots CH \cdots CH_2]^- + NH_3 .$$

Neutralization of each of these solutions gives a mixture of the two isomers.

Spectrophotometric measurements confirmed that on ionization, allyl- and pro-penylbenzenes are converted into the same ion $[C_6H_5CH \cdots CH \cdots CH_2]^-$ (λ_{max} = = 420 mμ and log ε_{max} = 4.34).

It is certain that a free carbanion is not always formed during the isomerization. It is most likely that in most cases there is partial protonization of the hydrogen and its synchronous transfer, for example, in a termolecular complex in the transition state.

B. Isomerization of Hydrocarbons by Bases

A. E. Favorskii [152] brought about the migration of the triple bond in a dialkyl-acetylene by catalysis with potassium hydroxide. Bourgeul [153] produced the same result by heating the acetylenic hydrocarbon with sodamide. Potassamide in liquid ammonia produces isomerization of alkenes (1-acetene into 2-octene [154] and 1-pen-tene into 2-pentene [107, 111]). Under these conditions, hydrocarbons with two iso-lated double bonds are converted into compounds with conjugated double bonds (di-allyl → dipropylene [111]). This is shown by a comparison of the constants of hydro-carbons before and after an experiment with literature data (Table 43).

As B. A. Kazanskii and I. V. Gostunskaya [144-147] showed, similar isomerizations of olefins and dienes occur under comparatively mild conditions when the hydrocarbon is boiled with solid calcium amide (for 1-2 hr) or when the vapor of the hydrocarbon is passed through a tube filled with lumps of the amide. It was shown that if deuterated calcium or potassium amide is used, together with isomerization of the hydrocarbon with migration of the double bond, there is hydrogen exchange in it [118]. A compari-son of the isomerization of some hexenes in contact with calcium amide at 80° showed [151a, b] that hydrocarbons with an allyl hydrogen in a methyne group (2,3-dimethyl-1-butene and 3-methyl-1-pentene) isomerize with greater difficulty than 2-methyl-1-pentene, 4-methyl-1-pentene, and 1-hexene, which have an allyl hydrogen in a methylene group. This is in accordance with the higher lability of hydrogen in a CH_2 group in comparison with a CH group (see pp. 116, 118, and 121).

The migration of a double bond in olefins and diolefins is catalyzed by organoal-kali compounds, hydrides, and alkoxides in addition to amides. Ipatieff and Pines and their co-workers [155-158] observed the isomerization of 1-butene into 2-butene (cis and trans), 1-pentene into 2-pentene (cis and trans), and 1-decene into a mixture of isomers in the presence of metallic sodium and a promoter (for example, chlorotoluene) which is readily converted into an organoalkali compound. Terpenes (limonene and others) have been isomerized by this method. Morton [159] compared the efficiency of different carbanions in the isomerization of 1-pentene into 2-pentene and found the following sequence: pentenyl > benzyl > octenyl > amyl > phenyl > triphenylmethyl > > fluorenyl > diallyl.

C. Rules of Isomerization

By investigating the isomerization of many unsaturated hydrocarbons, B. A. Kazan-skii and I. V. Gostunskaya were able to determine its rules and these were in agreement with the scheme presented. As was expected, the isomer in which the allyl position contained the greatest number of hydrogen atoms always reacted most rapidly in ac-

TABLE 43. Examples of the Isomerization of Hydrocarbons by a Solution of KNH_2 in Liquid NH_3

Hydrocarbon	B.p., °C	n_{20}^D
1-Pentene		
before experiment.........	29.5	1.3719
after experiment..........	36.0	1.3810
2-Pentene (literature data)		
trans-.................	37.0	1.3822
cis-..................	36.3	1.3790
1-Hexene		
before experiment.........	63.5	–
after experiment..........	67.5	–
2-Hexene (literature data)	68.0	–
3-Hexene (literature data)	66.8	–
1-Octene		
before experiment.........	121.5	1.4095
after experiment..........	123	1.4145
2-Octene (literature data)	124.4	1.4140
Diallyl		
before experiment.........	59	1.4040
after experiment..........	82	–
Dipropylene (literature data)	82	–

cordance with the general rule that as hydrogen atoms are replaced by electron-donor alkyl groups, the lability of the remaining hydrogen atom falls. For example, 2-methyl-1-butene (I) isomerizes into 2-methyl-2-butene (II) more rapidly than 3-methyl-1-butene (III):

$$H_2C = C - CH_2 - CH_3 \rightarrow H_3C - C = CH - CH_3 \leftarrow H_3C - HC - CH = CH_2 .$$
$$\qquad | \qquad\qquad\qquad\qquad | \qquad\qquad\qquad\qquad\qquad |$$
$$\qquad CH_3 \qquad\qquad\qquad\qquad CH_3 \qquad\qquad\qquad\qquad\qquad CH_3$$
$$\qquad I \qquad\qquad\qquad\qquad\quad II \qquad\qquad\qquad\qquad\qquad III$$

1-Hexene (IV) isomerizes into 2-hexene (V) more readily than 2,3-dimethyl-1-butene (VI) changes into 2,3-dimethyl-2-butene (VII).

$$CH_3 - CH_2 - CH_2 - CH_2 - CH = CH_2$$
$$IV$$
$$CH_3 - CH_2 - CH_2 - CH = CH - CH_3$$
$$V$$

$$CH_3 - CH - C = CH_2 \qquad CH_3 - C = C - CH_3$$
$$\qquad | \qquad | \qquad\qquad\qquad\quad | \qquad |$$
$$\qquad CH_3 \ CH_3 \qquad\qquad\qquad CH_3 \ CH_3$$
$$\qquad VI \qquad\qquad\qquad\qquad\quad VII$$

Morton [159] reported that the isomerization of hydrocarbons with a branched chain of atoms is slower than that of hydrocarbons with a straight chain.

7. ALKYLATION BY OLEFINS

A. Michael and Analogous Reactions

A reaction between weak acids (of the malonic ester type) and olefinic systems has been known for a comparatively long time under the name of the Michael reaction [160]. Such a weak acid is, for example, alcohol and a molecule of a substance with conjugated double bonds (butadiene, styrene, and acrylonitrile)added to it in an alkaline medium. A base converts the weak acid into an anion:

$$R - OH + B \rightarrow RO^- + BH^+,$$

which then participates in the reaction according to the following scheme:

$$RO^- + \overset{|}{C} = \overset{|}{C} - \overset{|}{C} = \overset{|}{C} \rightarrow RO - \overset{|}{C} - \overset{\ominus}{\underset{|}{C}} - \overset{|}{C} = \overset{|}{C}$$

$$RO - \overset{|}{\underset{|}{C}} - \overset{\ominus}{\underset{|}{C}} - \overset{|}{C} = \overset{|}{C} + HA \rightarrow RO - \overset{|}{\underset{|}{C}} - CH - \overset{|}{C} = \overset{|}{C} + A^-$$

(HA is an acid and A⁻ is its conjugation base). In recent years it has been possible to effect analogous reactions with even weaker acids than alcohols, namely, with amines and even hydrocarbons. It was then found that not only hydrocarbons with conjugated double bonds, but even simple olefins could participate in the reaction.

The reaction between olefins and amines proceeds according to the following scheme:

$$RNH_2 + B \rightarrow RNH^- + BH^+$$

$$RNH^- + \overset{|}{C} = \overset{|}{C} \rightarrow RNH - \overset{|}{\underset{|}{C}} - \overset{|}{\underset{|}{C}}{}^{\ominus}$$

$$RNH - \overset{|}{\underset{|}{C}} - \overset{|}{\underset{|}{C}}{}^{\ominus} + HA \rightarrow RNH - \overset{|}{\underset{|}{C}} - \overset{|}{\underset{|}{C}} - H + A^-.$$

The acid HA may be a second amine molecule.

Hydrocarbons in which a hydrogen atom is protonized in the presence of a strong base such as carbanion react with olefins. Let us denote a molecule of an aromatic hydrocarbon by ArH and the base by the symbol B, then

$$ArH + B \rightarrow Ar^- + BH^+$$

$$Ar^- + \overset{|}{C} = \overset{|}{C} \rightarrow Ar - \overset{|}{\underset{|}{C}} - \overset{|}{\underset{|}{C}}{}^{\ominus}$$

$$Ar - \overset{|}{\underset{|}{C}} - \overset{|}{\underset{|}{C}}{}^{\ominus} + HA \rightarrow Ar - \overset{|}{\underset{|}{C}} - \overset{|}{\underset{|}{C}} - H + A^-.$$

Normally HA is a second hydrocarbon molecule. The carbanion A⁻ is regenerated and the reaction continues. Ipatieff and Pines discovered alkylations of aliphatic-aromatic hydrocarbons and dimerizations of olefinic hydrocarbons which proceed according to this scheme.

B. Alkylation of Amines and Ammonia by Olefins

Let us examine examples of the alkylation of amines.

When sodamide is added to aniline, methylaniline, and o-toluidene, the latter are partly converted into sodium derivatives, which are alkylated by ethylene [161]:

$$C_6H_5NH_2 + NaNH_2 \rightarrow C_6H_5NH^\ominus Na^\oplus + NH_3$$
$$C_6H_5NH^\ominus Na^\oplus + CH_2 = CH_2 \rightarrow C_6H_5NHCH_2CH_2^\ominus Na^\oplus$$
$$C_6H_5NHCH_2CH_2^\ominus Na^\oplus + C_6H_5NH_2 \rightarrow C_6H_5NHC_2H_5 + C_6H_5NH^\ominus Na^\ominus.$$

By using a special method for preparing the sodium derivatives of aliphatic and alicyclic amines [162], the same authors ethylated dibutylamine, piperidene, and hexylamine under a pressure of 2-4 atm. Under much more drastic conditions (a pressure up to 1000 atm), it was possible to aminate simple olefins (ethylene, propylene, butylene, and hexene) with ammonia and butylamine in the presence of metallic sodium and sodium hydride [163].

C. Alkylation of Aromatic Hydrocarbons

The development of the reaction examined reached its logical conclusion when Ipatieff and Pines [164] proposed the alkylation of the weakest acids, namely, hydrocarbons, by a similar mechanism.

Monoalkylbenzenes (toluene, ethylbenzene, isopropylbenzene, and cyclohexylbenzene), xylenes, and mesitylene and also other aromatic hydrocarbons with an aliphatic substituent, such as diphenylmethane, are alkylated by simple olefins at 250-300° and a pressure of 50-150 atm. The reaction is promoted by substances (chlorobenzene and dimethylmercury) which form an organoalkali derivative in the presence of an alkali metal (Na, K, and Li).

$$C_6H_4(CH_3)Cl + 2Na \rightarrow C_6H_4Na(CH_3) + NaCl$$

$$C_6H_4Na\,(CH_3) \overset{heat}{\rightarrow} C_6H_5CH_2Na$$

$$(CH_3)_2Hg + 2Na \rightarrow 2CH_3Na + Hg.$$

The organoalkali compound is regarded as an undissociated ion pair, whose carbanion participates in an acid−base reaction with the alkylaromatic hydrocarbon.* In the latter, the most labile hydrogen atoms are those in the α-position (p. 101). One of these is first protonized and transferred to the carbanion obtained by the reaction above. For example, according to P. P. Shorygin, toluene reacts in the following way:

---

* The alkylation of alkylbenzenes, in the side chain is also initiated by sodium hydride and alkoxide [165, 166]. A summary of the work of Pines and his co-workers is given in the article of H. Pines and L. A. Schaap in Advances in Catalysis, Vol. 12, p. 117. Academic Press, New York and London, 1960.

$$C_6H_5CH_3 + R^\ominus Na^\oplus \rightleftarrows C_6H_5CH_2^\ominus Na^\oplus + RH.$$

The equilibrium is displaced further to the right, the greater the difference in the strengths of the acids (RH and the hydrocarbon metallated). Then, as in the reactions with amines described above, the olefin adds to the benzyl anion:

$$C_6H_5CH_2^\ominus Na^\oplus + H_2C = CH_2 \rightarrow C_6H_5CH_2CH_2CH_2^\ominus Na^\oplus.$$

A carbanion with a longer chain of carbon atoms is formed. It reacts with a second toluene molecule as shown in the following equation:

$$C_6H_5CH_2CH_2CH_2^\ominus Na^\oplus + C_6H_5CH_3 \rightleftarrows$$
$$\rightleftarrows C_6H_5CH_2CH_2CH_3 + C_6H_5CH_2^\ominus Na^\oplus.$$

The benzyl anion may then attack a new olefin molecule.

Thus, the reaction proceeds in several stages, of which two are protolytic reactions. Three different carbanions participate in them: I) the carbanion of the promoter ($R^\ominus$), II) a benzyl anion ($C_6H_5CH_2^\ominus$), and III) the carbanion $C_6H_5CH_2CH_2CH^\ominus$), which is formed as an intermediate reaction product. It is obvious that the hydrocarbon formed, whose alkyl group contains a larger number of carbon atoms than that of the original hydrocarbon, may react in its turn with a carbanion present in the system, for example, according to the following equation:

$$C_6H_5CH_2C_2H_5 + C_6H_5CH_2^\ominus Na^\oplus \rightarrow C_6H_5CHC_2H_5^\ominus Na^\oplus + C_6H_5CH_3.$$

With excess olefin, there is the reaction:

$$C_6H_5\overset{\ominus}{C}H C_2H_5 \overset{\oplus}{Na} + H_2C{=}CH_2 \rightarrow C_6H_5{\Big\langle}\begin{matrix} C_2H_5 \\ CH_2CH_2^\ominus Na^\oplus, \end{matrix}$$

and then

$$C_6H_5CH{\Big\langle}\begin{matrix} C_2H_5 \\ CH_2CH_2^\ominus Na^\oplus \end{matrix} + C_6H_5CH_3 \rightarrow C_6H_5CH(C_2H_5)_2 + C_6H_5CH_2^\ominus Na^\oplus$$

etc.

Experiments with unsymmetrical olefins (propylene and isobutylene) show that the intermediate carbanion normally has a primary structure in accordance with the higher stability of primary as compared with secondary and tertiary carbanions. The rule on the relative stability of anions (p. 116), which has already been mentioned several times, is again confirmed. This is evident from the following equations:

$$C_6H_5\overset{|}{\underset{|}{C}}{}^\ominus + H_3CCH{=}CH_2 \rightarrow C_6H_5\overset{|}{\underset{|}{C}} - \overset{CH_3}{\underset{H}{\overset{|}{\underset{|}{C}}}} - CH_2^\ominus$$

$$C_6H_5 \overset{|}{\underset{|}{C}}{}^{\ominus} + (H_3C)_2C{=}CH_2 \rightarrow C_6H_5\overset{|}{\underset{|}{C}} - \overset{CH_3}{\underset{CH_3}{\overset{|}{\underset{|}{C}}}} - CH_2^{\ominus}.$$

Further evidence in favor of the proposed mechanism of alkylation of aliphatic-aromatic hydrocarbons in the side chain was provided by Hart [167]. He showed that heating a mixture of $C_6H_5CHDCH_3$ and $C_6H_5CH(CH_3)C_2H_5$ at 165° with a small amount of metallic potassium or potassium hydride results in the transfer of deuterium from the α-CH bond of one hydrocarbon to the same position in the molecule of the second alkylbenzene. This is explained by the formation of an organoalkali compound, which was confirmed by a positive Gilman test [168] and the reddish color of the reaction mixture (p. 90). A deuteron is transferred from the α-carbon atom of the deuterated alkylbenzene molecule to the carbanion:[*]

$$Ar - \overset{|}{\underset{|}{C}}{}^{\ominus} + D - \overset{|}{\underset{|}{C}} - Ar_1 \rightarrow Ar - \overset{|}{\underset{|}{C}} - D + \overset{|}{\underset{|}{C}}{}^{\ominus} - Ar_1 .$$

On the basis of subsequent work [169], the hypothesis was put forward that the purely carbanion scheme is accurate only in the first approximation. The point is that the yield of the alkylation of toluene by propylene depends on the nature of the alkali metal used for the reaction. This would not have been the case had the course of the reaction been determined wholly by the formation of the benzyl carbanion. The authors of the work therefore assumed that there is another stage in the reaction in which the metal ion adds to the double bond of the propylene molecule to form a π-complex and the latter then reacts with the benzyl carbanion. The structure of this complex is determined by the dimensions of the cation and the energies and direction of the process depend on the latter.

From this example we can see that when reaction schemes involving free carbanions are used, in most cases these schemes are approximate. In actual fact, the carbanions are bound to oppositely charged ions in ionic pairs and there may also exist more complex aggregates of ions and complexes, including those with the solvent.

In addition, many reactions do not generally proceed as far as the formation of free carbanions and there is only a displacement of electrons in the direction of anionization of the molecule in the transition state. Below, in a more detailed discussion of the mechanism of acid—base reactions, we demonstrate the complexity of such phenomena which are normally described as a simple proton transfer, accompanied by ionization of the substance.

In recent years in the literature there has been an accumulation of facts indicating that in reactions which are normally illustrated by purely carbanion schemes, a very important part is actually played by the oppositely charged ion and the solvent.

It is also possible to alkylate an aromatic ring by a carbanion mechanism. Benzene (in the presence of phenlysodium) reacts with ethylene to give ethylbenzene [170] and with isobutylene to give a mixture of tert-butyl- and isobutylbenzene in a ratio

---

[*] This is deuterium exchange between a hydrocarbon and a carbanion, which is examined in more detail in the section "Deuterium exchange between hydrocarbons" (p. 114).

of 1 : 5. A by-product of the reaction is biphenyl, which is obtained as a result of the addition of a phenyl anion to a benzene molecule. This process is described by the following equations:

$$C_6H_6 + C_6H_5^{\ominus} \rightarrow C_6H_5{-}C_6H_5 + H^{\ominus}$$

$$\underset{H_3C}{\overset{H_3C}{\diagdown}}C = CH_2 + H^{\ominus} \rightarrow \underset{H_3C}{\overset{H_3C}{\diagdown}}CH - CH_2^{\ominus}$$

$$C_6H_6 + \underset{H_3C}{\overset{H_3C}{\diagdown}}CH - CH_2^{\ominus} \rightarrow \underset{H_3C}{\overset{H_3C}{\diagdown}}CH - CH_2C_6H_5 + H^{\ominus}.$$

The reaction between an olefin and tert-butylbenzene has also been studied. A mixture of tert-butylethylbenzenes was obtained, but as an alkyl group decreases the lability of the hydrogen atoms in an aromatic ring (cf., metallation, p. 93), and hydrogen exchange, p. 100), the product yield was lower than in the reaction with benzene.

D. Dimerization of Olefins

Olefins are readily metallated and converted into carbanions (p. 94). The carbanion of an olefin may add to an olefin molecule in the same way as carbanions of aliphatic-aromatic or aromatic hydrocarbons. The rule stating that the intermediate product is predominantly a primary carbanion holds. In accordance with this rule, propylene reacts according to the following equations [171]:

$$H_2C = CHCH_3 + R^{\ominus}Na^{\oplus} \rightarrow H_2C = CHCH_2^{\ominus}Na^{\oplus} + RH$$

$$H_2C = CHCH_2^{\ominus} + CH_3CH = CH_2 \rightarrow H_2C = CH - CH_2 - \underset{\underset{CH_3}{|}}{CH} - CH_2^{\ominus}$$

$$H_2C = CH - CH_2 - \underset{\underset{CH_3}{|}}{CH} - CH_2^{\ominus} + H_3C - CH = CH_2 \rightarrow$$

$$\rightarrow H_2C = CH - CH_2 - \underset{\underset{CH_3}{|}}{CH} - CH_3 + H_2C - CH = CH_2^{\ominus}.$$

At the same time, the carbanions present in the reaction scheme may isomerize the alkene with migration of the double bond (p. 117). Hydrogenation of the dimerization product yields 2-methylpentane. The dimerization of isobutylene proceeds analogously:

$$H_2C = \underset{\underset{CH_3}{|}}{C} - CH_2^{\ominus} + \underset{\underset{CH_3}{|}}{C} = CH_2 \rightarrow H_2C = \underset{\underset{CH_3}{|}}{C} - CH_2 - \underset{\underset{CH_3}{|}}{C} - CH_2^{\ominus}$$

$$H_2C = \underset{\underset{CH_3}{|}}{C} - CH_2 - \underset{\underset{CH_3}{|}}{\overset{\overset{CH_3}{|}}{C}} - CH_2^{\ominus} + H_3C - \underset{\underset{CH_3}{|}}{C} = CH_2 \rightarrow$$

$$\rightarrow H_2C = \underset{\underset{CH_3}{|}}{C} - CH_2 - \underset{\underset{CH_3}{|}}{\overset{\overset{CH_3}{|}}{C}} - CH_3 + H_2C - \underset{\underset{CH_3}{|}}{C} = CH_2^{\ominus}.$$

Hydrogenation of the mixture of isomeric trimethylpentenes gives 2,2,4-trimethyl-pentane.

The products from the reactions of cyclohexene with ethylene, propylene, and iso-butylene are ethyl-, isopropyl-, and tert-butylcyclohexenes, whose yields decrease in the given sequence. This is explained by the fact that the propylene and especially the isobutylene molecules add a carbanion with greater difficulty than the ethylene molecule as the electron density at the carbon atom of the methylene group of the first two hydrocarbons is low in comparison with ethylene (hyperconjugation effect, pp. 82 and 107).

The group of alkylations of aromatic hydrocarbons and dimerizations of olefins described in this section shows the same relations of reactivity to hydrocarbon structure as were found in metallation, hydrogen exchange, and isomerization. This identity of the rules is explained by the fact that all the reactions listed proceed through the formation of a carbanion by the reaction of an acid hydrocarbon with a strong base.

8. SUMMARY

1. As in other acids, in hydrocarbons the hydrogen may be replaced by a metal (normally an alkali metal) to form organoalkali compounds. This requires different conditions and reagents depending on the structure of the hydrocarbons and the degree of their acidity. Many organoalkali compounds are like salts: the cation is a metal ion while the anion (carbanion) contains trivalent, negatively charged carbon.

2. Another method of determining and comparing the strengths of hydrocarbons as acids is the measurements of the rate of hydrogen exchange with a strong base such as a solution of potassamide in liquid ammonia. The hydrocarbons may react as polybasic acids. Quantitative data on the exchange of hydrogen in them give valuable information on the structure of hydrocarbons and the intereffect of atoms in their molecules. The peculiar characteristics of hydrogen exchange in unsaturated compounds with a solution of potassamide in liquid ammonia are explained by the fact that the amide catalyzes isomerization with migration of the multiple bond.

3. The isomerization of unsaturated hydrocarbons, alkylation of hydrocarbons by olefins, and dimerization of olefins catalyzed by bases are caused by the acidity of hydrocarbons. Like metallation and hydrogen exchange, they proceed through the formation of a carbanion from the hydrocarbon. This explains the fact that the structure of the hydrocarbons has an analogous effect on all the reactions listed. Carbanions with a primary carbon atom are most stable and ions with a tertiary carbon atom are least stable. Free carbanions are not always formed in the reactions listed.

LITERATURE CITED

1. K. A. Kocheshkov and T. V. Talalaeva, Synthetic Methods in the Field of Organometallic Compounds of Lithium, Sodium, Potassium, Cesium, and Rubidium [in Russian] (Izd. AN SSSR, Moscow, 1949).
2. A. Morton, Behavior of organoalkali compounds, Chem. Rev. 35, 1-49 (1944).
3. H. Gilman, Organometallic Compounds In Organic Chemistry (Ed. H. Gilman) Vol. I (1938).
4. J. Schmidt, Organometallic Compounds [Russian translation] (Khimteoret., Leningrad, 1947) pp. 11-116.
5. G. E. Coates, Organometallic Compounds in the First Three Periodic Groups, Quart. Rev. 4, 217 (1950).

6. V. Grignard, G. Dupont, and R. Pacquin, Traité de chimie organique, Vol. V (Masson).

7. E. Krause and A. Grosse, Die Chemie der metallorganischen Verbindungen, Chap. 2. Berlin (Bonnhaeger, 1937).

8. H. Gilman and A. Morton, Organic Reactions (Edited by Adams) [Russian translation] (IL, Moscow, 1956) Vol. VIII, p. 333.

9. R. A. Benkeser, D. J. Foster, D. M. Sauve, and J. F. Nobis, Chem. Rev. 57, 867 (1957).

10. W. C. Fernelius and G. W. Watt, Reactions of Solutions of Metals in Liquid Ammonia, Chem. Rev. 20, 219-223 (1937).

11. G. W. Watt, Reactions of Organic and Organometallic Compounds with Solutions of Metals in Liquid Ammonia, Chem. Rev. 46, 317 (1950).

12. A. J. Birch, The reduction of organic compounds by metal ammonia solutions, Quart. Rev. 4, 69 (1950).

13. W. C. Fernelius and G. W. Bowman, Ammonolysis in Liquid Ammonia, Chem. Rev. 26, 35 (1940).

14. T. F. Rutledge, J. Org. Chem. 22, 649 (1957).

15. J. Thiele, Ber. 34, 68 (1901).

16. K. Ziegler, H. Froitzheim-Külhorn, and K. Hafner, Ber. 89, 434 (1956).

17. R. Weissgerber, Ber. 34, 1659 (1901); 41, 2913 (1908); 42, 569 (1909); 43, 3521 (1910).

18. J. Nieuwland and R. Vogt, Chemistry of Acetylene [Russian translation] (Inoizdat, Moscow, 1947) ch. 2.

19. Beilstein, Band 5. p. 700.

20. A. A. Morton and H. E. Ramsden, J. Am. Chem. Soc. 70, 3132 (1948).

21. J. de Postis, Compt. rend. 222, 398 (1946); see also Chem. Abstr. 40, 3104 (1946).

22. C. E. Claff and A. A. Morton, J. Org. Chem. 20, 440, 981 (1955).

23. E. A. Bried and G. F. Hennion, J. Am. Chem. Soc. 59, 1310 (1937).

24. T. H. Vaughn, D. F. Hennion, R. R. Vogt, and J. A. Niewland, J. Org. Chem. 2, 1 (1937).

25. H. Gilman and R. D. Gorsich, J. Org. Chem. 23, 550 (1958).

26. F. G. Cotrell, J. Chem. Phys. 18, 85 (1914).

27. H. Gilman and J. C. Bailie, J. Am. Chem. Soc. 65, 267 (1943).

28. C. B. Wooster and N. W. Mitchell, J. Am. Chem. Soc. 52, 688 (1930).

29. H. Rappoport and G. Smolinsky, J. Am. Chem. Soc. 80, 2810 (1958); 82, 934 (1960).

30. H. Gilman, A. L. Jacoby, and H. Ludeman, J. Am. Chem. Soc. 60, 2336 (1938).

31. A. A. Morton, C. E. Claff, and H. P. Kagen, J. Am. Chem. Soc. 76, 4556 (1954).

32. Ch. A. Kraus and R. Rosen, J. Am. Chem. Soc. 76, 4556 (1954).

33. A. I. Shatenshtein and E. N. Zvyagintseva, Doklady Akad. Nauk SSSR 117, 852 (1957).

34. A. M. Monoszon, E. N. Gur'yanova, and A. I. Shatenshtein, Zhur. Khim. Prom. 11, 389 (1935).

35. V. A. Pleskov, Zavodskaya Lab. 6, 178 (1937).

35a. I. V. Gostunskaya, M. I. Rozhkova, and B. A. Kazanskii, Doklady Akad. Nauk SSSR 114, 545 (1957); 118, 299 (1958); M. I. Rozhkova, Nauchnye Doklady Vysshei Shkoly. Khimiya i Khimicheskaya Tekhnologiya 1, 117 (1958).

35b. A. A. Vainshtedt, Zhur. Obshchei Khim. 4, 875 (1934).

35c. V. Boekelheide and C. E. Larrabee, J. Am. Chem. Soc. 72, 1245 (1950).

36. F. W. Bergstrom, W. C. Fernelius, and R. Levine, The Chemistry of the Alkali Amides, Chem. Rev. 12, 86-92 (1933); 20, 431-432 (1937); 54, 467 (1954).

37. H. Pines and V. Mark, J. Am. Chem. Soc. 78, 4316 (1956).

38. P. P. Shorygin, Investigation of Organometallic Compounds of Sodium [in Russian] (Moscow, 1910).

39. F. Hein, Angew. Chem. 51, 503 (1938).

40. S. V. Anantakrishnan, Proc. Indian Acad. Sci. 34, 299 (1951).

41. W. Schlenk and E. Markus, Ber. 47, 1664 (1914).

42. K. Ziegler and H. Vollschitt, Ann. 479, 123 (1930).

43. E. Swift, J. Am. Chem. Soc. 60, 1403 (1938).

44. N. V. Keevil and H. E. Bent, J. Am. Chem. Soc. 60, 193 (1938).

44a. D. G. Hill, J. Burkus, S. M. Luck, and C. R. Hauser, J. Am. Chem. Soc. 81, 2787 (1959).

45. Ch. A. Kraus and W. H. Kahler, J. Am. Chem. Soc. 55, 3537 (1933).

46. V. A. Pleskov and A. M. Monoszon, Zhur. Fiz. Khim. 3, 221 (1932).

47. E. Masdupuy and F. Gallais, Compt. rend. 232, 1935 (1951); Chem. Abstr. 45, 7400 (1951).

48. F. Hein, E. Petzschner, K. Wagler, and F. A. Segitz, Z. Anorg. Chem. 141, 161 (1924).

49. F. Hein and H. Pauling, Z. Phys. Chem. 165, 338 (1933).

50. G. Jander and L. Fischer, Z. Elektrochem. 62, 971 (1958).

51. F. Hein and H. Pauling, Z. Elektrochem. 38, 25 (1932).

52. C. B. Wooster, Chem. Rev. 11, 1 (1932).

53. J. E. Leffler, The Reactive Intermediates of Organic Chemistry (New York, London, 1956).

54. N. S. Wooding and W. C. E. Higginson, J. Chem. Soc. (1952) pp. 760, 774.

55. E. A. Izrailevich, D. N. Shigorin, I. V. Astaf'ev, and A. I. Shatenshtein, Doklady Akad. Nauk SSSR 111, 617 (1956).

56. I. V. Astaf'ev and A. I. Shatenshtein, Zhur. Optika i Spektroskopiya 6, 631 (1959).

57. I. V. Astar'ev, Absorption Spectra of Carbanions, Dissertation [in Russian] (Karpov Physicochemical Institute, Moscow, 1958).

58. L. C. Anderson, J. Am. Chem. Soc. 57, 1673 (1935).

59. W. West (Ed.), Chemical Applications of Spectroscopy (New York, 1956) p. 690.

60. H. C. Longuet-Higgins and J. A. Popple, Proc. Phys. Soc. 68A, 591 (1955).

61. W. Bingel, Z. Naturforsch. 10a, 462 (1955).

62. W. Schlenk and E. Bergmann, Ann. 463, 192 (1928).

63. J. B. Conant and G. W. Wheland, J. Am. Chem. Soc. 54, 1212 (1932).

64. W. K. McEwen, J. Am. Chem. Soc. 58, 1125 (1936).

65. R. D. Kleene and G. W. Wheland, J. Am. Chem. Soc. 63, 3321 (1941).

66. R. S. Stearns and G. W. Wheland, J. Am. Chem. Soc. 69, 2025 (1947).

67. D. Bryce-Smith, J. Chem. Soc. (1954) p. 1079.

68. A. A. Morton, J. T. Massengale, and M. L. Brown, J. Am. Chem. Soc. 67, 1620 (1945).

69. H. Gilman and L. Tolman, J. Am. Chem. Soc. 68, 522 (1946).

70. A. A. Morton and E. L. Little, J. Am. Chem. Soc. 71, 487 (1949).

71. A. A. Morton and E. J. Lanpher, J. Org. Chem. 23, 1636 (1958).

72. A. A. Morton, M. L. Brown, M. E. T. Holden, R. L. Letsinger, and E. E. Magat, J. Am. Chem. Soc. 67, 2224 (1945).

73. A. A. Morton, E. L. Little, and W. O. Strong, J. Am. Chem. Soc. 65, 1339 (1943).

74. K. A. Kocheshkov and T. V. Talalaeva, Doklady Akad. Nauk SSSR 77, 621 (1951).
75. A. A. Morton, J. B. Davidson, T. R. P. Gibb, E. L. Little, E. F. Clarke, and A. G. Green, J. Am. Chem. Soc. 64, 2250 (1942).
76. H. Gilman and R. L. Bebb, J. Am. Chem. Soc. 61, 109 (1939).
77. H. Gilman and A. L. Jacobi, J. Org. Chem. 3, 108 (1938).
78. A. A. Morton, F. D. Marsh, R. D. Coombs, A. L. Lyons, S. E. Penker, H. E. Ramsden, V. B. Baker, E. L. Little, and R. L. Letsinger, J. Am. Chem. Soc. 72, 3785 (1950).
79. A. A. Morton and M. L. Brown, J. Am. Chem. Soc. 69, 160 (1947).
80. A. A. Morton, M. L. Brown, and E. Magat, J. Am. Chem. Soc. 69, 161 (1947).
81. A. A. Morton and M. E. T. Holden, J. Am. Chem. Soc. 69, 1675 (1947).
82. A. A. Morton, Ind. and Eng. Chem. 42, 1488 (1950).
83. D. C. Pepper, Ionic Polymerization, Quart. Rev. 8, 88 (1954).
84. A. A. Morton, G. M. Richardson, and A. T. Hallowell, J. Am. Chem. Soc. 63, 327 (1941).
85. H. Gilman, F. W. Moore, and O. Baine, J. Am. Chem. Soc. 63, 2482 (1941).
86. A. D. Petrov, Synthesis and Isomeric Conversions of Aliphatic Hydrocarbons [in Russian] (Izd. AN SSSR, Moscow, 1947).
87. P. D. Bartlett, S. Friedman, and M. Stiles, J. Am. Chem. Soc. 75, 1771 (1953).
88. A. I. Brodskii, Chemistry of Isotopes [in Russian] (Izd. AN SSSR, Moscow, 1957).
89. G. P. Miklukhin, Uspekhi Khimii 17, 663 (1948).
90. K. F. Bonhöffer, K. H. Geib, and O. Reitz, J. Chem. Phys. 7, 664 (1939).
91. M. S. Kharasch and W. G. Brown, J. Org. Chem. 2, 36 (1937); 4, 442 (1939).
92. A. I. Brodskii, L. L. Chervyatsova, and G. P. Miklukhin, Zhur. Fiz. Khim. 24, 968 (1950).
93. R. Breslow, J. Am. Chem. Soc. 79, 1762 (1957).
94. R. C. Lord and W. D. Philips, J. Am. Chem. Soc. 74, 2429 (1952).
95. L. Reyerson and B. Gillespie, J. Am. Chem. Soc. 57, 779, 2250 (1935); 58, 282 (1936); 59, 900 (1937).
95a. P. Ballinger and F. A. Long, J. Am. Chem. Soc. 81, 3148 (1959).
95b. H. B. Charman, G. W. D. Tiers, M. M. Kreevoy, and G. Filipovich, J. Am. Chem. Soc. 81, 3149 (1959).
96. M. Koizumi, Bull. Chem. Soc. Japan 14, 491 (1939).
96a. O. Reitz, Handbuch der Katalyse, Vol. II (Springer, 1940).
97. D. N. Kursanov and Z. N. Parnes, Doklady Akad. Nauk SSSR 109, 315 (1956).
98. D. N. Kursanov and Z. N. Parnes, Doklady Akad. Nauk SSSR 91, 1125 (1953).
99. Z. N. Parnes and D. N. Kursanov, Doklady Akad. Nauk SSSR 99, 265 (1954).
100. D. N. Kursanov and Z. N. Parnes, Doklady Akad. Nauk SSSR 103, 847 (1955).
101. Z. N. Parnes, Ukr. Khim. Zhur. 22, 40 (1956).
102. D. N. Kursanov, Ukr. Khim. Zhur. 22, 35 (1956).
103. D. N. Kursanov, M. E. Vol'pin, and Z. N. Parnes, Khim. Nauk. i Prom. 3, 159 (1958).
104. A. I. Shatenshtein and Yu. P. Vyrskii, Doklady Akad. Nauk SSSR 70, 1029 (1950).
105. A. I. Shatenshtein and Yu. P. Vyrskii, Zhur. Fiz. Khim. 25, 1206 (1951).
106. A. I. Shatenshtein, N. M. Dykhno, E. A. Izrailevich, L. N. Vasil'eva, and M. Faivush, Doklady Akad. Nauk SSSR 79, 479 (1951).
107. A. I. Shatenshtein, L. N. Vasil'eva, N. M. Dykhno, and E. A. Izrailevich, Doklady Akad. Nauk SSSR 85, 381 (1952).
108. A. I. Shatenshtein and E. A. Izrailevich, Zhur. Fiz. Khim. 28, 3 (1954).

109. N. M. Dykhno and A. I. Shatenshtein, Zhur. Fiz. Khim. 28, 11 (1954).

110. N. M. Dykhno and A. I. Shatenshtein, Zhur. Fiz. Khim. 28, 14 (1954).

111. A. I. Shatenshtein, L. N. Vasil'eva, and N. M. Dykhno, Zhur. Fiz. Khim. 28, 193 (1954).

112. A. I. Shatenshtein and E. A. Izrailevich, Doklady Akad. Nauk SSSR 94, 923 (1954).

113. A. I. Shatenshtein and L. N. Vasil'eva, Doklady Akad. Nauk SSSR 95, 115 (1954).

114. A. I. Shatenshtein and E. A. Yakovleva, Doklady Akad. Nauk SSSR 105, 1024 (1955).

115. A. I. Shatenshtein and E. A. Izrailevich, Doklady Akad. Nauk SSSR 108, 294 (1956).

116. A. I. Shatenshtein and E. N. Zvyagintseva, Doklady Akad. Nauk SSSR 117, 852 (1957).

117. A. I. Shatenshtein and E. A. Yakovleva, Zhur. Obshchei Khim. 28, 1713 (1958).

118. A. I. Shatenshtein, Yu. R. Dubinskii, E. A. Yakovleva, I. V. Gostunskaya, and B. A. Kazanskii, Izvest. Akad. Nauk SSSR, Otdel. Khim. Nauk (1958) p. 104.

119. A. I. Shatenshtein and A. V. Vedeneev, Collection: "Isotopes and Radiation in Chemistry" [in Russian] (Izd. AN SSSR, Moscow, 1958) p. 7.

120. A. I. Shatenshtein and A. V. Vedeneev, Zhur. Obshchei Khim. 28, 2644 (1958).

121. A. I. Shatenshtein and E. A. Izrailevich, Zhur. Obshchei Khim. 28, 2939 (1958).

122. A. I. Shatenshtein and E. A. Izrailevich, Zhur. Fiz. Khim. 32, 2711 (1958).

123. A. I. Shatenshtein, E. A. Yakovleva, M. I. Rikhter, M. Yu. Lukina, and B. A. Kazanskii, Izvest. Akad. Nauk SSSR, Otdel. Khim. Nauk (1959) p. 1805.

124. E. N. Yurygina, P. P. Alikhanov, E. A. Izrailevich, P. N. Manochkina, and A. I. Shatenshtein, Zhur. Fiz. Khim. 34, 587 (1960).

125. A. I. Shatenshtein, A. P. Talanov, and Yu. I. Ranneva, Zhur. Obshchei Khim. 30, 583 (1960).

126. Yu. G. Dubinskii, E. A. Yakovleva, and A. I. Shatenshtein, Zhur. Obshchei Khim. (in press).

127. A. I. Shatenshtein, Uspekhi Khimii 21, 914 (1952).

128. A. I. Shatenshtein, Collection: "Problems in Chemical Kinetics, Catalysis, and Reactivity" [in Russian] (Izd. AN SSSR, Moscow, 1955) p. 699.

129. A. I. Shatenshtein, Uspekhi Khimii 24, 377 (1955).

130. A. I. Shatenshtein, Ukr. Khim. Zhur. 22, 3 (1956).

131. A. I. Shatenshtein, E. N. Zvyagintseva, E. A. Yakovleva, E. A. Izrailevich, Ya. M. Varshavskii, M. G. Lozhkina, and A. V. Vedeneev, Collection: "Isotopes in Catalysis" [in Russian] (Izd. AN SSSR, Moscow, 1957) p. 218.

132. A. I. Shatenshtein, Collection: "Problems in Physical Chemistry" [in Russian] (Goskhimizdat, Moscow, 1958), No. 1, p. 202; Zhur. Fiz. Khim. 34, No. 3 (1960).

133. A. I. Shatenshtein, Uspekhi Khimii 28, 3 (1959).

133a. A. I. Shatenshtein, V. R. Kalinachenko, E. N. Yurygina, and V. M. Basmanova, Zhur. Obshchei Khim. 29, 849 (1959).

134. G. E. Hall, R. Piccolini, and J. D. Roberts, J. Am. Chem. Soc. 77, 4540 (1955).

135. V. L. Broude, E. A. Izrailevich, A. L. Liberman, M. I. Onoprienko, O. S. Pakhomova, A. F. Prikhot'ko, and A. I. Shatenshtein, Zhur. Optika i Spektroskopiya 5, 113 (1958).

136. P. H. Hermans, Introduction to Theoretical Organic Chemistry (Amsterdam, New York, London, 1954) p. 410.

137. G. S. Landsberg, A. I. Shatenshtein, G. V. Peregudov, E. A. Izrailevich, and L. A. Novikova, Izvest. Akad. Nauk SSSR, Ser. Fiz. No. 6, 669 (1954); Zhur. Optika i Spektroskopiya 1, 34 (1956).

138. A. I. Shatenshtein, G. V. Peregudov, E. A. Izrailevich, and V. R. Kalinachenko, Zhur. Fiz. Khim. 32, 146 (1958).

139. A. D. Walsh, Trans. Farad. Soc. 45, 179 (1949).

140. R. Ya. Levina, Synthesis and Catalytic Conversions of Unsaturated Hydrocarbons [in Russian] (Izd. MGU, 1949).

141. R. Ya. Levina et al., Vestnik Mosk. Univ. No. 6, 115 (1949); No. 10, 115 (1949); No. 2, 77 (1951); No. 2, 87 (1952); Zhur. Obshchei Khim. 23, 562 (1953).

142. E. Vogel, Fortschr. Chem. Forsch. 3, 430 (1955).

143. V. T. Aleksanyan, Kh. E. Sterin, M. Yu. Lukina, L. G. Sal'nikova, and I. L. Safonova, Transactions of the Tenth Conference on Spectroscopy [in Russian] (L'vov, 1957) p. 64.

143a. G. A. Razuvaev, G. G. Petukhov, M. A. Shubenko, and V. A. Voitovich, Ukr. Khim. Zhur. 22, 45 (1956).

143b. H. Hart and R. E. Crocker, J. Am. Chem. Soc. 82, 418 (1960).

143c. A. I. Shatenshtein, E. A. Yakovleva, and E. S. Petrov, Zhur. Obshchei Khim. 32, No. 4 (1962).

143d. E. A. Yakovleva, E. S. Petrov, S. P. Solodovnikov, V. V. Voevodskii, and A. I. Shatenshtein, Doklady Akad. Nauk SSSR 133, 645 (1960).

143e. A. I. Shatenshtein, E. A. Yakovleva, and E. S. Petrov, Doklady Akad. Nauk SSSR 136, 882 (1961).

144. I. V. Gostunskaya and B. A. Kazanskii, Zhur. Obshchei Khim. 25, 1995 (1955).

145. I. V. Gostunskaya, N. I. Tyun'kina, and B. A. Kazanskii, Doklady Akad. Nauk SSSR 108, 473 (1956).

146. B. A. Kazanskii, A. L. Liberman, M. Yu. Lukina, and I. V. Gostunskaya, Khim. Nauk i Prom. 2, 172 (1957).

147. I. V. Gostunskaya, Calcium Hexammoniate in the Reduction of Mono- and Di-olefinic Hydrocarbons, Dissertation [in Russian] (MGU, 1955).

148. R. A. Benkeser, D. J. Foster, D. M. Sanoe, and J. N. Nobis, Chem. Rev. 57, 867 (1957).

148a. W. O. Haag and H. Pines, J. Am. Chem. Soc. 82, 387 (1960).

148b. L. Reggel, S. Friedman, and J. Wekder, J. Org. Chem. 23, 1136 (1958).

148c. E. A. Rabinovich, I. V. Astaf'ev, and A. I. Shatenshtein, Zhur. Obshchei Khim. 32 (1962).

149. H. Levy and A. C. Cope, J. Am. Chem. Soc. 66, 1684 (1944).

150. T. W. Campbell and W. G. Young, J. Am. Chem. Soc. 69, 688, 3066 (1947).

151. H. F. Herbrandsen and D. S. Mosney, J. Am. Chem. Soc. 79, 5809 (1957).

151a. I. V. Gostunskaya, N. B. Dobroserdova, M. P. Berednikova, and B. A. Kazanskii, Kinetika i Kataliz 1, 612 (1960).

151b. I. V. Gostunskaya, N. I. Tyun'kina, and B. A. Kazanskii, Izvest. Akad. Nauk SSSR, Otdel. Khim. Nauk (1960) p. 132.

152. A. E. Favorskii, Zhur. Russ. Fiz. Khim. Obshchest. 19, 553 (1887).

153. M. Bourgeul, Ann. chim. (10), 3, 325 (1925).

154. A. J. Birch, J. Chem. Soc. (1947) p. 1642.

155. H. Pines, J. A. Vesely, and V. N. Ipatieff, J. Am. Chem. Soc. 77, 347 (1955).

156. H. Pines and H. E. Eschinazi, J. Am. Chem. Soc. 77, 6314 (1955); 78, 1178, 5950 (1956).

157. H. Pines and L. Schaap, J. Am. Chem. Soc. 79, 2956 (1957).

158. H. Pines and W. O. Haag, J. Org. Chem. 23, 328 (1955); J. Am. Chem. Soc. 82, 387 (1960).

159. A. A. Morton and E. J. Lanpher, J. Org. Chem. 20, 839 (1955).

160. Bruson, Organic Reactions (Edited by Adams) [Russian translation] (IL, Moscow) Vol. V.

161. R. G. Closson, J. P. Napolitano, G. G. Ecke, and A. J. Kolka, J. Org. Chem. 22, 646 (1957).

162. J. D. Danoforth, Chem. Abstr. 45, 3870 (1951).

163. F. W. Howk, E. L. Little, S. L. Scott, and G. M. Witzman, J. Am. Chem. Soc. 76, 1899 (1954).

164. H. Pines, J. A. Vesely, and V. N. Ipatieff, J. Am. Chem. Soc. 77, 554 (1955).

165. H. Pines and V. Mark, J. Am. Chem. Soc. 78, 4316 (1955).

166. H. Pines and L. Schaap, J. Am. Chem. Soc. 79, 2956, 4967 (1957).

167. H. Hart, J. Am. Chem. Soc. 78, 2619 (1956).

168. H. Gilman and F. Schulze, J. Am. Chem. Soc. 47, 2002 (1925).

169. R. M. Schramm and G. E. Langlois, J. Am. Chem. Soc. 82, 4912 (1960).

170. H. Pines and V. Mark, J. Am. Chem. Soc. 78, 4316 (1956).

171. V. Mark and H. Pines, J. Am. Chem. Soc. 78, 5946 (1956).

Chapter 2. HYDROCARBONS AS BASES

1. INTRODUCTION

Some hydrocarbons react with very strong acids as normal Brønsted bases and add a proton to form the conjugate acids, namely, carbonium ions. However, carbonium ions are formed not only according to an acid–base scheme, but also as a result of rupture of bonds between carbon atoms and other atoms (halogen and oxygen). A general idea of this type of reaction should be given.

Acid hydrogen exchange is caused by the affinity of a hydrocarbon for a proton (deuteron) and therefore the appropriate literature is naturally given in this chapter on hydrocarbons as bases.

Like other bases, hydrocarbons can participate in equilibria with aprotic acidlike substances. The range of molecular compounds of the latter with hydrocarbons is vast and a detailed review is beyond the scope of this book, though it would not be correct to ignore these compounds completely and thus impoverish the characterization of hydrocarbons as bases. In addition, equilibrium reactions of hydrocarbons with electrophilic reagents (for example, halogens) are intermediate stages in irreversible electrophilic replacement of hydrogen (for example, halogenation). This explains the great similarity in the rules of these chemical reactions and acid hydrogen exchange in hydrocarbons, which we will discuss later in the last section of the book on the mechanisms of hydrogen replacement.

2. CARBONIUM IONS

Carbonium ions contain carbon whose electronic envelope has a structure $1s^2 2s 2p^2$. On ionization, carbon changes from the state of sp^3-hybridization to the state of sp^2-hybridization. The valence bonds, which were previously directed toward the corners of a tetrahedron, lie in one plane. Optically active compounds are therefore racemized if a carbonium ion is formed (see [1-6] for the voluminous literature on carbonium ions).

Carbonium ions are obtained in the following reactions:

1) Rupture of the bond of carbon with an electronegative atom or group;

2) Elimination of a hydride ion (H^-) from a hydrocarbon;

3) An exchange between a carbonium ion and a hydrocarbon;

4) The addition of a proton (H^+) or a carbonium ion to an unsaturated hydrocarbon.

In addition, V. L. Tal'roze [7, 8], working in V. N. Kondrat'ev's laboratory, discovered the addition of a proton to a saturated hydrocarbon molecule. The ions formed in this way (for example, the methonium ion CH_5^+) are also called carbonium ions, though they do not contain trivalent carbon and their structure has not been established as yet.

A. Rupture of a Bond Between Carbon and an Electronegative Atom or Group

The conception of carbonium ions arose at the beginning of the twentieth century as a result of the work of Walden [9] and Gomberg [10], who discovered that a solution of triphenylmethyl chloride ($(C_6H_5)_3CCl$ in liquid sulfur dioxide has a yellow color and conducts an electric current. When $SnCl_4$ was added, the intensity of the color and the electrical conductivity increased. The latter approached that of a typical electrolyte, namely, potassium iodide in liquid sulfur dioxide. In the heterolytic rupture of the bond ($\equiv C : Cl \rightarrow \equiv C^+ + Cl^-$), the pair of electrons is transferred to the chlorine atom. The energy required for the rupture of the bond is imparted as a result of solvation of the chlorine anion by sulfur dioxide molecules. The addition of stannic chloride displaces the ionization equilibrium as a result of the addition of a chlorine ion to the coordinationally unsaturated tin atom:

$$(C_6H_5)_3CCl + SnCl_4 \rightleftarrows (C_6H_5)_3C^+ + SnCl_5^-$$

$$(C_6H_5)_3CCl + SnCl_5^- \rightleftarrows (C_6H_5)_3C^+ + SnCl_6^{--}.$$

The conclusions of the authors were confirmed by experiments on the electrolysis of solutions in liquid sulfur dioxide [11], etc. A detailed study was made later [12-14] of the effect of ionization of the nature of the substituents and their position in the aromatic rings of triphenylmethyl chloride. The formation of triarylcarbonium ions is promoted by electropositive groups (NH_2, OCH_3, etc.) and hindered by electronegative substituents (NO_2, etc.).

Triphenylcarbinol is ionized similarly in liquid sulfur dioxide:

$$(C_6H_5)_3COH + SO_2 \rightleftarrows (C_6H_5)_3C^+ + SO_3H^-.$$

The equilibrium is displaced to the right even more strongly when the carbinol is dissolved in sulfuric acid:

$$(C_6H_5)_3COH + 2H_2SO_4 \rightleftarrows (C_6H_5)_3C^+ + H_3O^+ + 2HSO_4^-.$$

Cryoscopic measurements [14-17] demonstrated the appearance of four osmotically active particles. Absorption spectra of solutions in liquid sulfur dioxide and in sulfuric acid are the same because the same cations are formed in the two solvents [18-22]. As the position of the ionization equilibrium depends on the acidity of the solution, it was suggested that spectrophotometric measurements of the concentration of carbonium ion may be used to determine the acidity of concentrated acids (see p. 51 for details on the acidity function J_0). Hart and Fish [23] were able to prepare a stable, doubly charged carbonium ion by solution of trichloromethylpentamethylbenzene in 100% sulfuric acid. The solution had an intense red color and its spectrum contained the following maxima (λ in mμ): 545 (1031), 393 (16160), 265 (3860) and 235 (5120). The extinction coefficients (ε) are given in brackets. The authors considered that two molecules of HCl were formed in the reaction:

$$C_6(CH_3)_5CCl_3 + 2H_2SO_4 \rightarrow H_3C \underset{CH_3\ CH_3}{\overset{CH_3\ CH_3}{\langle\!\!+ \rangle}} = C^+ - Cl + 2HCl + 2HSO_4^-.$$

In actual fact, cryoscopic measurements confirmed that the Van't Hoff factor i equaled 5 (it did not change after 24 hr). When dry nitrogen was passed through the solution, two moles of HCl were absorbed in a receiver with alkali even after 15 min and the amount of HCl did not increase in the next 12 hr. In accordance with the hypothesis on the formation of a double charged ion, the molecular electrical conductivity of the solution practically equaled that of o-$C_6H_4(NH_2)_2$, which gives a double charged ion in sulfuric acid (Table 44).

TABLE 44. Molecular Electrical Conductivity of Solutions in 100% H_2SO_4 (at 25°)

M	$C_6(CH_3)_5CCl_3$	o-$C_6H_4(NH_2)_2$
0.05	304	304
0.10	213	212
0.20	155	164

The same ions are produced if a large excess of aluminum chloride is added to a solution of trichloromethylpentamethylbenzene in nitromethane.

Aliphatic carbonium ions can also be produced by rupture of a C—Hal bond, for example: $RHal + AlHal_3 \rightarrow R^+ + AlHal_4^-$, but this reaction is less favored energetically. Even in 1902, V. A. Plotnikov [24] observed an increase in the electrical conductivity of ethyl bromide when aluminum bromide was dissolved in it (see also [25, 26]). The number of free ions in such a solvent with a low dielectric constant is small. Measurements of the dielectrical polarization [27] of alkyl halides (C_2H_5Br, tert-C_4H_9Br, etc.) in cyclohexane show that ion pairs of the type $C_2H_5^+Al_2Br_7^-$ exist in the presence of $AlBr_3$, $SnBr_4$, and BBr_3. The lability of the C—Hal bond observed when electrophilic reagents like $AlBr_3$ are added is confirmed by the exchange of bromine between $AlBr_3$ and alkyl bromides, which was first discovered by S. Z. Roginskii and his co-workers [28-30]. Similar work was carried out by Fairbrother [31].

The amount of heat (Q) absorbed during the ionization of an alkyl halide in solution is made up of several terms: D, the heat of dissociation of the R—X bond, I, the ionization potential of the alkyl group, E, the affinity of the halogen atom X for an electron, and, finally, S^+ and S^-, the heats of solvation of the ions [2]:

$$Q = D + I - E - S^+ - S^-.$$

The ionization potentials of different alkyl groups (in the gas phase) taken from [3] are given below:

Alkyl group	CH_3	C_2H_5	$n\text{-}C_3H_7$	$iso\text{-}C_3H_7$	$tert\text{-}C_4H_9$
I, kcal/mole	230	201	183	171	159

Primary are less stable than secondary aliphatic carbonium ions. Tertiary carbonium ions are most stable: $CH_3^+ < C_2H_5^+ < CH(CH_3)_2^+ < C(CH_3)_3^+$. In accordance with this rule, the electrical conductivity of the complex of tert-buryl fluoride with boron tri-fluoride (general formula RBF_4) is 200 times that of the electrical conductivity of a similar complex with methyl fluoride [32].

B. Exchange Between a Carbonium Ion and a Hydrocarbon

The reaction between a carbonium ion and a hydrocarbon

$$R^+ + R_1H \rightleftarrows R_1^+ + RH$$

formally corresponds to the reaction between a carbanion and a hydrocarbon examined above:

$$R^- + R_1H \rightleftarrows R_1^- + RH.$$

The reaction essentially consists of the abstraction of H^- by the ion R^+ (see below). An example is the formation of a tropylium ion from cycloheptatriene and $(CH_3)_3C^+$ [33]. It is considered that the formation of a carbonium ion R_1^+ is part of the mechanism of hydrogen exchange between saturated hydrocarbons containing a tertiary carbon atom and deuterosulfuric acid. A carbonium ion is obtained by oxidation of a methyne group and the reaction between the carbonium ion and the hydrocarbon then proceeds by a chain mechanism (p. 184).

C. Elimination of a Hydride Ion

Whitmore [34] envisaged the formation of a carbonium ion from a hydrocarbon by the elimination of a hydride ion (H^-) from it by an electrophilic reagent (E). The reaction is represented schematically by the following equation:

$$RH + E \xrightarrow{-H^-} R^+ + EH^-.$$

It is probable that polar complexes R^+EH^- are usually formed rather than free ions.

Such reactions occur, for example, when cycloheptatriene is treated with electrophilic reagents (including acids) and are used to prepare tropylium salts.

For example, with sulfuric acid the reaction proceeds according to the equation:

$$\text{(cycloheptatriene with CH}_2\text{)} + 2H_2SO_4 \longrightarrow \text{(tropylium)} HSO_4^- + 2H_2O + SO_2$$

and with aluminum chloride:

$$\text{(cycloheptatriene with CH}_2\text{)} + AlCl_3 \longrightarrow \text{(tropylium)} HAlCl_3^- .$$

The hydride transfer leading to the conversion of cycloheptatriene to the tropylium ion in an acid medium is reversible. This is demonstrated by the fact that in the reaction of deuterocycloheptatriene with tropylium hydrobromide in formic or acetic acid, the deuterium is uniformly distributed between the tropylium and cycloheptatriene [33a].

D. N. Kursanov, M. E. Vol'pin, and their co-workers, who studied such reactions, made a valuable contribution to the chemistry of tropylium [33, 35] (see the review [36]). These authors consider that the formation of a tropylium ion, a nonbenzenoid aromatic carbonium ion, by this scheme has much in common with reactions in which aliphatic carbonium ions are obtained by the action of concentrated sulfuric acid on saturated hydrocarbons with a tertiary carbon atom (p. 184). The elimination of a hydride ion from cycloheptatriene and the conversion of the latter into a tropylium ion are produced by reaction with sulfuric acid and by oxidation with CrO_3 and SeO_2 in an acid medium. Thus, this provides a link between acid−base and oxidation−reduction reactions. In the opinion of D. N. Kursanov, it is possible that there may be a connection here between the mechanisms of hydrogen exchange in saturated hydrocarbons with sulfuric acid, which is simultaneously an oxidizing agent, and with liquid hydrogen halides ($HBr + AlBr_3$ and $HF + BF_3$), which do not have an oxidizing action (see p. 182 for more details). A detailed review of reactions of hydrocarbons involving the transfer of a hydride ion was published by Nenitzescu [37].

D. Addition of a Proton to an Unsaturated Hydrocarbon

The hypothesis that aliphatic carbonium ions are formed in the reactions of unsaturated hydrocarbons catalyzed by acids was put forward by Whitmore [33, 38] more than twenty-five years ago. Thus, a proton adds to the double bond of an olefin through its electrons:

$$> C::C + H^+ \rightarrow > \overset{\oplus}{C}:\underset{\overset{..}{H}}{C} < \cdot$$

Aprotonic acidlike substances react analogously, for example:

$$> C::C< + AlCl_3 \rightarrow > \overset{\oplus}{C}:\underset{\overset{\ominus}{\ddot{A}lCl_3}}{C} < .$$

A pair of electrons of the double bond fills the deficit in the electron cloud of the atom of the electrophilic reagent. Below we will examine the already numerous studies which demonstrate the capacity of aromatic hydrocarbons to react with acids (and with acidlike substances) like Brønsted bases. The summarization and discussion of this work represents a considerable part of this chapter.

There are cases where a carbonium ion may be obtained in two ways with the same reagent [3]:

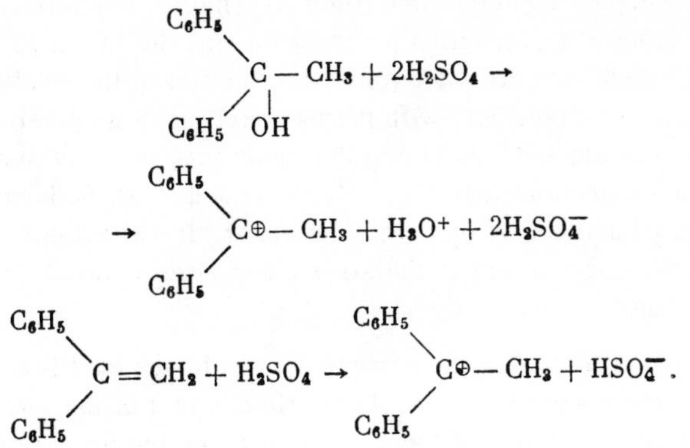

$$\begin{array}{c} C_6H_5 \\ \diagdown \\ \diagup \quad \diagup \\ C_6H_5 \quad OH \end{array} C - CH_3 + 2H_2SO_4 \rightarrow$$

$$\rightarrow \begin{array}{c} C_6H_5 \\ \diagdown \\ \diagup \\ C_6H_5 \end{array} C\oplus - CH_3 + H_3O^+ + 2HSO_4^-$$

$$\begin{array}{c} C_6H_5 \\ \diagdown \\ \diagup \\ C_6H_5 \end{array} C = CH_2 + H_2SO_4 \rightarrow \begin{array}{c} C_6H_5 \\ \diagdown \\ \diagup \\ C_6H_5 \end{array} C\oplus - CH_3 + HSO_4^-.$$

When an acid reacts with diphenylmethylcarbinol, the hydroxyl group is eliminated, while in the reaction of the same acid with diphenylethylene, a proton is added to the double bond.

Other examples are given in [38a]. The identity of the carbonium ions formed from a phenylated olefin and an aromatic carbon was demonstrated by the coincidence of the absorption spectra.

E. Addition of a Carbonium Ion to an Aromatic Hydrocarbon

We should also mention that a carbonium ion is formed if an aliphatic carbonium ion such as CH_3^+ instead of a proton is added to the molecule of an aromatic hydrocarbon. The methylation of hexamethylbenzene (and other methylbenzenes) by the Friedel-Crafts method gives a heptamethylbenzonium ion $C_6(CH_3)_7^+$ [39]. Its spectrum is very similar to that of the ion $C_6(CH_3)_6H^+$:

$C_6(CH_3)_7^+$: 287 mμ (log ε = 3.83); 397 mμ (log ε = 3.93)
$C_6(CH_3)_6H^+$: 283 mμ (log ε = 4.31); 401 mμ (log ε = 4.42),

(see also pp. 141, 150, 159, and 160) on the spectra of ions obtained by the addition of a proton). The problem of the nomenclature of carbonium ions is discussed in [39].

F. Detection of a Carbonium Ion by Deuterium Exchange

In the chemical literature [4], the idea of unstable carbonium ions is often used to explain reaction mechanisms without adequate grounds. An elegant method of demonstrating the formation of a carbonium ion during a reaction is the racemization of an optically active substance as a result of a change in the type of valence hybridization of carbon (p. 130) [40, 41].

D. N. Kursanov, Z. N. Parnes, V. N. Setkina, S. V. Vitt, and E. V. Bykova [40-44] used hydrogen exchange as an indicator of the formation of carbonium ions. They started from the hypothesis that aliphatic carbonium ions readily participate in hydrogen exchange. This idea was checked on a series of heterolytic reactions in which it was assumed that carbonium ions were formed in the interpretation of their mechanism given in the literature. These reactions were the pinacolin rearrangement [40], the dimerization of 1-methyl-1-cyclohexene under the action of phosphoric acid, the alkylation of resorcinol [41], and replacements of iodine in alkyl iodides [43]. As an example, we will consider the last paper.

Ingold found that the hydrolysis of tertiary alkyl halides is described by a kinetic equation of the first order as this reaction proceeds through the intermediate formation of carbonium ions, whose rate also determines the kinetics of the whole process. The carbonium ion reacts instantaneously with the hydroxyl ion of an alkali or with a water molecule to form an alcohol. Assuming that there may be rapid exchange of hydrogen for deuterium in a carbonium ion, D. N. Kursanov and V. N. Setkina studied the hydrolysis of tertiary butyl iodide by deuterium oxide. It was actually possible to show [43] that all the hydrogen atoms in the alkyl group were replaced by deuterium in the tert-butanol obtained.

It was established [41] that the possibility of hydrogen exchange in halogen derivatives depends on the magnitude of the dielectric constant of the solvent. A high dielectric constant facilitates rupture of the C−Hal bond and the formation of a carbonium ion, in which the exchange reaction occurs. Thus, for example, 1-methyl-1-chlorocyclohexane undergoes deuterium exchange with anhydrous formic acid (DC = 58.5), but does not undergo exchange with deuteroacetic acid, whose dielectric constant is a factor of ten less (DC = 6).

According to the data of the same authors [44], there is no exchange between tertiary butyl chloride and acetic acid enriched in deuterium in the carboxyl group, but dueterium exchange proceeds readily when a number of acidlike substances are added. D. N. Kursanov and V. N. Setkina consider that the reason for this phenomenon is heterolysis of the C−Cl bond, whereupon hydrogen exchange is caused by the formation of a carbonium ion as a result of heterolysis of the C−Cl bond analogous to the equation given on p. 133. By breaking the C−Cl bond with various halogen compounds, the authors were able to estimate their relative electrophilicity, which is represented by the following series:

$$FeCl_3 \simeq SbCl_5 > SnCl_4 > ZnCl_2 > HgCl_2.$$

3. EQUILIBRIUM REACTIONS OF HYDROCARBONS WITH ACIDS

A. Introduction

Below we give a summary of equilibrium reactions of hydrocarbons (predominantly aromatic, but partly ethylenic) with acids to which the following scheme can be applied:

$$ArH + HA \rightleftarrows ArH_2^+ + A^-, \qquad (I)$$

where a molecule of an aromatic hydrocarbon ArH adds a proton to form a carbonium ion [45]. Let us first consider the nature of these equilibria: This makes it possible to demonstrate that there are no qualitative differences between hydrocarbons and any other bases. There are quantitative differences as most hydrocarbons are very weak bases and for this reason their reactions with acids, more frequently than the reactions of other bases, do not proceed to ionization, but are limited to the formation of a molecular complex:

$$ArH + HA \rightleftarrows HAr \cdots HA. \tag{II}$$

A knowledge of the protolytic reactions of hydrocarbons provides further evidence for the need for the development of the concepts of acids and bases.

Solutions of hydrocarbons in liquid hydrogen fluoride and in sulfuric are considered first. The strong protogenicity of these solvents together with their high dielectric constant promotes the ionization of hydrocarbons. This is indicated by data on electrical conductivity, vapor pressure, and absorption spectra. The relative strengths of hydrocarbons as bases can be compared by means of data on the distribution between an acid and an inert solvent.

The low dielectric constants of the other hydrogen halides (HCl, HBr, and HI) are the reason why the reaction proceeds predominantly according to scheme (II) when hydrocarbons are dissolved in them. The formation of molecular compounds is indicated by breaks on the freezing curves of the systems. The deviation of the solubility of hydrogen halides from the ideal (according to Henry's law) indicates a weak interaction between the hydrocarbon and the acid, which consists of the formation of π-complexes (p. 156).

The addition of an aluminum halide to the system consisting of an aromatic hydrocarbon and a hydrogen halide polarizes the H−Hal bond so strongly that carbonium salts of haloaluminic acids are formed readily and these are colored and conduct an electric current. Other electrophilic reagents such as gallium halides have an action similar to that of aluminum halides.

A few hydrocarbons which are among the strongest bases in this class can be converted to carbonium salts even by solution in formic and acetic acids.

B. Liquid Hydrogen Fluoride

a. SOLUBILITY OF HYDROCARBONS

In contrast to saturated hydrocarbons (alkanes and cycloalkanes), many aromatic hydrocarbons (benzene, toluene, xylenes, mesitylene, hexamethylbenzene, anthracene, and diphenylethylene) are soluble in liquid HF [46-48]. Klatt [46] reported that the solubility of aromatic hydrocarbons is increased appreciably by the addition of salts (for example, $Hg(CN)_2$, AgN_3, AgF, and TlF) which give complex salts with the F^- ion. Hammett [6, p. 239] explained this phenomenon by the displacement of the equilibrium between the hydrocarbon and HF:

$$C_6H_6 + HF \rightleftarrows C_6H_7^+ + F^-$$

to the right in the presence of these salts as their cations bind fluorine ions, for example:

$$Hg^{++} + 4F^- \rightleftarrows HgF_4^{--}.$$

It was shown subsequently [48, 49] that the solubility of hydrocarbons is increased considerably by the addition of BF_3 and other substances which are capable of adding fluorine ions as a result of the reaction:

$$C_6H_6 + HF + BF_3 \rightleftarrows C_6H_7^+ + BF_4^-.$$

Naphthalene and phenanthrene, which are sparingly soluble in liquid HF, become appreciably soluble in the presence of BF_3 [48]. The solubility of anthracene [50] at 0°
is 0.18 mole/1000 g of liquid HF. The addition of one mole of BF_3 to the same amount
of HF increases the solubility to 1.05 mole/1000 g of HF. On the other hand, the addition of one mole of NaF reduces the solubility of anthracene to 0.04 mole/1000 g of
HF as it increases a factor in the ionic product $L_{ArH} = m_{ArH^+} m_{F^-}$, where m is the
molarity of the ions. A qualitative estimate of the acidity of solutions of fluorides in
liquid HF gave the following series [51]:

$$SbF_5 > AsF_5 > BF_3 > SnF_4.$$

The solubility of α-olefins in liquid HF at temperatures from -80 to 0° is less than
1%. If BF_3 is passed into a heterogeneous system consisting of equimolecular amounts
of an olefin and HF, up to one mole of it is absorbed vigorously, while BF_3 is sparingly
soluble in the components of the mixture separately. When BF_3 is dissolved in a mixture of an olefin and HF, the system acquires a yellow color. The experiments were
carried out with 2-methylbutene, 1-butene, and propylene [51a]. The phenomena described are explained by the formation of a complex, similar to those formed by aromatic hydrocarbons:

$$RCH = CH_2 + HF + BF_3 \rightarrow RCH - CH_3^+ BF_4^-.$$

b. ELECTRICAL CONDUCTIVITY OF SOLUTIONS

As early as 1935 Klatt [52] observed that a solution of anthracene in liquid hydrogen fluoride conducts an electric current as well as a solution of potassium fluoride: at
a solution molarity of 0.12, i.e., 0.12 mole of anthracene in 1000 g of HF, the equivalent electrical conductivity $\lambda = 300$ ohm^{-1} cm^2, while the electrical conductivity of KF
solution equaled 255 ohm^{-1} cm^2. This is evidently due to the fact that liquid hydrogen
fluoride, in addition to being an acid solvent, has an exceptionally high dielectric constant. The measurements of Kilpatrick and Luborsky [53, 54] showed that even a solution of benzene in liquid HF conducts a current, though very weakly. The ionization
constant of benzene as a base in liquid hydrogen fluoride is very low: $K_i^{HF} = 10^{-8}$ at

20°; the ionization constant increases progressively, i.e., the degree of ionization increases with an increase in the number of methyl groups in an alkylbenzene and reaches
$K_i^{HF} = 5 \cdot 10^{-2}$ for hexamethylbenzene. Table 45 gives λ_1, the equivalent electrical

conductivity of a solution at a hydrogen concentration of the order of 0.1 mole/liter,
K_i^{HF}, the ionization constant of the hydrocarbon in liquid hydrogen fluoride at 20°, and

λ_2, the equivalent electrical conductivity when excess boron trifluoride is added. The
latter is close to the electrical conductivity of a solution of a binary salt such as potassium fluoride in the same solvent.

When boron trifluoride is added to a solution of a hydrocarbon in liquid hydrogen
fluoride, the electrical conductivity gradually increases. There is a break on the electrical conductivity curve at an equimolecular ratio of the hydrocarbon and boron trifluoride. This confirms that they react with each other in a ratio of a mole to a mole
according to the equation:

$$ArH + HF + BF_3 \rightleftarrows ArH_2^+ + BF_4^-.$$

If the alkylbenzene is weakly ionized in pure hydrogen fluoride, the addition of boron trifluoride promotes ionization and the electrical conductivity of the solution increases;

TABLE 45. Electrical Conductivity of Solutions of Benzene and Methylbenzenes in Liquid HF (at 20°)

Hydrocarbon	λ_1	λ_2	K_i^{HF}
Benzene	1	—	$1 \cdot 10^{-8}$
Methylbenzene (toluene).	5	200	$2 \cdot 10^{-7}$
1,4-Dimethyl-(p-xylene)	6	208	$3 \cdot 10^{-7}$
1,2-Dimethyl-(o-xylene)	6	240	$4 \cdot 10^{-7}$
1,3-Dimethyl-(m-xylene).	28	270	$8 \cdot 10^{-6}$
1,2,4-Trimethyl-(pseudocumene) .	45	240	$2 \cdot 10^{-5}$
1,2,3-Trimethyl-(hemimellitene).	50	230	$2 \cdot 10^{-5}$
1,3,5-Trimethyl-(mesitylene) . . .	270	250	$4 \cdot 10^{-3}$
1,2,4,5-Tetramethyl-(durene) . . .	—	—	$2 \cdot 10^{-4}$
1,2,3,4-Tetramethyl-(prehnitene).	102	250	$1 \cdot 10^{-4}$
1,2,3,5-Tetramethyl-(isodurene). .	255	230	$4 \cdot 10^{-3}$
Pentamethylbenzene	300	—	$9 \cdot 10^{-3}$
Hexamethylbenzene.	320	340	$5 \cdot 10^{-2}$

when the equilibrium (I) is displaced far to the right (as in solutions of polymethylbenzenes), the addition of boron trifluoride hardly changes the number of ions, but leads to replacement of the F^- ions by BF_4^- ions. As a result of the lower mobility of the latter in comparison with fluorine ions, the electrical conductivity of the solution decreases. In both cases, the electrical conductivity of the solution measured with excess BF_3 (Table 45, the value λ_2) corresponds to the maximum ionization of the hydrocarbon, i.e., its complete conversion into carbonium ions. An analogous decrease in the electrical conductivity as a result of the formation of the complex ion BF_4^- was observed when BF_3 was passed into a solution of KF [54]. Olah [55-57] prepared individual complexes* of toluene, m-xylene, mesitylene, and isodurene with $HF-BF_3$, which consisted of equimolecular amounts of the hydrocarbon, HF, and BF_3. The complexes were intensely colored (see below for details of their spectra). They were stable at low temperatures and conducted a current well. Their melting points (°C) and specific electrical conductivities (ohm^{-1} cm^{-1}) are given in Table 46.

TABLE 46. Properties of Protonized Complexes of Methylbenzene Tetrafluoborates

Complex	M.p., °C	$\varkappa$
$[C_6H_5CH_3]^+BF_4^-$	—65	$0.8 \cdot 10^{-2}$
$[m-C_6H_5(CH_3)_2]^+ BF_4^-$	—55	$2.0 \cdot 10^{-2}$
$[1,3,5-C_6H_4(CH_3)_3]^+ BF_4^-$	—15	$1.6 \cdot 10^{-2}$
$[1,2,3,5-C_6H_2(CH_3)_4]^+BF_4^-$	—10	$0.6 \cdot 10^{-2}$

---

*If DF was present in the complex with toluene, decomposition of the complex yielded toluene deuterated in the ortho- and para-positions.

c. VAPOR PRESSURE OF SOLUTIONS

A clear idea of the relative strengths of alkylbenzenes as bases in liquid HF and the composition of complexes with BF_3 is given by measurements of the vapor pressure of ternary systems, made by McCaulay and Lien [49]. The ratio between the number of moles of HF and the hydrocarbon was set (10:1) and the ratio between the amount of hydrocarbon and BF_3 varied.

In Fig. 16, the first line corresponds to the vapor pressure system $HF-BF_3$, while the last line is reproduced from literature data for the vapor pressure of a solution of KF in liquid HF. The curves obtained for the systems with hexamethylbenzene and mesitylene coincide with the line for $KF-HF$ approximately up to the point corresponding to a ratio of hydrocarbon to BF_3 of 1:1. It is evident that these hydrocarbons, which conduct a current well in liquid HF even without BF_3 and are completely ionized when it is present (p. 141), form complexes containing a mole of the hydrocarbon and a mole of BF_3.

The identical lowering of the vapor pressure of liquid hydrogen fluoride when equimolecular amounts of the hydrocarbons complex or a salt, namely, potassium fluoride, are dissolved in it demonstrates that the complex consists of two ions like potassium fluoride. In a solution containing BF_3, alkylbenzenes are present in the form of an ionized complex, whose composition corresponds to the general formula ArH_2^+, BF_4^- (ArH is a molecule of the aromatic hydrocarbon).

The vapor pressure curves show that the stability of the carbonium salt depends both on the degree of methylation of the benzene and on the position of the methyl group in the hydrocarbon molecule; the strengths of alkylbenzenes as carbo bases decrease in the sequence: hexamethylbenzene > mesitylene > > m-xylene > o-xylene > p-xylene > toluene.* This is shown by the fact that in the region of molar ratios of BF_3 and the hydrocarbon close to 1:1, the vapor pressure of the system with hexamethylbenzene is greater than that with mesitylene. The curve for m-xylene approaches the curve for the first two substances only when more than an equimolecular amount of BF_3 is added. The complexes with o- and p-xylene are still less stable. An interaction between toluene and BF_3 is indicated by a change in the Henry constant in comparison with the value for the binary system $HF-BF_3$ (compare also the melting points of the complexes, above which they begin to decompose. The complex with toluene has the lowest melting point, Table 46).

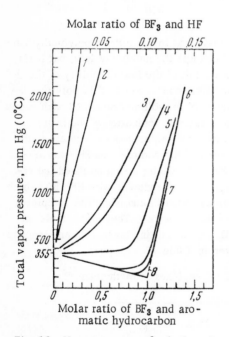

Molar ratio of BF_3 and HF

Total vapor pressure, mm Hg (0°C)

Molar ratio of BF_3 and aromatic hydrocarbon

Fig. 16. Vapor pressure of solutions in liquid HF. 1) $HF-BF_3$; 2) toluene; 3) p-xylene; 4) o-xylene; 5) m-xylene; 6) mesitylene; 7) hexamethylbenzene; 8) KF.

_____

* It was established later [58] that some alkylbenzenes are isomerized in HF + BF_3.

The form of the curve for the binary system $HF-BF_3$ indicates that fluoboric acid HBF_4 does not exist in a free state. Electrochemical measurements [59] indicate the presence of the equilibrium

$$2HF + BF_3 \rightleftarrows H_2F^+ + BF_4^- ,$$

whose constant is very low $(1.4 \cdot 10^{-9})$. Antimony pentafluoride reacts similarly with HF [60]:

$$2HF + SbF_5 \rightleftarrows H_2F^+ + SbF_6^- .$$

The conclusions of the work described [49] are in agreement with the results obtained by Mackor and Hofstra [50]. They found that with equimolecular ratios of pyrene and BF_3 in liquid HF* the lowering of the solution vapor pressure is the same as for an equivalent solution of $NaBF_4$. It should be noted that the basicity constant of pyrene is greater than that of mesitylene (p. 147).

d. ABSORPTION SPECTRA OF SOLUTIONS

Klatt [46] described the color of solutions of some hydrocarbons in liquid HF. More complete information was obtained in the Isotopic Reactions Laboratory [48]: anthracene gives a yellow-green solution; m-xylene, diphenylmethane, and triphenylmethane give yellow solutions; mesitylene and hexamethylbenzene give bright yellow solutions; diphenylethylene gives a bright orange solution. The addition of BF_3 not only increases the solubility of benzene, naphthalene, anthracene, and phenanthrene, but also leads to a color. Solutions of benzene, toluene, p-xylene, chlorobenzene, and even methylcyclohexane have a yellow color. A solution of biphenyl is yellow-green, one of naphthalene, red, and one of phenanthrene, cherry. If solutions in HF are colored, the addition of BF_3 increases the intensity of the color.

The spectra of carbonium ions obtained by the reaction of benzene, naphthalene, anthracene, phenanthrene, naphthacene, and pyrene, and also toluene, mesitylene, and hexaethylbenzene with a solution of boron trifluoride in liquid hydrogen fluoride were first described by Reid [61] in an article entitled "The Aromatic Carbonium Ion." The measurements, which were difficult experimentally, were made at $-78°$. The spectra of the carbonium ions were divided into two groups with absorption maxima in the region of 4000 and 4800 A. Here it is particularly interesting to note that the spectrum of a solution of anthracene in the system $HF-BF_3$ is almost completely identical with the spectrum of a solution of anthracene in anhydrous sulfuric acid [62]. This is a convincing demonstration of the fact that anthracene is actually present in the form of the carbonium ion $C_{14}H_{11}^+$ in both solutions. The spectra of solutions of benzene and toluene are close to the theoretically calculated spectra of the corresponding carbonium ions [63, 64].

Thorough investigations of the spectra of solutions of hydrocarbons in liquid hydrogen fluoride were made by Mackor and his co-workers [65-67].

These authors first measured the spectra of carbonium ions of anthracene and 1,2-benzanthracene.

---

*It was assumed [50] that in concentrated solutions of pyrene (P) there is an equilibrium of a more complex form: $6P + 2HF \rightleftharpoons P_6H_2^{++} + 2F^-$.

A proton adds to the anthracene molecule most readily in a meso-position (position 9 or 10). In the 1,2-benzanthracene molecule, positions 9 and 10 are unsymmetrical. Since a proton can add to each of them, the spectrum of a solution of this hydrocarbon and its methylated derivatives is a combination of the spectra of the two ions.

Table 47 gives the position of the absorption maxima of aromatic ions (in $m\mu$), determined experimentally (λ_{exp}) and calculated theoretically by the molecular orbital method (λ_{calc}).

TABLE 47. Long-Wave Absorption Maxima of Carbonium Ions of Some Polycyclic Hydrocarbons

Ion	λ_{exp}	λ_{calc}
9-Anthracenium	~445	420
	~377	401
	408	398
9-Naphthacenium	592	549
	448	410
	~425	397
1,2-Benz-9-anthracenium	520	563
	444	415
1,2-Benz-10-anthracenium	520	495
10-Methyl-1,2-benz-9-anthracenium	520	563
	444	415
	408	396
3-Methyl-1,2-benz-10-anthracenium	592	549
	448	410
	~425	397

The spectra of solutions of some isomers of methyl-1,2-benzanthracene were also measured [65] and the spectral data used to calculate the equilibrium constants of the ionization

$$ArH + HF \rightleftharpoons ArH_2^+ + F^-.$$

The equilibrium constant

$$K = \frac{m_{ArH_2^+} m_{F^-}}{m_{ArH}} \cdot \frac{f_{\pm}^2}{f_{ArH}} ,$$

where m is the concentration of mole/1000 g of HF (molarity); $f_{\pm}$ and f_{ArH} are the activity coefficients of the ion pair and the hydrocarbon molecule for an infinitely dilute solution in liquid HF.

The logarithms of the constants are given in brackets: 1,2-benzanthracene (2.6); 2'-methyl-(3.2); 3'-methyl-(3.3); 4'-methyl-(3.0); 3-methyl-(4.0); 4-methyl-(3.1); 5-methyl-(3.2); 6-methyl-(3.9); 7-methyl-(3.8); 8-methyl-(3.1); 9-methyl-(5.9); 10-methyl-(4.6); 9,10-dimethyl-1,2-benzanthracene (6.2). The methyl derivatives of benzanthracene are quite strong bases (cf., the constants for methylbenzene and a series of other hydrocarbons with condensed rings on p. 141 *). In an article published later [66], the authors noted that while the values of the constants are largely accurate, they still require refining.

In other work, Dallinga, Mackor, and Stuart [67] measured the absorption spectra of solutions of aromatic hydrocarbons in oxygen-free liquid HF.† The weaker bases (marked with an asterisk) are wholly converted into carbonium ions by the addition of BF_3. All the spectra have strong characteristic ionization bands in the visible and ultraviolet regions (see Table 48); λ is the wavelength in mμ and ϵ the absorption coefficient.

The spectrum of the benzene cation contains two strong bands at 370 and 260 mμ and a weaker one in the region of 200 mμ. The spectrum of the anthracene cation contains an intense band at 408 mμ and a series of weaker bands in the region of shorter wavelengths. Beer's law was obeyed with the spectra studied.

The positions of the bands in the absorption spectra have been calculated for a large number of carbonium ions (including isomeric ions with a proton added to different carbon atoms) by the molecular orbital method. The calculated maxima agree well with experimental values. In some cases where nonequivalent carbon atoms have a similar affinity for a proton, the calculated spectra helped in the assignment of the bands. Alkyl derivatives of the corresponding hydrocarbons were also used for this purpose. The introduction of an alkyl group has little effect on the spectrum, but strongly increases the affinity of a carbon atom for a proton (precisely which carbon atom depends on the position of the alkyl group). For example, in 1,2-benzanthracene, the carbon atoms in positions 9 and 10 are almost equivalent. Their relative basicities may be changed by almost an order by replacement of various hydrogen atoms in the molecule of this hydrocarbon by CH_3 groups. The spectra of the carbonium ions of 10-methyl- and 3-methyl-1,2-benzanthracene differ very strongly because the proton adds al-

---

*For comparison, the constants given here were obtained analogously to those on p. 141. The constants determined from spectral data differed comparatively little.

†Oxygen very strongly changes the spectra of some ions, in particular, perylene. Its spectrum approaches that of a solution in concentrated sulfuric acid, in which there are evidently oxidation processes with the formation of ion-radicals, which have paramagnetic properties (see [68, 69, 244] and the literature there).

most exclusively to position 9 in the former and to position 10 in the latter hydrocarbon. Analysis of the spectra shows that most of them correspond to the formation of

TABLE 48. Maxima in the Spectra of Carbonium Ions of Aromatic Hydrocarbons

Hydrocarbon	Band A		Band B		Band C		Band D	
	λ	ε	λ	ε	λ	ε	λ	ε
Mesitylene*	355	11100	256	8800	—	—	—	—
1,2,3,5-Tetra-methylbenzene	365	7500	261	5350	—	—	—	—
Pentamethyl-benzene	377	9800	272	4300	—	—	—	—
Bimesityl	362	5800	258	3900	—	—	—	—
Naphthalene*	410	—	390	10900	280	14800(?)	254	13000(?)
1,4-Dimethyl-naphthalene	—	—	381	18700	269	8900	252	7600
2,3-Dimethyl-naphthalene	—	—	381	11200	270	6600	253	—
Acenaphthene	420	2800	354	12200	264	5000	255	5000
9-Methylphen-anthrene	510	6300	320	10250	259	26900	—	—

* BF_3 added.

one ion. However, some curves, for example that of phenanthrene, are evidently obtained by the superposition of the spectra of several ions (the proton in positions 9, 4, and possibly 1).

Mesitylene and hexamethylbenzene are so strongly protonized in liquid hydrogen fluoride that they can be used as Hammett indicators (see Section II, p. 53). To determine the acidity function of liquid hydrogen fluoride and solutions in it, Kilpatrick and Hyman [70] recorded two maxima at 260 and 360 mμ for a solution of mesitylene, while the absorption maximum for hexamethylbenzene (completely ionized in liquid HF) lay at λ = 395 mμ. Mackor used anthracene as an indicator to determine the acidity function of solutions in liquid HF [71].

e. DISTRIBUTION OF AN AROMATIC HYDROCARBON BETWEEN HF AND AN INERT SOLVENT

This method was used by McCaulay and Lien [49, 72] for the first quantitative estimate of the strengths of methylbenzenes as bases. Their distribution between liquid hydrogen fluoride and heptane was studied in the first series of experiments. Equal volumes of a solution of the aromatic hydrocarbon in n-heptane and liquid hydrogen fluoride were shaken and the amount of methylbenzene extracted by the acid was determined. The distribution coefficient α = N_{HF}/N_{hept}, where N_{HF} and N_{hept} are the mole fractions of the hydrocarbons in liquid HF, and heptane, respectively.

Hydrocarbon	$\alpha \cdot 10^4$
p-Xylene	4
m-Xylene	12
Mesitylene	100
Hexamethylbenzene	1100

The solubility of the hydrocarbon in liquid HF increases with methylation, especially when the methyl groups are meta to each other.

As was pointed out above (p. 137), the solubility of aromatic hydrocarbons in HF is increased by the addition of BF_3. In the second series of experiments, the authors mentioned dissolved a mixture of two alkylbenzenes (normally 1:1) in heptane and added liquid HF containing 0.5 mole of BF_3 for each mole of alkylbenzene mixture. When equilibrium had been reached in the system, the amount of each of the hydrocarbons in the acid layer and in heptane was determined. The results were expressed in the form of a separation factor α', which equaled the quotient of the ratios of the mole fractions of the two components (1 and 2) in each of the phases:

$$\alpha' = \frac{N_2}{N_1} \bigg/ \frac{N_2'(h)}{N_1'(h)}.$$

$N_{(h)}'$ denotes the heptane layer.

TABLE 49. Comparison of Relative Basicities of Alkylbenzenes

Hydrocarbon	$\alpha'/\alpha'_{\text{p-xylene}}$	$K/K_{\text{p-xylene}}$
Methylbenzene (toluene).........	0.01	0.63
1,4-Dimethyl-(p-xylene)........	1	1.0
1,2-Dimethyl-(o-xylene)........	2	1.1
1,3-Dimethyl-(m-xylene)........	20	26
1,2,4-Trimethyl-(pseudocumene)...	40	63
1,2,3-Trimethyl-(hemimellitene)...	40	69
1,2,4,5-Tetramethyl-(durene).....	120	140
1,2,3,4-Tetramethyl-(prehnitene)...	170	400
1,3,5-Trimethyl-(mesitylene).....	2800	13,000
1,2,3,5-Tetramethyl-(isodurene)....	5600	16,000
Pentamethylbenzene...........	8700	29,000
Hexamethylbenzene (mellitene)....	89,000	97,000

Table 49 gives the values of α' relative to p-xylene, for which it was arbitrarily taken that $\alpha' = 1$. For comparison, we also give the ratios of the ionization constants of the same hydrocarbons to the ionization constant of p-xylene $K/K_{\text{p-xylene}}$ from electrical conductivity measurements summarized in Table 45.

This method* was used to study the complex forming power of various fluorides in liquid HF [73]. It was found that like BF_3, titanium, tantalum, and niobium fluorides bind fluorine ions:

$$2TiF_4 + F^- \rightleftharpoons Ti_2F_9^-$$
$$TaF_5 + F^- \rightleftharpoons TaF_6^-$$
$$NbF_5 + F^- \rightleftharpoons NbF_6^-$$

---

*A mixture of p- and m-xylene was dissolved in n-heptane and extracted with liquid HF with the addition of the fluoride tested. The complex forming power of the fluoride was determined from the ratio of the concentrations of the two isomers in the starting mixture and in the extracted products.

PF_5 is a much weaker complex former than the substances given above. SiF_4, BaF_2, PbF_2, CrF_3, ZrF_4, WF_6, and ZnF_2 do not participate in complex formation (see p. 141 on SbF_5).

The same procedure was developed into a strictly quantitative method by Mackor, Hofstra, and van der Waals [74-76], who used it to determine the basicity constants K (p. 183) of twenty-seven aromatic hydrocarbons. The basicity constants covered a range of seven orders.

The basicity constants of strong bases were determined from data on their distribution between a solution of sodium fluoride in HF and n-heptane; NaF is a base in liquid HF (it reduces the acidity of the solution, as is shown by direct determinations of the acidity function described in Section II, p. 52). The distribution coefficient of the hydrocarbon between two phases is α:

$$\alpha = \frac{m_{ArH_2^+} + m_{ArH}}{m'_{ArH}},$$

where m_{ArH} is the concentration of the aromatic hydrocarbon in HF; m_{ArH_2} is the concentration of its cationic form; m'_{ArH} is the concentration (in mole per 1000 g of solvent) of the hydrocarbon in the heptane layer.

The hydrocarbon is in the molecular state in the nonpolar solvent. As has been shown by conductimetric measurements (p. 139), a series of aromatic hydrocarbons are very weakly ionized in liquid HF and therefore for them it is sufficiently accurate to assume that

$$\alpha_0 = \frac{m_{ArH}}{m'_{ArH}}.$$

At low concentrations there is a linear relation between $\log \alpha_0$ and the molar volume of the hydrocarbon, which may be used to determine the value of α_0 for bases which are appreciably ionized in liquid HF. On the other hand, the following equation applies to hydrocarbons which are completely ionized in liquid HF:

$$\alpha = \frac{m_{ArH_2^+}}{m'_{ArH}}.$$

In this work, Mackor derived an equation relating the basicity constant K (p. 143) to α/α_0 and this had the form

$$\log K \alpha_0 = \log \alpha m_{NaF} + 2\log f_{\pm},$$

where m_{NaF} is the concentration of NaF dissolved in the liquid HF.

The standard state is an infinitely dilute solution in liquid HF. The authors assumed that within the limits of accuracy of the measurements, the following equation applies for all aromatic hydrocarbons:

$$\log f_{\pm} = -0.53 \sqrt{m_{NaF}} + 0.24\, m_{NaF}.$$

To determine the basicity constant K of strong bases, they measured α at several NaF concentrations. The basicity constants of weak bases were determined by measuring the distribution coefficient of the hydrocarbon between a solution of $NaBF_4$ in liquid HF and CCl_4 at a known partial pressure of BF_3 vapor. As already stated several times above (pp. 138-139), in the ternary system ArH + HF + BF_3 there is the equilibrium

$$ArH + BF_3 + HF \rightleftarrows ArH_2^{+} + BF_4^{-} .$$

The constant of this equilibrium is denoted by K^+. It is given by the equation

$$K^+ = \frac{m_{ArH_2^+} \cdot m_{BF_4^-}}{m_{ArH} f_{ArH} m_{BF_3} f_{BF_3}} f_{\pm}^{\circ}.$$

By using hydrocarbons which are bases of medium strength (for example, 2-methylnaphthalene), the authors determined the values of K^+ and K for them and thus found the conversion factor (log C = log K^+/K = 6.60), which made it possible to reduce measurements made under different conditions to a single scale of K.

TABLE 50. Basicity Constants of Aromatic Hydrocarbons

Hydrocarbon	log K	n	log K'
Benzene	− 9.4	6	− 10.2
Toluene	− 6.3	3	− 6.8
p-Xylene	− 5.7	4	− 6.3
Biphenyl	− 5.5	6	− 6.3
Triphenylene	− 4.6	6	− 5.4
Naphthalene	− 4.0	4	− 4.6
Phenanthrene	− 3.5	4	− 4.1
m-Xylene	− 3.2	3	− 3.7
Fluorene	− 2.4	2	− 2.7
Chrysene	− 1.7	2	− 2.0
1-Methylnaphthalene	− 1.7	1	− 1.7
2-Methylnaphthalene	− 1.4	1	− 1.4
Mesitylene	− 0.4	3	− 0.9
Hexamethylbenzene	1.4	6	0.6
Acenaphthene	1.6	2	1.3
Hexaethylbenzene	2.0	6	1.2
Pyrene	2.1	4	1.5
1,2:5,6-Dibenzanthracene	2.2	2	1.9
1,2-Benzanthracene	2.3	2	2.0
Anthracene	3.8	2	3.5
Perylene	4.4	4	3.8
2-Methylanthracene	4.7	1	4.7
9-Ethylanthracene	5.4	1	5.4
9-Methylanthracene	5.7	1	5.7
Naphthacene	5.8	4	5.2
9,10-Dimethylanthracene	6.4	2	6.1
3,4-Benzopyrene	6.5	1	6.5

Without dwelling on the details of the derivation of the calculation formulas and the methods of determining the activity coefficients, which are described in the article [74], we give in Table 50 a summary of the values of log K, arranging the hydrocarbons in order of increasing strength as bases. The structural formulas of the hydrocarbons are given in Table 51.

TABLE 51. Basicity Constants of Aromatic Hydrocarbons in Liquid HF

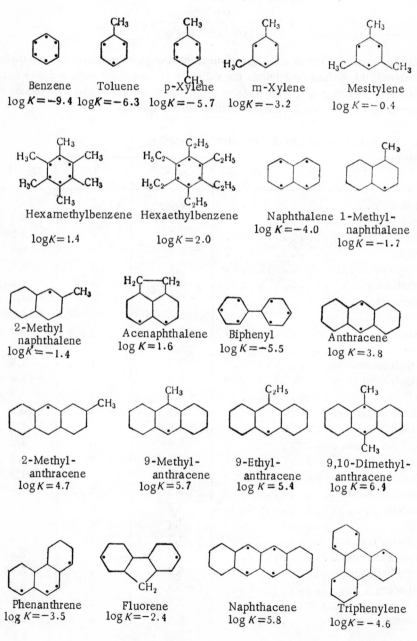

Benzene   Toluene   p-Xylene   m-Xylene   Mesitylene
$\log K = -9.4$ $\log K = -6.3$ $\log K = -5.7$ $\log K = -3.2$ $\log K = -0.4$

Hexamethylbenzene   Hexaethylbenzene   Naphthalene   1-Methyl-
$\log K = 1.4$ $\log K = 2.0$ $\log K = -4.0$ naphthalene
                                                        $\log K = -1.7$

2-Methyl   Acenaphthalene   Biphenyl   Anthracene
naphthalene $\log K = 1.6$ $\log K = -5.5$ $\log K = 3.8$
$\log K = -1.4$

2-Methyl-   9-Methyl-   9-Ethyl-   9,10-Dimethyl-
anthracene   anthracene   anthracene   anthracene
$\log K = 4.7$ $\log K = 5.7$ $\log K = 5.4$ $\log K = 6.9$

Phenanthrene   Fluorene   Naphthacene   Triphenylene
$\log K = -3.5$ $\log K = -2.4$ $\log K = 5.8$ $\log K = -4.6$

TABLE 51 (continued)

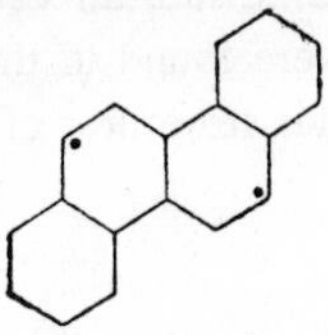

Chrysene
$\log K = -1.7$

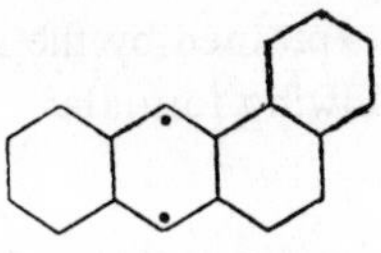

1,2-Benzanthracene
$\log K = 2.3$

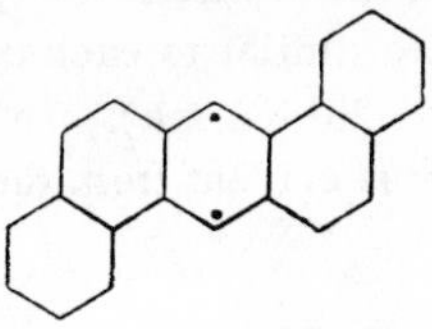

1,2 : 5,6-Dibenzanthracene
$\log K = 2.2$

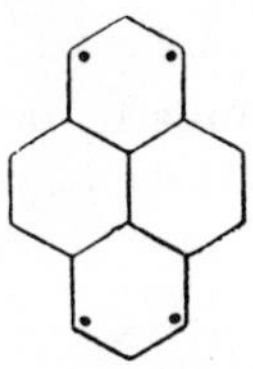

Pyrene
$\log K = 2.1$

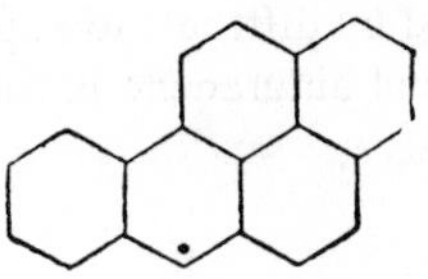

3,4-Benzpyrene
$\log K = 6.5$

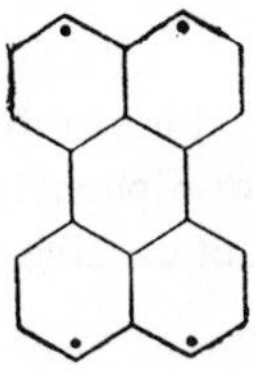

Perylene
$\log K = 4.4$

The basicity constants of the hydrocarbons given refer to the carbon atoms in the molecule to which a proton may be added when the hydrocarbon is dissolved in liquid HF. In table 51, they are marked by dots. Table 50 gives the number of them n. The same table also gives the values of log K' = log K — log n. It should be remembered that in some hydrocarbons (for example, in toluene and biphenyl) the carbon atoms marked as equivalent actually differ in their capacity to add a proton. This may be demonstrated, for example, by comparing the partial rate factors of deuterium exchange with liquid HBr for the ortho- and para-hydrogen atoms in these substances (pp. 170 and 177). However, in the case of solutions in liquid HF, the mark should be regarded as correct since the differences in the strengths of bases are very much reduced in this solvent.

C. Sulfuric Acid

As a result of its strong protogenicity and high dielectric constant, sulfuric acid is quite suitable for the preparation of carbonium ions by the addition of a proton to a hydrocarbon, but as it is extremely active chemically, it often brings about side reactions such as the oxidation and sulfonation of hydrocarbons. This is probably the reason why the volume of work with sulfuric acid is less than with liquid hydrogen fluoride, although the former is much more convenient as regards experimental techniques.

a. ABSORPTION SPECTRA OF SOLUTIONS

Gold and his co-workers [62] published a series of articles under the general heading of "Basicity of Hydrocarbons." By cryoscopic measurements with solutions of 1,1-diphenylethylene in 100% sulfuric acid, it was shown that the molecule of the ethylenic hydrocarbon is converted into two ions, which confirms the proposed reaction:

$$(C_6H_5)_2C=CH_2 + H_2SO_4 \rightleftarrows (C_6H_5)_2CCH_3^+ + HSO_4^- .$$

The ultraviolet spectra of solutions of 1,1-diphenylethylene, triphenylethylene, and anthracene in anhydrous sulfuric acid were measured. The measurements were made under conditions during which irreversible processes were excluded. All the spectra differed sharply from the spectra of solutions of the same substances in an inert solvent and were similar to each other (absorption maxima were found in the regions of 305-315 and 425-435 mμ). This is explained by the similar structures of the carbonium ions, as is evident from the following formulas:

$$(C_6H_5)_2\overset{\oplus}{C}CH_3; \ (C_6H_5)_3\overset{\oplus}{C}CH_2C_6H_5; \ C_6H_5 \diagup^{\displaystyle CH_2^{\displaystyle \cdot}}_{\diagdown \underset{\oplus}{CH} \diagup} \diagdown C_6H_5 \ .$$

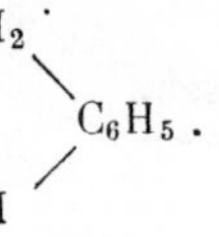

Leffler [1] noted that it would be difficult to explain the similarity in the spectra of solutions of diphenylethylene and anthracene in sulfuric acid if the ions had a non-classical structure like π-complexes:

$$\overset{\displaystyle H^+}{\underset{\displaystyle (C_6H_5)_2C \, = \, CH_2}{\vdots}} \ .$$

The results of Grace and Symons [76a] are in agreement with this conclusion. Having completely confirmed the measurements of the spectra of solutions of anthracene and phenylated ethylenes in sulfuric acid, the authors established that when the same substances are dissolved in a mixture of benzene and trichloroacetic acid or in a mixture of sulfuric and acetic acids, absorption of light is observed in the region of 610-660 mμ. It was considered that in this case the absorption was caused by the formation of a π-complex (see also [76b]).

Like anthracene, perylene

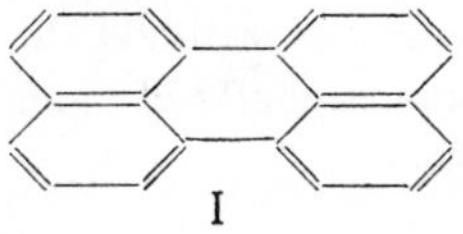

I

reacts reversibly with anhydrous sulfuric acid and may be isolated from a solution in it in an almost unchanged state [77]. The spectrum of a solution in sulfuric acid differs sharply from a spectrum of a solution in heptane, as is shown by the positions of the maxima on the absorption curve. The following wave numbers are expressed in mm^{-1} and the logarithms of the absorption coefficients are given in brackets: spectrum in heptane: 2302 (4.55); 2445 (4.37); 3962 (4.62); spectrum in 100% sulfuric acid: 1820 (4.47); 3120 (3.91). See also [78] on the spectrum of perylene.

Azulene, a hydrocarbon isomeric with naphthalene, is a base, to whose molecule a proton can be added even from an aqueous solution of sulfuric acid:

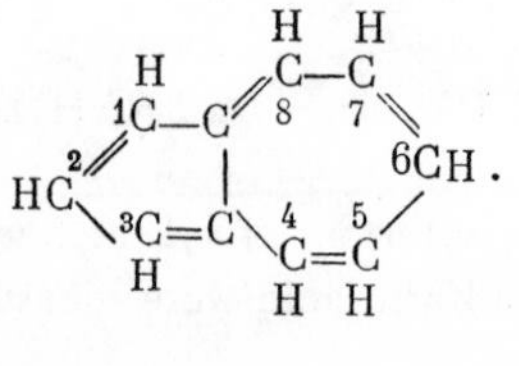

A stable carbonium ion is formed:

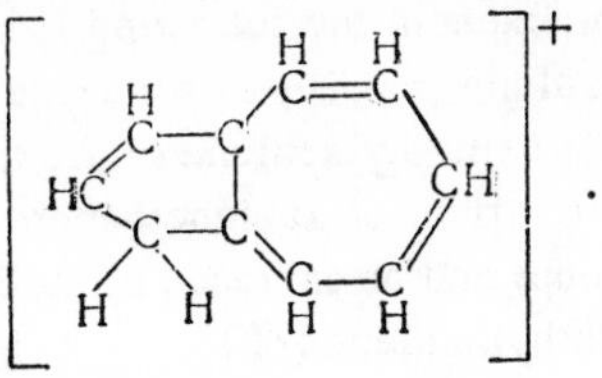

The color of the solution thereupon changes from blue to yellow [79]. The long-wave maximum is displaced by 200-350 mμ. Below we give the positions of the absorption maxima (mμ) in the spectra of solutions of azulene and its homologs in ligroin and in 50% sulfuric acid:

	Molecule	Ion	$\Delta\lambda$
Azulene	580	352	228
Guaiazulene*.	603	357	346
2-Methylazulene.	566	369	197

The changes in the spectra are reversible. When a sulfuric acid solution is diluted with alcohol, the color changes from yellow to blue and back again to yellow when sulfuric acid is added again.

The absorption spectra of the cations of azulene and seventeen of its methyl, di-methyl, and other alkyl derivatives have been measured [80]. The position of the first maximum varies over the range of 350-370 mμ, while it lies in the region of 560-610 mμ in the spectra of the starting hydrocarbons. The similarity in the spectra of the cations indicates that they have the same structure. The proton always adds to the carbon atom in position 1 or 3. The observed displacement of the band is close to that calculated by the molecular orbital method. The structure of the ion is discussed in [81] and the behavior of azulene and its homologs as bases are reviewed in [36, 82].

b. ELECTRICAL CONDUCTIVITY OF SOLUTIONS

The approximate value λ = 78 ohm^{-1} cm^2 was obtained for the molar electrical conductivity of guaiazulene in sulfuric acid monohydrate, and this is close to the molar electrical conductivity of solutions of ammonia, water, and alcohol in the same solvent [79].

The electrical conductivity of a 10^{-2} M solution of perylene in sulfuric acid is 6% higher than that of sulfuric acid. Experiments on current transference revealed the existence of positively charged ions of the hydrocarbon [78]. It is probable that in this case the positively charged ion was obtained not as a result of the addition of a proton, but because of the elimination of an electron, i.e., through an oxidation reaction, as was indicated by the formation of SO_2 [68] (see footnote on p. 143).

c. DISTRIBUTION OF A HYDROCARBON BETWEEN SULFURIC ACID AND AN INERT SOLVENT

Azulene and its homologs are extracted from a solution in ligroin by an aqueous solution of sulfuric acid [83]. The coefficient of the distribution of a substance between two solvents is the ratio of its concentration in these solvents, $\alpha = C_1/C_2$.

---

* 7-Isopropyl-1,4-dimethylazulene.

There is a linear relation between the solubility of azulene in an acid and its acidity function H_0. The value of H_0 at which $\alpha = 1$ is specific to each azulene derivative. The first figure in brackets in the following list is H_0 and the second, the acid concentration in percent: azulene (-3.00; 47.8); 1-methylazulene (-2.96; 47.3); 2-methylazulene (-2.69; 44.7); 5-methylazulene (-2.30; 40.3); and 6-methylazulene (-2.66; 44.3). The alkyl derivatives of azulene become weaker bases with the successive introduction of alkyl groups and an increase in the length of the latter, but azulene is a weaker base than methylazulene [82].

The coefficient for the distribution of anthracene between sulfuric acid and carbon tetrachloride also depends on the acid concentration (and on the acidity function H_0) [84]:

H_2SO_4, %	$-H_0$	$\alpha = \dfrac{C_{H_2SO_4}}{C_{CCl_4}}$
84.0	7.47	0.0003
85.0	7.62	0.0006
85.9	7.72	0.0014
87.8	7.94	0.012
90.0	8.16	0.07

The polycyclic hydrocarbon 3,4-benzopyrene is extracted from a solution in a mixture of benzene and ligroin by sulfuric acid [85].

On the basis of experiments on the distribution of phenylated ethylenes between an aqueous solution of sulfuric acid and cyclohexane it was concluded [62] that they may be arranged in the following sequence with respect to their strength as bases:

$$(C_6H_5)_2C=CH_2 > \underset{C_{10}H_7}{\overset{C_6H_5}{>}}C=CH_2 > (C_6H_5)_2C=CHC_6H_5 >$$

$$> \begin{cases} C_6H_5CH=CHC_6H_5 \\ (C_6H_5)_2C=C(C_6H_5)_2 \end{cases}$$

d. SOLUBILITY OF HYDROCARBONS

Olefinic hydrocarbons dissolve in sulfuric acid and, moreover, react with it. The ratio of the rates of solution of ethylene, propylene, and isobutylene is 1 : 300 : 54,000. Thereupon, the chemical compounds $CH_3CH_2HSO_4$, $CH_3CH(HSO_4)CH_3$, and $(CH_3)_2C(HSO_4)CH_3$. It is considered that the reaction proceeds in two stages. The rate of the reaction is determined by the addition of a proton and after this, a bisulfate ion adds rapidly. The stability of the carbonium ion increases in the sequence given above (i.e., from primary to secondary and tertiary) [86] (p. 133).

D. Hydrogen Chloride and Bromide

a. FREEZING POINT OF HYDROCARBON-HYDROGEN HALIDE SYSTEM

Freezing curves of solutions of aromatic and unsaturated aliphatic hydrocarbons in liquid hydrogen bromide were determined first by Maass and his co-workers [87-90]. Similar measurements with solutions in HCl and HBr were made recently by Cook and Schneider [91-93] and Terres [94].

Let us first consider systems with HCl. Table 52 gives the maxima, minima, and breaks on freezing curves. The hydrocarbon concentration is given in molar percents and the corresponding temperature is given in brackets.

TABLE 52. Maxima, Minima, and Breaks on the Freezing Curves of Hydro-carbon-Hydrogen Chloride Systems

Hydrocarbon	Maximum	Minimum	Break
Toluene	34 (—84.5°)	13,5 (—132.5°); 83 (—111°)	30—47 (—115°)
Ethylbenzene	56 (—101)	14,6 (—136.5); 75 (—110.5)	41.5 (—104.5)
n-Propylbenzene	50 (—106)	15,0 (—137.0)	41.5 (—104.5)
Isopropylbenzene	50 (—84)	11.5 (—124.5); 81.5 (—102)	—
tert-Butylbenzene	50 (—71)	13.5 (—125.5); 61.0 (—88.5)	—
m-Xylene	45—48 (—77)	9 (—124.5); 58.0 (—78)	—
Pseudocumene	—	7.5 (—120.0);	42 (—90.5)
Mesitylene	50 (—63)	6 (—117.0); 72.5 (—72.5)	—
p-Cymene	32 (—76.5)	2.7 (—115.0); 52.6 (—100.5)	—
Cyclohexene	50 (—109.8)	17.5 (—135); 75 (—127)	—
1-Octene	—	15.0 (—124.5) —	39 (—116.5)
1-Nonene	—	4.0 (—114.5) —	42 (—98.5)
1-Decene	—	—	35 (—90.0)

The freezing curve of a solution of an alkylbenzene has a maximum. Its position corresponds to the formation of a molecular compound with a 1:1 composition. According to Cook and Schneider [91], toluene also gives a second compound $C_6H_5CH_3 \times$ $\times$ 2HCl. The melting points of the molecular compounds according to the determinations of these authors are as follows: benzene (— 110°); toluene (— 112 and — 91°); o-xylene (— 64°); m-xylene (— 78°); p-xylene (— 93°); mesitylene (— 66°). The compound of mesitylene is the most stable, while the complexes of m- and o-xylene are more stable than those of toluene and benzene [91]. Pseudocumene reacts with HCl less vigorously than mesitylene and m-xylene [94]. The stability of the complexes of monoalkylbenzenes falls in the following sequence: tert-butylbenzene > isopropylbenzene > propylbenzene > ethylbenzene > toluene [94].

Terres stated that the compounds of alkylbenzenes with HCl are yellow, orange, or pink, while Schneider reported that these compounds are colorless,* which is in agreement with observations on the color of solutions of the same hydrocarbons in liquid HBr in the Isotopic Reactions Laboratory [95].

---

* The compound of HCl with trans-stilbene is blue, while that with anthracene is yellow.

According to Maass [87, 89], hydrogen bromide gives compounds with the following compositions: $2C_6H_6 \cdot HBr$ (− 86.5°); $2C_6H_5C_2H_5 \cdot HBr$ (− 103.8°); $C_6H_5C_2H_5 \cdot HBr$ (− 105.5°); $1,3,5\text{-}C_6H_3(CH_3)_3 \cdot HBr$. The data of Terres are given in Table 53. In contradiction to Maass, Terres found that with the exception of p-cymene, which reacts with both HBr and HCl to give a compound with the composition $C_6H_4(CH_3)CH(CH_3)_2 \times \times 2HBr$; all the complexes have a 1:1 composition.

TABLE 53. Maxima, Minima, and Breaks on the Freezing Curves of Hydrocarbon-Hydrogen Bromide Systems

Hydrocarbon	Maximum	Minimum	Break.
Ethylbenzene	50 (—106.25°)	21.5 (—125°); 66.5 (—113°)	—
Isopropylbenzene	50 (—93.0)	19.0 (—126); 76.0 (—108)	—
m-Xylene	50 (—66.5)	14.0 (—108.5); 60.0 (—72)	—
Pseudocumene	—	12.0 (—113)	46 (—82°)
p-Cymene	34 (—61)	6.0 (—92.5); 64 (—86)	—

As regards alkenes and alkynes, the measurements of Maass [88, 90] showed that acetylene does not form a molecular compound with hydrogen bromide at low temperature, while methylacetylene (allylene) gives a molecular compound with a 1:1 composition. The reaction of hydrogen bromide with propylene is much more marked than that with ethylene. When the temperature is raised, the addition of the hydrogen halide to the double bond begins.

Cook and Schneider [91] concluded that there are molecular compounds of alkenes with HCl with 1:1 and 1:2 compositions. Addition products with a 1:4 composition also are obtained with alkynes (only a 1:1 compound was found with acetylene). The molecular compounds normally melt at temperatures below − 130°.

Of the olefins listed in Table 52, only cyclohexene forms a stable complex (1:1).[*] Only weakly expressed inflexions were observed on the freezing curves of the other substances. An increase in the number of carbon atoms in an alkene with an unbranched chain is accompanied by a fall in basicity:

Cyclohexene > 1-octene > 1-nonene > 1-decene

b. SOLUBILITY OF HYDROGEN HALIDES IN HYDROCARBONS

Hydrogen chloride. To compare the basic properties of aromatic hydrocarbons, Brown and Brady [96] measured the solubility of hydrogen chloride at a low temperature (− 78.5°) in heptane containing a definite concentration of the substances studied. The solubility of gas in the liquid (N_{HCl}) is directly proportional to the partial pressure of the gas (p, mm): N = kp. The proportionality factor (k) is called Henry's constant.

---

[*]A melting point of − 116.5° was reported in [93].

Henry's constant for a solution of hydrogen chloride in heptane at $-78.5°$ equals 4520. As column 5 in Table 54 shows, the values of Henry's constant decrease with an increase in the basic properties of the aromatic hydrocarbon present in the solution. The authors also calculated the dissociation constant of the complex (ArH.HCl) which is formed between the hydrocarbon and the hydrogen chloride:

$$K_{diss} = \frac{N_{ArH} \, p_{HCl}}{N_{ArH \cdot HCl}} \, .$$

As k for pure heptane is known, it is possible to calculate the concentration of hydrogen chloride in solution at a given partial pressure which corresponds to no interaction between HCl and ArH. The difference between this value and the actual HCl concen-

TABLE 54. Solutions of HCl in Toluene and Heptane Solutions of Aromatic Hydrocarbons at $-78.51°$

Substance	Toluene		Heptane		K'_{ass}
	K_{diss}	k	K_{diss}	k	
Benzotrichloride	—	332	3180	4520	—
Chlorobenzene	—	318	1500	4220	—
Benzene	706	308	700	4000	0.61
Toluene	466	299	460	3500	0.92
p-Xylene	430	294	460	3170	1.00
o-Xylene	382	286	360	2980	1.13
m-Xylene	340	278	240	2550	1.26
Pseudocumene	316	272	—	—	1.36
Hemimellitene	294	265	—	—	1.46
Mesitylene	270	254	—	—	1.59
Durene	—	—	—	—	—
Prehnitene	264	250	—	—	1.63
Isodurene	258	246	—	—	1.67
Pentamethylbenzene	—	—	—	—	—
Hexamethylbenzene	—	—	—	—	—

tration in the presence of an aromatic hydrocarbon corresponds to the concentration of the complex. Consequently, the concentration of the hydrocarbon not bound in the complex equals its initial concentration minus the concentration of the complex. Thus, all the data are obtained for calculating the dissociation constants and these are given in column 4 in Table 54. The same sequence for the changes in Henry's constant for hydrocarbons was obtained when toluene was used as the solvent instead of heptane, (columns 2 and 3).

The last column in Table 54 gives the relative association constants for molecular compounds of hydrogen chloride with methylbenzenes:

$$K'_{ass} = \frac{K_{ass}}{K_{ass(p\text{-}xylene)}}, \text{ where } K_{ass} = \frac{1}{K_{diss}} \, .$$

The authors compared data on the basicity of hydrocarbons relative to p-xylene determined from the solubility of HCl (I) with corresponding results obtained by study-ing the distribution of a hydrocarbon between hexane and a solution of BF_3 in liquid HF (II) (p. 145), and also from the rate of halogenation (III) (p. 165). Brown and Brady noted several cases of discrepancies between the series established in this way. The first of them concerns m-xylene, whose basic properties differ little from those of the other isomers according to their determination, while judging by other data, it is a much stronger base:

I. p-xylene < o-xylene < m-xylene

II. p-xylene < o-xylene << m-xylene

III. p-xylene < o-xylene << m-xylene.

A similar deviation was observed with mesitylene. Its strength as a base determined from the solubility of hydrogen chloride is somewhat higher than that of the other tri-methylbenzenes, but less than that of prehnitene, while methods II and III indicate that it is a stronger base than prehnitene:

I. pseudocumene < hemimellitene < mesitylene < prehnitene

II. pseudocumene < hemimellitene < prehnitene << mesitylene

III. pseudocumene < hemimellitene < prehnitene << mesitylene .

For monoalkylbenzenes, Brown and Brady found a sequence (toluene < ethylben-zene < isopropylbenzene < tert-butylbenzene) which is directly opposite to the sequence found by measuring the halogenation rates.

Brown and Brady explained all these differences by the fact that the comparatively weak interaction between HCl and a hydrocarbon corresponds to the formation of a π-complex, while a σ-complex is formed in the strong interaction of a hydrocarbon with a solution of BF_3 in HF and in halogenation (see p. 15).

In Fig. 17, the values of the relative basicity of hydrocarbons determined by Brown and Brady are plotted along the ordinate axis and the relative reactivity of the same hydrocarbons in halogenation along the abscissa axis. In Fig. 18, the same values as in the previous graph are plotted along the abscissa axis, while the relative basicity of benzene and methylbenzenes measured by McCaulay and Lien [49] by the method de-scribed on p. 140 are plotted along the ordinate axis. A comparison of these two figures clearly illustrates the deviations which, according to Brown are explained by the diff-erences in the structures of π- and σ-complexes between hydrocarbon-bases and acids of different strengths. Under the conditions of Brown and Brady's experiments, HCl does not react irreversibly with alkenes. There is a weak interaction through the π-electrons, which increases in the sequence:

$$Cl_2C = CCl_2 < RCH = CH_2 < R_2C = CH_2, RCH = CHR < R_2C = CHR,$$

i.e., as the hydrogen atoms are replaced by alkyl groups.

Hydrogen bromide. In parallel with the experiments on the solubility of hydrogen chloride, measurements were made on the solubility of hydrogen bromide in aromatic hydrocarbons at 0° (the measurements with benzene were made at 5.7°) [97]. The second column of the table gives the ratio of Henry's constants k_{obs}/k_{ideal}, i.e., that found experimentally at 0° and that calculated for no reaction between hydrogen bromide and the hydrocarbon, while the third column gives similar values for HCl at $-78.5°$.

Substance	k_{obs}/k_{ideal}	
	HBr (0°)	HCl (− 78.5°)
Benzene . . .	0.500	0.774
Toluene . . .	0.441	0.701
m-Xylene . .	0.371	0.650
Mesitylene. .	0.335	0.564

The observed Henry coefficient in the case of HBr differs from the calculated one even more than in the solution of hydrogen chloride, though in the experiments with the latter, the temperature was almost 80° lower. This is the result of the higher acidity of hydrogen bromide.

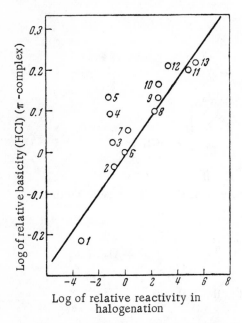

Fig. 17. Relation between the stability of π -complexes of methylbenzenes with HCl and the relative rate of halogenation. 1) Benzene; 2) toluene; 3) ethylbenzene; 4) isopropylbenzene; 5) tert-butylbenzene; 6) p-xylene; 7) o-xylene; 8) m-xylene; 9) pseudocumene; 10) hemimellitene; 11) mesitylene; 12) prehnitene; 13) isodurene.

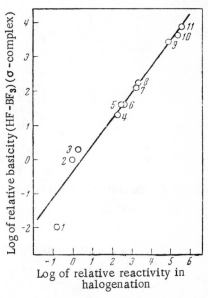

Fig. 18. Relation between the stability of σ-complexes of methylbenzenes with HF − BF₃ and the relative rate of halogenation. 1) Toluene; 2) p-xylene; 3) o-xylene; 4) m-xylene; 5) pseudocumene; 6) hemimellitene; 7) durene; 8) prehnitene; 9) mesitylene; 10) isodurene; 11) pentamethylbenzene.

c. ABSORPTION SPECTRUM

F. F. Cheshko [Zhur. Fiz. Khim., 33, 745 (1959)] showed that when hydrogen chloride is dissolved in benzene, the ultraviolet spectrum of the latter changes. This is explained by the formation of a molecular compound of the hydrocarbon with the acid.

E. Carbonium Salts of Haloaluminic Acids

a. TENSIMETRIC MEASUREMENTS

As Brown and his co-workers [97, 98] demonstrated, carbonium ions of aromatic hydrocarbons are formed in ternary systems consisting of an aromatic hydrocarbon, an aluminum halide, and a hydrogen halide; these are carbonium salts of complex halo-aluminic acids, which do not exist in the free state [99, 100] (see also [101-104]).

Aluminum chloride does not dissolve appreciably in toluene at $-80°$, but dissolves when excess HCl is present. This yields a clear solution with an intense green color. The decrease in the hydrogen chloride pressure indicates that one mole of hydrogen chloride is consumed per mole of aluminum chloride. The reaction is reversible and evaporation of the hydrogen chloride precipitates aluminum chloride, which redissolves when hydrogen chloride is added again. This operation may be repeated several times without producing irreversible changes. As already pointed out, the reaction consists of the formation of a carbonium salt of the acid $HAlCl_4$, which does not exist in the free state:

At $-45.4°$, one mole of HCl is added per mole of Al_2Cl_6 to give a yellow solution.

Al_2Cl_6 dissolves readily in the system obtained and this is explained by the formation of carbonium salts of higher complex acids of the general formula $[ArH_2]^+ [AlX_4 \times n AlX_3]^-$, in the same way as the solution of sulfur trioxide in sulfuric acid yields higher acids: $H_2S_2O_7$, $H_2S_3O_{10}$, $H_2S_4O_{13}$, etc.

The saltlike nature of the complex examined is confirmed by its high electrical conductivity and the solubility in it of polar substances such as alkali metal halides. A detailed study was made of analogous ternary systems which contained aluminum bromide and hydrogen bromide together with benzene and its homologs (toluene, m-xylene, and mesitylene [97]). Most attention was paid to the ratio of the amounts of hydrogen bromide and aluminum bromide in the complex. It was only in experiments with toluene that the amount of the aromatic hydrocarbon was measured also to determine its role in the complex.

In the case of benzene, toluene, m-xylene, and mesitylene, one mole of hydrogen bromide per mole of Al_2Br_6 and mole of hydrocarbon is absorbed, i.e., the reaction is expressed by the equation

$$ArH + HBr + Al_2Br_6 \rightleftarrows ArH_2^+ + Al_2Br_7^- .$$

The solubility of hydrogen bromide in a mixture of an aromatic hydrocarbon and aluminum bromide increases in the sequence: benzene < toluene < m-xylene [104a]. The stability of the complex increases with the methylation of the aromatic ring.

If hydrogen bromide is passed into the complex from m-xylene or mesitylene and $AlBr_3$, one mole is absorbed to yield a solid crystalline substance with the composition $ArH_2^+ \cdot AlBr_4^-$. This reaction proceeds at $0°$. In contrast to m-xylene and mesitylene, toluene, which is a weaker carbo base, forms an analogous, but unstable compound only at a low temperature $(- 80°)$.

Like HBF_4 (p. 141), $HAlCl_4$ (p. 158), $HAlBr_4$, and HAl_2Br_7 do not exist in a free form [97, 105], though their organic and inorganic salts are stable.

It should be noted that the complexes normally have a more complicated composition than that given above. It is considered that the carbonium ion is solvated by strongly polarized molecules of the aromatic hydrocarbon.

b. ELECTRICAL CONDUCTIVITY

V. A. Plotnikov [106, 107] was a pioneer in the measurement of electrical conductivity of binary and ternary systems containing an aromatic hydrocarbon and an aluminum halide. Analogous work was published later by Wertyporoch [108]. Solutions of $AlBr_3$ in benzene, toluene, and xylene do not conduct a current, but become electrically conducting when a hydrogen halide is passed into the solution.

When hydrogen bromide is introduced into the system $C_6H_6 - Al_2Br_6$, the solution acquires a yellow color and as the concentration of HBr is increased, the specific electrical conductivity of the system increases; it reaches $\varkappa = 1.9 \cdot 10^{-5}$ $ohm^{-1}cm^{-1}$. With further addition of HBr, the system separates into two phases. If still more hydrogen bromide is added, the specific electrical conductivity of the upper layer remains at the given value $(1.9 \cdot 10^{-5}$ $ohm^{-1}cm^{-1})$, while the specific electrical conductivity of the lower layer reaches $1 \cdot 10^{-2}$ $ohm^{-1}cm^{-1}$ [103, 109]. It has also been established that the complex $1,3,5-C_6H_3(C_2H_5)_3 \cdot Al_2Br_6 \cdot HBr$ conducts a current [110, 111].

Experiments on the transference of current showed that the positive ion contains carbon, while the negative ion contains aluminum and a halogen. The analyses corresponded to the following simplest formula: $[C_6H_3(C_2H_5)_3H]^+ [Al_2Br_7]^-$.

c. ABSORPTION SPECTRA

The light absorption maximum of the system $C_6H_6 - Al_2Br_6 - HBr$ lies at a wavelength of $\lambda = 330$ $m\mu$, while for the binary system $C_6H_6 - Al_2Br_6$, its position corresponds to $\lambda = 278.5$ $m\mu$ [112]. The electronic spectra of binary and ternary systems consisting of benzene, toluene, or methylnaphthalene, aluminum bromide or chloride, and hydrogen bromide or chloride were measured by Luther [113]. These measurements showed that no molecular compound is formed in a binary system consisting of a hydrocarbon and an aluminum halide. No molecular compounds is formed either in the ternary system $C_6H_6 - Al_2Cl_6 - HCl$, but one is formed if toluene is used instead of benzene or aluminum bromide instead of aluminum chloride.

F. Carboxylic Acids

Carboxylic acids have much lower strengths than mineral acids. Consequently, only hydrocarbons which are among the strongest bases in this class of organic substances form carbonium salts with carboxylic acids.

a. ABSORPTION SPECTRA

The absorption spectra of solutions of azulene in 50 and 96% sulfuric acid and in 99% formic acid are practically the same. Consequently, azulene is completely ionized in HCOOH [79]. The equilibrium constants (K) of the reactions between azulene and tri- and di-chloroacetic acids were determined by spectrophotometric measurements. These constants equal 0.08 and 0.01, respectively, at room temperature.

Carotene, a hydrocarbon whose molecule contains 10 conjugated double bonds, very readily gives carbonium salts with acids, including carboxylic acids [114, 115]. The addition of an acid is accompanied by the disappearance of the light absorption maximum in the region of 450 mμ and the appearance of a new long-wave maximum. The equilibrium constants of the reactions of carotene with HCl, CCl_3COOH, $CHCl_2COOH$, $CH_2ClCOOH$, $C_6H_2(NO_2)_3OH$, and dodecylbenzenesulfonic acid were determined spectrophotometrically. A linear relation was found between the equilibrium constants of the reactions with chloroacetic acids and the dissociation constants of the latter in aqueous solution. Judging by the fact that the ratio of the equilibrium constants of reactions of acids with carotene and with azulene is close to unity, carotene is a base of approximately the same strength as azulene.

Like β-carotene, lycopene, a hydrocarbon with the empirical formula $C_{40}H_{58}$ and containing 13 conjugated double bonds, adds a proton on solution in trichloroacetic acid. The ionization is accompanied by a sharp change in the spectrum. The spectrum of the molecule is characterized by two narrow bands in the region of 500 mμ, while the spectrum of the ion has a wide band in the region of 900 mμ [115a].

Zethrene, the hydrocarbon synthesized by Clar,

is capable of reacting reversibly even with acetic acid. Zethrene is sparingly soluble in neutral solvents, but dissolves readily in acetic acid and is liberated unchanged on neutralization of the latter. A solution in acetic acid and other acids is violet, while zethrene itself is green. Zethrene is the most strongly basic aromatic hydrocarbon known as yet. Mackor [74] roughly estimated the basicity constant of zethrene in liquid HF. It was found that log pK = 8.

b. ELECTRICAL CONDUCTIVITY

The strength of azulene as a base was compared with other bases by measurements of the electrical conductivity of solutions in anhydrous formic acid [79].

The molecular electrical conductivities of solutions extrapolated to zero concentration of the substance in HCOOH and the ionization constants of the bases in water are given below (Table 55).

In the transference of current in solutions of monoacidic bases, formate ions are responsible for four-fifths of the electrical conductivity. Therefore, the electrical conductivities of different formates in formic acid differ little. Two formate ions add to the last two substances given in the table, i.e., to the diamines. A comparison of the electrical conductivities shows that only one proton is transferred to the azulene molecule. The carbonium salts of azulene and carotene with tri- and dichloroacetic acids conduct a current even in benzene solution [114, 115].

TABLE 55. Molecular Electrical Conductivity of Solutions

Substance	λ, ohm^{-1} cm^2	$K_i^{H_2O}$
Azulene	39	—
p-Nitroaniline	50	—
Aniline...............	55	$5 \cdot 10^{-10}$
Guaiazulene	58	—
Triethylamine	58	$6 \cdot 10^{-4}$
Ethylenediamine...........	101	$9 \cdot 10^{-5}$
Phenylenediamine..........	107	$3 \cdot 10^{-10}$

4. EQUILIBRIUM REACTIONS OF BASIC HYDROCARBONS WITH ACIDLIKE SUBSTANCES

A. Introduction

Bases, i.e., substances with an affinity for a proton, are generally capable of participating in equilibrium reactions with not only hydrogen acids, but also many substances which do not contain hydrogen, but have an affinity for an electron. Such electrophilic reagents are called acidlike substances (p. 4).

It was shown above that many hydrocarbons participate in equilibrium reactions with hydrogen acids, thereupon fulfilling the function of a base. They differ from most normal bases in a lower affinity for a proton so that these reactions can be achieved only with the strongest acids. Being bases, hydrocarbons are also capable of reacting with acidlike substances. There is a vast literature, whose volume has grown particularly rapidly in recent years, on the preparation and properties of molecular compounds of acidlike substances with aromatic hydrocarbons (see the reviews of Andrews [104] and A. N. Terenin [116]). Measurements of the stability constants of complexes give an idea of the relative strengths of hydrocarbons as bases in their reactions with acidlike substances. This sequence is found to be very close to that established for equilibrium reactions for the same hydrocarbons with protonic acids.

We will limit ourselves to a brief account of facts on reactions in which there are formed molecular compounds of aromatic hydrocarbons with sulfur dioxide or with halogens.* We consider sulfur dioxide because the concept of acidlike substances arose [118] in a study of the behavior of acid-base indicators in liquid sulfur dioxide [119]; the equilibrium reactions between aromatic hydrocarbons and halogens are interesting because the molecular compounds formed are intermediate products in the electrophilic replacement of hydrogen, to the discussion of whose mechanism we return later (Section V).

B. Sulfur Dioxide

Sulfur dioxide and a basic indicator (an aminoazo compound) participate in an equilibrium reaction. At a low temperature, the equilibrium is displaced toward the

---

* The account is based on the review [117], which was compiled before the publication of the articles [104] and [116].

formation of a molecular compound. The change in the absorption spectrum of the solution of the indicator in liquid sulfur dioxide produced by this is completely identical with that observed when hydrogen chloride is introduced into the same solution at room temperature [120-121]. Consequently, sulfur dioxide and hydrogen chloride form molecular compounds of the same structure with a basic indicator. Typical bases such as amines react vigorously with sulfur dioxide; their solutions are yellow, orange, and red in color. Many molecular compounds have been isolated (see [122]).

An interaction between aromatic hydrocarbons and sulfur dioxide is indicated by the high solubility of the former in the latter [123]. The solutions obtained are colored; thus, for example, a solution of pseudocumene is yellow and one of mesitylene, orange-yellow [124]. The spectrum of a solution of anthracene in liquid sulfur dioxide has been measured. It loses the fine structure characteristic of solutions of anthracene in inert solvents and the absorption maximum is displaced toward longer wavelengths [125].

By measuring the freezing curves of binary systems of sulfur dioxide and an aromatic hydrocarbon (benzene, toluene, ethylbenzene, mesitylene, and pseudocumene), Carli [124] detected the following crystalline compounds whose melting points are given in brackets: $C_6H_6 \cdot SO_2$ ($-15°$); $C_6H_6 \cdot 2SO_2$ ($-40°$); $C_6H_6 \cdot 3SO_2$ ($-52°$); $C_6H_5CH_3 \times 2SO_2$ ($-85.5°$); $C_6H_5CH_3 \cdot 3SO_2$ ($-80°$); $1,3,5\text{-}C_6H_3(CH_3)_3 \cdot SO_2$ ($-49.4°$); $1,2,4\text{-}C_6H_3(CH_3)_3 \cdot SO_2$ ($-60°$).

The association between molecules of aromatic hydrocarbons and sulfur dioxide acting as a solvent is also observed in measurements of the viscosity of solutions; the degree of interaction increases as the hydrogen atoms in benzene are replaced by methyl groups [126]. The same rule was found by Andrews and Keefer [127], who determined the stability of molecular compounds formed by sulfur dioxide and benzene or its homologs in carbon tetrachloride. The equilibrium constant was calculated in the same way as in the work of the same authors described on p. 163. The absorption maximum lies in the region of 280-305 mμ and with progressive methylation of benzene, the maximum is displaced toward longer wavelengths. The reciprocals of the equilibrium constants, i.e., the stability constants K_{stab}* are given in Table 56.

C. Halogens

It has been known for a long time that there are iodine solutions of two colors, namely, violet and brown (and colors transitional between them) [104, 116, 128]. Solutions of iodine in inert solvents such as carbon tetrachloride and aliphatic hydrocarbons are violet, while brown solutions are obtained in alcohols, ethers, and other solvents of the electron-donor type. Hildebrand put forward the hypotheses (see footnote in [128]) that these peculiarities of solutions of iodine in aromatic hydrocarbons are connected with the formation of molecular compounds of the acid—base type between iodine and the hydrocarbon molecule. This hypothesis was substantiated in more detail in the work of Hildebrand and Benesi [129] on the absorption spectra of solutions of iodine in benzene and its homologs. The spectra of the solutions have two maxima, including the characteristic maxima in the region of 300 mμ, whose presence is ascribed to the formation of a complex with an equimolecular composition (ArH $\cdot$ I_2). By

---

* The equilibrium $ArH + E \rightleftharpoons ArH \cdot E$, is considered, where ArH is an aromatic hydrocarbon and E is an electrophilic reagent. The equilibrium constant $K = (ArH \cdot E)/(ArH)(E)$. The stability constant $K_{stab} = (ArH)(E)/(ArH \cdot E)$.

measuring the spectra of solutions containing different concentrations of iodine and the hydrocarbon in carbon tetrachloride, the authors calculated the equilibrium constant of the formation of the molecular compounds of iodine with benzene and with mesitylene and confirmed the expected higher stability of the second complex. The temperature coefficient of the equilibrium constant was determined later [130].

Hildebrand's work gave rise to a series of analogous studies. Outstanding among these are the investigations of Keefer and Andrews [104], who systematically measured the spectra of carbon tetrachloride solutions of mixtures of benzene and its homologs with halogens: chlorine [131], bromine [132], iodine [133, 129] (see also [136]), and iodine chloride [133-135]. There are also data on molecular compounds of naphthalene

TABLE 56. Stability Constants of Complexes with Alkylbenzenes

Substance	SO_2 [127]*	Cl_2 [1321]	Br_2 [132]	I_2 [129]	I_2 [133]	I_2 [136]	ICl [135]	ICl [133]
Benzene	0.47	0.33	1.04	1.72	1.55	1.60	4.76	5.56
Toluene	0.79	—	1.44	—	1.65	—	7.97	8.96
p-Xylene	1.34	—	2.26	—	3.19	—	13.4	15.5
o-Xylene	1.65	—	2.29	—	3.19	—	15.4	13.1
m-Xylene	1.49	0.62	2.16	—	2.78	—	16.0	14.3
Mesitylene	2.11	—	—	7.2	8.45	5.96	—	47.3
Durene	—	—	—	—	6.49	—	—	43.8
Pentamethylbenzene	—	—	—	—	9.06	9.72	—	66.2
Hexamethylbenzene	—	—	—	—	13.9	15.2	—	233
Naphthalene	—	—	2.38**	—	2.58	2.66**	—	14.3
Biphenyl	—	—	—	—	4.74	—	—	15.2

* The numbers in brackets are literature references.
** See [137].

with iodine and bromine [137]. In all these studies it was established that there is a spectral band characteristic of the molecular compound obtained as a result of the reaction:

$$ArH + Hal_2 \rightleftarrows ArH \cdot Hal_2.$$

Table 56 compares the stability constants K_{stab}, which equal the equilibrium constants K of the given reaction, known at the present time.

$$K_{stab} = \frac{(ArH) \cdot (Hal_2)}{(ArH)(Hal_2)}.$$

To calculate the equilibrium constants and the molecular absorption coefficient ε of the complex in solution, the authors used the relation:

$$\frac{(Hal_2) l}{d} = \frac{1}{K\varepsilon} \cdot \frac{1}{N_{ArH}} + \frac{1}{\varepsilon},$$

where l is the layer thickness; d is the maximum optical density of the solution in the region of the absorption maximum; (Hal_2) is the molecular concentration of the halogen; N_{ArH} is the concentration of the aromatic hydrocarbon in mole fractions.

The values of $(Hal_2)l/d$ were plotted against $1/N_{ArH}$ on a graph to give straight lines, whose interaction with the ordinate axis gave the value of $1/\varepsilon$, while the slope corresponded to $1/K\varepsilon$. In the original papers, the values of the constants given were different from those in columns 5 and 8 in Table 56. This is explained by the fact that the hydrocarbon concentration was expressed as a molar concentration and not in mole fractions. (See [136] on the conversion of one form of the constant to the other).

The relative stability constants of the complexes with aromatic hydrocarbons are compared in Table 57 (the stability constant of the complex with benzene is taken as unity).

The relative stability constants of the complexes with aromatic hydrocarbons are compared in Table 57 (the stability constant of the complex with benzene is taken as

TABLE 57. Relative Stability Constants of Complexes with Alkylbenzenes

Hydrocarbon	SO_2 [127]*	Cl_2 [131]	Br_2 [132]	I_2 [129]	I_2 [133]	I_2 [136]	I Cl [135]	ICl [133]	HCl [96]
Benzene	1.00	1.00	1.00	1.00	1.00	1.00	1.00	1.00	1.00
Toluene	1.68	—	1.39	—	1.34	—	1.67	1.61	1.51
p-Xylene	2.85	—	2.17	—	2.07	—	2.82	2.78	1.64
o-Xylene	3.51	—	2.20	—	1.80	—	3.24	2.30	1.85
m-Xylene	3.17	1.88	2.08	—	2.07	—	3.36	2.58	2.06
Mesitylene	4.49	—	—	2.39	5.47	3.72	—	8.50	2.39
Durene	—	—	—	—	4.19	—	—	7.85	2.74
Pentamethylbenzene	—	—	—	—	5.87	6.08	—	10.9	—
Hexamethylbenzene	—	—	—	—	9.00	9.50	—	42.0	—
Naphthalene	—	—	—	—	1.63	—	—	2.57	—

* The numbers in brackets are the literature references.

unity). An examination of this table shows that the stability of the molecular compound increases with respect to the halogen in the sequence:

$$Cl_2 < Br_2 < I_2 < ICl.$$

This sequence corresponds to the increase in the "acidity" of a halogen from iodine to chlorine [138]. There is also a relation between the long-wave absorption maximum of the complex of a hydrocarbon with iodine and the ionization potential of the electron-donor molecule [139]. The latter increases in the direction: mesitylene < m-xylene < toluene < benzene. A fall in the ionization potential corresponds to higher basicity of the hydrocarbon, an increase in the stability of the molecular compound, and a higher stability constant. The stability of the molecular compound increases with a change from benzene to hydrocarbons with two rings, i.e., naphthalene and biphenyl. As regards benzene homologs, the stability of the molecular compound increases with an increase in the number of methyl groups, which corresponds to the increase in the strength of the hydrocarbon as a base.

A linear relation between the free energy of formation of the complex and the ionization potential of the aromatic hydrocarbon has been found not only in reactions with iodine, but in reactions with other electron-acceptor acidlike substances (poly-nitro compounds [140, 141], tetrachloroquinone [142], etc.) and with proton donors[143]. This is in agreement with general theoretical views on a donor-acceptor interaction (p. 12).

The molecular compounds of aromatic hydrocarbons with halogens existing in so-lution are polar, as follows from measurements of the dielectric polarization [144]. A clear idea of the degree of interaction between the halogen and the electron-donor molecule is given by the magnitude of the deviation of the experimental values of the dielectric polarization of dissolved substances from the values calculated by addition. As the work of Ya. K. Syrkin and V. M. Kazakova [145] shows, with a change from aro-matic hydrocarbons to stronger nitrogen bases, a much greater effect is observed. A solution of iodine in pyridine conducts a current well [146, 147] (cf., p. 15).

5. IRREVERSIBLE REACTIONS OF BASIC HYDROCARBONS WITH ACIDLIKE SUBSTANCES

The existence of direct proportionality between the logarithm of the relative rate constant of chlorination of methylbenzenes and the logarithm of a value characterizing their relative basicity was first pointed out by Condon [148, 149] and Brown [96] simul-taneously. In particular, they were considering data obtained by McCaulay and Lien [49] on equilibria in the systems $ArH - HF - BF_3$ (p. 140), which characterize the strength of hydrocarbons as bases. This relation is quite understandable if it is assumed that the irreversible reaction of electrophilic replacement proceeds through the equilibrium of the formation of a molecular compound with the halogen [150], whereby the state of the equilibrium is determined by the strength of the hydrocarbon as a base, as we have already demonstrated. Several years after the publication of the work cited, there ap-peared a series of analogous investigations of Brown, de la Mare, and Andrews and Keefer [151-157].

Table 58 gives data on the relative rate of electrophilic replacement of hydrogen in benzene and its methylated derivatives. The rate constant of the reaction with ben-zene is arbitrarily taken as unity. The data refer to chlorination by molecular chlorine without a catalyst [151], bromination by Br_2 without a catalyst [152, 154] and with the addition of $ZnCl_2$ [153], iodination by ICl with the same catalyst [155], and mercura-tion [156] by mercuric acetate. The solvent was acetic acid. The benzoylation [157] was carried out with benzoyl chloride in nitrobenzene with $AlCl_3$ as the catalyst. (See [153a] on chlorination in different solvents.)

If the hydrocarbons are arranged in order of their relative reactivity, the sequence is found to be similar for all the electrophilic replacements of hydrogen listed above:

benzene < p-xylene < o-xylene < m-xylene < { pseudocumane / durene } < mesitylene <

< pentamethylbenzene.

Deviations are observed in iodination and these are explained [155] by steric hindrance, which appears particularly in reactions with highly methylated hydrocarbons.

A comparison of the relative strengths of the same compounds as bases shows that the sequence is very similar to that just mentioned. Thus, the sequence with respect to the electrical conductivity of solutions in liquid HF was found to be as follows (p. 139):

TABLE 58. Relative Rates of Electrophilic Replacement of Hydrogen in Benzene and Methylbenzene

Hydrocarbon	Cl_2 [151]*	Br_2 [152]	$Br_2 + ZnCl_2$ [153]	Br_2 [154]	$ICl + ZnCl_2$ [155]	C_6H_5COCl [157]	$(CH_3COO)_2Hg$ [156]
Benzene	1	1	1	1	1	1	1
Toluene	$3.44 \cdot 10^2$	$3.5 \cdot 10^2$	$1.5 \cdot 10^2$	$6.05 \cdot 10^2$	$1.4 \cdot 10^2$	$1.5 \cdot 10^2$	5
o-Xylene	$2.10 \cdot 10^3$	—	$1.1 \cdot 10^3$	$5.32 \cdot 10^3$	$1.3 \cdot 10^3$	$1.36 \cdot 10^3$	$1.6 \cdot 10^1$
p-Xylene	$1.85 \cdot 10^5$	$4.4 \cdot 10^5$	$3.4 \cdot 10^3$	$5.14 \cdot 10^5$	$2.3 \cdot 10^4$	$3.91 \cdot 10^3$	$3.5 \cdot 10^1$
m-Xylene	$2.08 \cdot 10^3$	—	$4.8 \cdot 10^2$	$2.52 \cdot 10^3$	$1.0 \cdot 10^3$	$1.42 \cdot 10^2$	8.2
Hemimellitene	—	—	—	$1.67 \cdot 10^5$	—	$1.33 \cdot 10^4$	$6.8 \cdot 10^1$
Pseudocumene	—	$1.4 \cdot 10^6$	$6.4 \cdot 10^3$	$1.52 \cdot 10^6$	$7.5 \cdot 10^4$	$7.60 \cdot 10^3$	$4.9 \cdot 10^1$
Mesitylene	—	—	$3 \cdot 10^5$	$1.89 \cdot 10^8$	$1.8 \cdot 10^6$	$1.25 \cdot 10^5$	$2.1 \cdot 10^2$
Prehnitene	—	—	—	$1.10 \cdot 10^7$	—	$3.55 \cdot 10^4$	$1.3 \cdot 10^2$
Isodurene	—	—	$6.5 \cdot 10^5$	$4.2 \cdot 10^8$	$1.6 \cdot 10^6$	$2.12 \cdot 10^5$	$2.6 \cdot 10^2$
Durene	—	$2.6 \cdot 10^6$	$3.1 \cdot 10^4$	$2.83 \cdot 10^6$	$2.3 \cdot 10^4$	$1.10 \cdot 10^4$	$3 \cdot 10^1$
Pentamethylbenzene	—	—	$1.4 \cdot 10^6$	$8.1 \cdot 10^8$	$7.4 \cdot 10^5$	$1.39 \cdot 10^5$	$2.2 \cdot 10^2$

* The figures in brackets are the literature references.

benzene < p-xylene < o-xylene < m-xylene < { $\begin{array}{l}\text{pseudocumene} \\ \text{hemimellitene}\end{array}$ < prehnitene <

< durene < mesitylene < isodurene < pentamethylbenzene

The sequence found from the distribution of hydrocarbons between hexane and a solution of BF_3 in HF (p. 145) was as follows:

benzene < p-xylene < o-xylene < m-xylene < hemimellitene <
< pseudocumene < durene < prehnitene < mesitylene < isodurene <
< pentamethylbenzene < hexamethylbenzene.

6. ISOTOPIC EXCHANGE OF HYDROGEN BETWEEN HYDROCARBONS AND ACIDS
A. Introduction

In 1936 Ingold and his co-workers [158-164] achieved for the first time isotopic exchange in benzene and its derivatives and also some saturated hydrocarbons containing a tertiary carbon atom, using concentrated deuterosulfuric acid. Exchange reactions with saturated hydrocarbons were studied by Burwell and Grodon and their co-workers [165-169], Beeck and Stevenson [170-172], D. N. Kursanov and V. N. Setkina [173-181], and others. The study of hydrogen exchange between aromatic hydrocarbons and sulfuric acid was also developed (Gold and Satchell [84], [182-186], Melander and Olsson [187, 188], and Eaborn and Taylor [188a]). Numerous experiments have also been carried out with chlorosulfonic, perchloric [71, 186, 189], hydrochloric [184], and phosphoric [184] acids.

Hydrocarbons are sparingly soluble in concentrated aqueous solutions of mineral acids and in anhydrous sulfuric acid. In addition, sulfuric, chlorosulfonic, and perchloric acids produce chemical side reactions (oxidation and sulfonation). It is difficult, therefore, to obtain quantitative data on the kinetics of exchange reactions with these acids.

Liquid deuterium bromide does not have these drawbacks: even saturated hydrocarbons are readily soluble in it and, as a rule, they do not react chemically with the solvent. It is true that comparatively complex experimental techniques are required as a result of the high vapor pressure of this solvent at room temperature. Work on deuterium exchange with liquid DBr was undertaken in the Isotopic Reactions Laboratory in 1950 [190] and continued vigorously in subsequent years. This was the beginning of a systematic study of deuterium exchange with liquid hydrogen halides [95, 190-216] (DF, DCl, DBr, and DI), which was contributed much to the understanding of the reactivity of hydrocarbons and their derivatives and to the elucidation of some problems in the mechanism of hydrogen exchange and the equilibrium distribution of hydrogen isotopes in bonds of hydrogen with various elements. Foreign authors (Olah [55-57, 217] and Mackor [71]) subsequently made individual observations on hydrogen exchange with liquid hydrogen fluoride.

In the search for a solvent capable of dissolving hydrocarbons and sufficiently acid to induce hydrogen exchange in hydrocarbons, investigators turned to perfluorocarboxylic acids, namely, trifluoroacetic acid [71, 218, 219, 223, 233] and heptafluorobutyric acid [189], C_3F_7COOH. Carbon tetrachloride was added to dissolve hydrocarbons of higher molecular weights and the acidity of the solution was varied by the addition of H_2SO_4, $HClO_4$, or HBF_3OH.

Glacial acetic acid has been used repeatedly [220-223], usually with its acidity increased by the addition of H_2SO_4 [71, 220, 221], HCl [224], and aprotic acidlike substances ($ZnCl_2$[225] and $SnCl_4$[226]), which form stronger complex acids with CH_3COOH. With rare exceptions, acetic acid has been found to be insufficiently acid for hydrogen exchange in hydrocarbons. It has been used to obtain data only on hydrogen exchange in phenol ethers and salts of carboxylic acids. Other carboxylic acids, namely, HCOOH, $CH_2ClCOOH$, and CCl_3COOH have been used [223].

A brief review of the results of studies of hydrogen exchange between acids and aromatic, alicyclic, and aliphatic hydrocarbons is given below.

B. Aromatic Hydrocarbons

a. BENZENE

Ingold and his co-workers [158-160] were the first to show that with a sulfuric acid concentration of 65 mol. % hydrogen exchange in benzene proceeds comparatively rapidly, but the benzene is sulfonated considerably at the same time. The chemical side reaction is practically absent at an acid concentration of 51-52 mol. %, but then the equilibrium of the exchange reaction is reached only after 3-4 days. When the acid concentration is reduced to 40 mol. %, hydrogen exchanges very slowly. Thus, even in the first work a relation was found between the rate of acid exchange and the acid concentration. Using benzene derivatives, Ingold followed the relation between the deuterium rate and the acid strength: $D_2SO_4 > D_2SeO_4 > D_3O^+ > C_6H_5OD > D_2O$. Ingold and his co-workers prepared fully deuterated benzene for the first time [161] and made a detailed study of its vibration spectra. Twenty years later Gold and Satchell [183] measured the rate of deuterium exchange in monodeuterobenzene with sulfuric acid of various concentrations and established that the logarithm of the deuterium exchange rate is proportional to the acidity function H_0 (p. 59). The reaction proceeded in homogeneous solution and 220 ml of acid was used with 0.15 g of benzene. When the H_2SO_4 concentration was raised from 66.5 to 83.2 wt. %, the rate constant increased by three orders. A linear relation was found between the rate constant k and the acidity function of the sulfuric acid.

H_2SO_4(wt. %)	66.5	68.7	73.5	75.8	79.2	83.2
$-H_0$	5.08	5.40	5.99	6.27	6.72	7.34
$k \cdot 10^7$(sec^{-1})	2.68	5.75	35.2	115	453	3450

Mackor [71] also obtained a straight line with a slope equal to one on a graph of log k against $-H_0$ for solutions of benzene in CF_3COOH with various amounts of H_2SO_4 (and also $HClO_4$) added.

Polanyi and his co-workers [227] discovered that the hydrogen in benzene exchanges for deuterium much more rapidly if sulfuric acid is replaced by gaseous hydrogen chloride in the presence of aluminum chloride. The equilibrium of the exchange reaction was reached after 3 hr at 25°. Klit and Langseth [228-230] confirmed the high activity of this reagent and obtained hexadeuterobenzene by passing DCl into dry benzene in which was dissolved freshly sublimed $AlCl_3$.

Aluminum bromide is an exceptionally active catalyst of deuterium exchange between hydrocarbons and DBr (see the work of the Isotopic Reactions Laboratory [190, 194, 195, 215, 216]). At 20°, the rate constant of hydrogen exchange in benzene k = $5 \cdot 10^{-8}$ sec^{-1} (see [199, 215]), but even an $AlBr_3$ concentration of 10^{-6} mole per mole of DBr, k = $1 \cdot 10^{-5}$ sec^{-1}. Hexadeuterobenzene was prepared by deuterium exchange with liquid DBr in the presence of AlBr$_3$ [205]. In other work in the Isotopic Reactions Labora-tory [215], a quantitative comparison was made of the catalytic activity of AlBr$_3$ and other bromides in deuterium exchange between liquid DBr and C$_6$H$_6$ (p. 42).

The hydrogen in benzene exchanges with liquid deuterium fluoride [202] four and a half orders faster than with liquid deuterium bromide and the rate constants (at 25°) are $1 \cdot 10^{-3}$ and $5 \cdot 10^{-8}$ sec^{-1}, respectively. The reason for the acceleration of the re-action was discussed above (p. 29) and it was also pointed out that the rate of hydrogen exchange with DF does not change when a base (NaF) is added, but is increased very considerably when BF$_3$ is added (see pp. 27 and 63).

b. POLYCYCLIC HYDROCARBONS

The rate of deuterium exchange with liquid DBr increases when the number of aro-matic rings is increased [199]:

benzene $\ll$ naphthalene $<$ phenanthrene $<$ anthracene

benzene $<$ biphenyl $\leq$ terphenyl.

The hydrogen in hydrocarbons with condensed rings exchanges more rapidly than that in hydrocarbons with the same number of rings connected linearly:

naphthalene $>$ biphenyl; anthracene $>$ p-terphenyl.

Table 59 gives the fraction of atoms exchanging at 25° in 30 min (in an hour for biphenyl and p-terphenyl) as a percentage of the total number in the hydrocarbon mole-cule. For comparison it should be noted that only two out of the six hydrogen atoms in benzene could be exchanged after a year.

TABLE 59. Hydrogen Exchange in Polycyclic Hydrocarbons

Hydrocarbon	Degree of exchange of hydrogen atoms,%			
	35°	25°	0°	−40°
Naphthalene	60	50	50	35
Phenanthrene	80	80	80	—
Anthracene	100	100	—	—
Pyrene	100	100	100	—
Biphenyl	—	14	—	—
p-Terphenyl	—	20	—	—

The hydrogen atoms of polycyclic hydrocarbons are nonequivalent as regards their reactivity. This is well known from their chemical reactions and, in particular, from equilibrium reactions with acids (p. 149). The nonequivalence of the hydrogen atoms is reflected in the exchange kinetics, producing a decrease in the rate constant with time. For example, for naphthalene at 0°, after 5 min, $k = 1 \cdot 10^{-3}$, while after 6 hr, $k = 9 \cdot 10^{-5}$; biphenyl, after 1 hr, $k = 1 \cdot 10^{-4}$, while after 12 hr, $k = 2 \cdot 10^{-5}$ (the rate con-stants are in sec^{-1}). When six hydrogen atoms in the biphenyl molecule and ten in p-terphenyl had exchanged, no further exchange of hydrogen with liquid deuterium bro-mide was observed even after 330 hr. This may be ascribed to the very low reactivity of the meta-atoms, in electrophilic replacement as the phenyl group is ortho-para-di-recting in the latter.

Conclusions drawn from a knowledge of the over-all kinetics of forward exchange were later confirmed by measurements of the rate of reverse exchange of deuterium for hydrogen in isomeric monodeuterated hydrocarbons [200, 211]. The rate of exchange of an α-deuterium atom in the naphthalene molecule is 40-60 times greater than that of a β-atom: $k_{25^\circ} = 2\text{-}3 \cdot 10^{-3}$ and $5 \cdot 10^{-5}$ sec^{-1}. In biphenyl, a para-deuterium atom exchanges approximately five times faster than an ortho-atom ($k_{25^\circ} = 1 \times 10^{-4}$ and $3 \cdot 10^{-5}$). There is a similar relation in the exchange with liquid HI [213] ($k_{25^\circ} = 4 \cdot 10^{-6}$ and $1 \cdot 10^{-6}$ sec^{-1}). The rate constant of exchange of a meta-atom with HBr is less than $2 \cdot 10^{-8}$ sec^{-1}.

Let us compare the rates of exchange of different hydrogen atoms in naphthalene and biphenyl with HBr with the rate of exchange of hydrogen in benzene. The ratio of the rate constant of exchange of each of these atoms to the rate constant of exchange of the hydrogen atoms in benzene is called the partial rate factor. Their approximate values are given below.

Naphthalene		Biphenyl		
α-	β-	para-	ortho-	meta-
$5 \cdot 10^4$	$1 \cdot 10^3$	$3 \cdot 10^3$	$5 \cdot 10^2$	$4 \cdot 10^{-1}$

According to Gold and Long [84], in hydrogen exchange between sulfuric acid and anthracene, the two hydrogen atoms in the meso-position (positions 9 and 10) in the molecule of the latter are particularly reactive and are replaced most readily in chemical reactions with electrophilic reagents.

The logarithm of the rate constant of exchange of a meso-atom is directly proportional to the acidity function H_0 of the sulfuric acid.

The following data were obtained on the relation of the rate constant k to the acidity function H_0.

H_2SO_4 (wt. %)	84.0	85.0	85.9	87.8	90.0
$-H_0$	7.47	7.62	7.72	7.94	8.16
$k \cdot 10^4$ (sec^{-1})	5.2	10	17	43	230

Dallinga, Mackor, et al. [219] measured the rates of hydrogen exchange in polynuclear hydrocarbons containing a deuterium atom in a definite position. The exchange was carried out in trifluoroacetic acid with the addition of CCl_4 (to increase the solubility of the hydrocarbon) and sulfuric acid (to vary the acidity).

In most experiments, one millimole of the hydrocarbon was dissolved in a mixture of 10.0-10.7 g of CF_3COOH, 40 g of CCl_4 [in two cases, the CCl_4 was replaced by p-$C_6H_4Cl(CH_3)$], and 0.007-0.2 g of H_2SO_4. The acidity function was measured in mixtures of $CF_3COOH-H_2SO_4$ and quite arbitrarily was assumed to be the same for solutions containing CCl_4. In most experiments, $H_0 = -4.40$. If H_0 differed from this value, then for convenient in comparison of the rate constants, they were converted by means of the formula : $\log k$ (at $H_0 = -4.40$) $= \log k_{exp} + 0.86 (H_0 + 4.40)$. This equation was established experimentally for naphthalene, but was also used in experiments with the other hydrocarbons, which naturally is not strictly correct.

The fourth column of Table 60 gives the values of $\log k$ obtained in this way for the nonequivalent hydrogen atoms whose positions in the hydrocarbon molecule are given in the second column of the table, while the number of them n is given in the third column. In addition, the table gives the logarithms of the basicity constants K' for the same hydrocarbons, determined in liquid HF as described above on p. 147. The last column of the table gives the logarithms of the partial rate factors for nitration f according to the data of Dewar and his co-workers [231].

TABLE 60. Comparison of Deuterium Exchange Rate Constants, Basicity Constants, and Nitration Partial Rate Factors of Hydrocarbons

Hydrocarbon	Position of atom in molecule	m	log k	log K'	log f
Benzene	—	6	~—4.5	—10.2	0.00
Naphthalene	1	4	—2.25	—4.6	2.67
	2	4	—3.13	—	1.70
Anthracene	1	4	—1.55	—	—
	2	4	—2.25	—	—
	9	2	+1.61	3.5	—
Biphenyl	2	4	—2.93	—6.3	1.48
	3	4	>—5	—	—
	4	2	—2.93	—	1.26
Pyrene	3	4	+0.43	1.5	4.23
	1	4	—	—	>2
1,2-Benzanthracene	9	1	1.1	2.0	—
	10	1	1.1	—	—
Perylene	3	4	+0.53	3.8	4.89
	1	4	+0.53	—	>2
Triphenylene	2	6	—3.01	—5.4	2.78
	6	6	—3.50	—	2.78
Chrysene	2	2	—1.23	—	3.54
	6	—	—1.87	—2.0	—

Table 61 gives the structural formulas of the same hydrocarbons and the partial rate factors for hydrogen exchange in them. The figures given are extremely approximate as the rate constant of hydrogen exchange in benzene has been determined inaccurately.

Eaborn and Taylor [231a] later measured the rate of exchange of tritium in positions 2, 3, and 4 in the biphenyl molecule and also in the α- and β-positions of naph-

TABLE 61. Partial Rate Factors for Hydrogen Exchange in Polycyclic Hydrocarbons

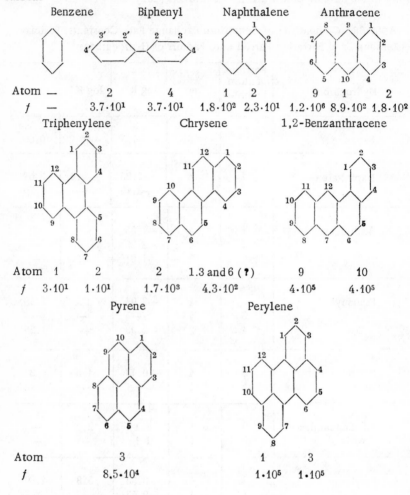

	Benzene	Biphenyl		Naphthalene		Anthracene		
Atom	—	2	4	1	2	9	1	2
f	—	$3.7 \cdot 10^1$	$3.7 \cdot 10^1$	$1.8 \cdot 10^2$	$2.3 \cdot 10^1$	$1.2 \cdot 10^6$	$8.9 \cdot 10^2$	$1.8 \cdot 10^2$

	Triphenylene		Chrysene		1,2-Benzanthracene	
Atom	1	2	2	1.3 and 6 (?)	9	10
f	$3 \cdot 10^1$	$1 \cdot 10^1$	$1.7 \cdot 10^3$	$4.3 \cdot 10^2$	$4 \cdot 10^5$	$4 \cdot 10^5$

	Pyrene	Perylene	
Atom	3	1	3
f	$8.5 \cdot 10^4$	$1 \cdot 10^5$	$1 \cdot 10^5$

thalene with acids of the following composition: I) CF_3COOH (95.31) – H_2O (2.21) – H_2SO_4 (2.48) and II) CF_3COOH (92.04) – H_2O (5.45) – $HClO_4$ (2.51). The concentrations in mol. % are given in brackets. The following rate constants ($k \cdot 10^7$ in sec^{-1}) were obtained for tritium exchange at 25°.

Substance	$CF_3CO_2H + H_2SO_4$	$CF_3CO_2H + HClO_4$
T-benzene	3.6	14.3
p-T-Toluene	2526	4480
2-T-Biphenyl	477	52
3-T-Biphenyl	—	9.8
4-T-Biphenyl	515	52
α-T-Naphthalene	3886	370
β-T-Naphthalene	456	62

The partial rate factors of tritium exchange were calculated.

The absolute values of the partial rate factors of exchange in the system $CF_3CO_2H - H_2SO_4$ are less than the corresponding factors for exchange with HBr (see p. 177). The reason for this is discussed on p. 280.

Let us compare the partial rate factors of deuterium and tritium exchange in biphenyl and in naphthalene with acids.

Partial Rate Factors of Deuterium and Tritium Exchange in Biphenyl

Acid	2-	3-	4-	Reference
$CF_3CO_2H - H_2SO_4 - CCl_4$	37	–	37	[219]
$CF_3CO_2H - HClO_4$	52	0.68	52	[231a]
$CF_3CO_2H - H_2SO_4$	133	–	143	[231a]
HBr	500	< 0.4	3000	[211]

Partial Rate Factors of Deuterium and Tritium Exchange in Naphthalene

Acid	α-	β-	Reference
$CF_3CO_2H - H_2SO_4 - CCl_4$	180	23	[219]
$CF_3CO_2H - HClO_4$	370	62	[231a]
$CF_3CO_2H - H_2SO_4$	1079	127	[231a]
HBr	50,000	1000	[200]

As has been stated, the replacement of liquid HBr by liquid HF in hydrogen exchange with benzene increases the exchange rate very considerably. Nonetheless, judging by the partial rate factors, hydrogen exchange with HBr in polycyclic hydrocarbons proceeds much more rapidly than in benzene. However, the conclusion that the rate of hydrogen exchange in polycyclic hydrocarbons with HF should be much greater than in benzene, which seems quite logical at first glance, actually is not borne out [193, 208]: the hydrogen in benzene exchanges with HF much more rapidly than in anthracene. Thus, in the first 5 min at 50° it is possible to replace two thirds of the total number of the hydrogen atoms by deuterium, while only two atoms are replaced in anthracene (probably instantaneously). The equilibrium of the exchange is not reached even after 2 hr (n = 9.3). The mean rate constant without allowance for the exchange of the most reactive meso-atoms $k = 4 \cdot 10^{-4}$, while for benzene $k = 4 \cdot 10^{-3}$ sec^{-1}. The reason for the retardation of hydrogen exchange in anthracene has already been discussed above (p. 31). The basicity constant of anthracene in liquid HF is 14 orders greater than that for benzene in the same solvent (p. 147). Anthracene is considerably ionized in HF as a result of protonization of the meso-carbon atom and hydrogen exchange in the cation is hampered. The addition of BF_3 completely ionizes not only anthracene, but also benzene. In this case, exchange in anthracene is limited to only the two meso-atoms even after 2 hr. All the hydrogen atoms in benzene are equivalent and therefore its ionization, on the contrary, results in the very rapid establishment of exchange equilibrium.

Azulene is ionized readily when dissolved in not only sulfuric acid, but even carboxylic acids (pp. 150 and 160). The proton thereupon adds in position 1. (Position 3 is symmetrical with respect to it). In accordance with this, in the reaction of D_2SO_4 with azulene, the two hydrogen atoms in these positions exchange readily and the exchange reaction is limited to this [231b, 231c] in the same way as in the reaction of anthracene with DF examined above.

c. ALKYLBENZENES AND ALKYLNAPHTHALENES

Acid hydrogen exchange in these hydrocarbons was achieved for the first time in the Isotopic Reactions Laboratory with liquid DBr [95, 190, 192]. The methylation of benzene and naphthalene, which increases their strength as bases (pp. 139 and 145), considerably accelerates hydrogen exchange in the aromatic ring. Within the limits of experimental error, the rates of hydrogen exchange in monoalkylbenzenes are the same, as is shown by the number of hydrogen atoms exchanging n (Table 62).

TABLE 62. Comparison of Numbers of Hydrogen Atoms in Monoalkylbenzenes Exchanging with Liquid DBr (at 25°)

Substance	Experiment time		
	2 hr	5 hr	9 hr
Toluene	1.5	2.3	2.5
Ethylbenzene	1.5	2.3	2.5
Isopropylbenzene	1.3	2.0	2.3
tert-Butylbenzene	1.4	2.2	2.5
n-Butylbenzene	1.5	2.3	2.6

Mackor [71] confirmed that the rate constants and activation energies of the exchange of a deuterium atom in the para-position relative to the alkyl group in toluene, ethylbenzene, and tert-butylbenzene with $CF_3COOH + H_2SO_4$ are also equal within the limits of experimental accuracy. The same conclusion was drawn in the work of Lauer and his co-workers [233].

As Table 62 shows, the first three hydrogen atoms in monoalkylbenzenes (ortho- and para-atoms) are exchanged for deuterium comparatively rapidly. From the change in the rate constant with time it was possible to calculate the approximate values of the rate constant of exchange (at 25°) of the para-atom ($k = 3 \cdot 10^{-4}$) and the ortho-atoms ($k = 5 \cdot 10^{-5}$ sec^{-1}). The rate of exchange of a meta-hydrogen atom in mono-alkylbenzenes is three orders lower ($k = 3 \cdot 10^{-7}$ sec^{-1}). The accurate constants are given below (p. 177).

The equilibrium of the exchange of five hydrogen atoms in toluene with liquid DF, in contrast to liquid DBr, is reached in a few minutes [202]. However, even with catalysis by BF_3 and an experiment time of 100-200 hr, no exchange of the hydrogen atoms of the methyl group was observed and likewise with catalysis by $AlBr_3$ in liquid DBr. No exchange of hydrogen in hexamethylbenzene was observed even after 140 hr in a solution of $DF + BF_3$ or in liquid DBr after a year. The hydrogen atoms of alkyl groups bound to an aromatic ring do not participate in acid hydrogen exchange. The same conclusion was reached by Olsson and Melander [187], who attempted to effect an exchange reaction between toluene tritiated in the methyl group and sulfuric acid.

The rate of isotopic exchange of different hydrogen atoms in the aromatic ring of toluene with acids was determined in subsequent work of the Isotopic Reactions Laboratory and of foreign authors by means of isomers of monoduetero- and monotritiotoluene (HBr [211], HI [213], $H_2SO_4 - H_2O$ [185, 187, 188], $HClO_4 - H_2O$ [189], C_3F_7COOH [189], $CF_3CO_2H - H_2O$ [232], $CF_3COOH - H_2SO_4$ [71].

The results of measurements of the relative rate of exchange of different hydrogen atoms in the toluene nucleus with the acids listed are given in Table 63.

TABLE 63. Relative Rate of Exchange of Different Hydrogen Atoms in Toluene

Medium (acid concentration in brackets)	Reaction conditions					Relative exchange rate			Literature reference
	isotope	temp., °C	system	DC	$-H_0$	para	ortho	meta	
HClO$_4$ – H$_2$O (70.8%)	D	22	ht		7.0	15.4	16.8	1	[189]
						0.92	1		
H$_2$SO$_4$ – H$_2$O (80.8%)	T	25	ht	95	6.9	37.2	45.4	1	[188]
						0.82	1		
H$_2$SO$_4$ – H$_2$O (80.8%)	D	25	ht	95	6.9	46.7	51.6	1	[188]
						0.91	1		
H$_2$SO$_4$ – H$_2$O (73.24%)	T	25	hm		–	50.9	–	1	[188a]
H$_2$SO$_4$ – H$_2$O (75.30%)	T	25	hm		6.2	48.5	–	1	[188a]
H$_2$SO$_4$ – H$_2$O (68%)	D		hm	94	5.3	44	44	1	[185]
						1	1	–	
C$_3$F$_7$COOH (100%)	D	78-119	hm		3.3	15.4	6.2	1	[189]
						2.4	1	–	
CF$_3$COOH + H$_2$SO$_4$ (0.25 M)	D		hm		3.3	100	70	1	[71]
						1.4	1	–	
CF$_3$COOH – H$_2$O	D	70	hm			105	62	1	[232]
						1.7	1		
HBr (100%)	D	25	hm	4	–	730	208	1	[211]
						3.5	1		
HI (100%)	D	25	hm	2	–	3.2	1	–	[213]

The data on the exchange rate in each of the systems are expressed in two forms: in the first line, the rate constant of exchange of the deuterium atom in the meta-position is taken as unity and in the second line, the ortho-position was the standard. The following abbreviations are used: D – monodeuterotoluene; T – monotritiotoluene; °C – temperature; ht – heterogeneous system; hm – homogeneous solution; DC – dielectric constant (this was interpolated for the sulfuric acid-water system from the values $DC_{H_2SO_4} = 101$ and $DC_{H_2O} = 80$); H_0 – acidity function.

If a concentrated aqueous solution of a strong acid (HClO$_4$ or H$_2$SO$_4$) is used in the exchange, then the para-atom exchanges somewhat more slowly than the ortho-atom. On the other hand, when toluene is dissolved in perfluorocarboxylic acids of liquid hydrogen halides, the para-position is found to be much more reactive than the ortho-position. The relative reactivity of the para- and meta-positions may be varied over a wide range (from 15:1 to 730:1). It is noteworthy that these changes accompany a decrease in the dielectric constant of the solvent. For example, in experiments with sulfuric and trifluoroacetic acids, the latter differ by a factor of more than ten (96 and

8), while the values of H_0 are almost the same (-6.9 and -6.4).[*] The difference in the degree of polarity of the bond in the activated state is probably of importance.

As regards steric hindrance to deuterium exchange in monoalkylbenzenes, it has much less effect on the rate than in nitration, for example (p. 268). Table 64 gives a comparison of the partial rate factors for the exchange of deuterium atoms in different positions in the molecules to toluene, ethylbenzene, isopropylbenzene, and tert-butylbenzene with trifluoroacetic acid containing heavy water [233].

TABLE 64. Partial Rate Factors of Deuterium Exchange in Monoalkyl-benzenes

Hydrocarbon	f_o	f_m	f_p	f_o / f_m	f_o / f_m	f_p / f_m
Benzene	1	1	1	1	1	1
Toluene	253	3.5	421	0.60	72	120
Ethylbenzene	257	5	449	0.57	51	92
Isopropylbenzene	259	—	493	0.53	—	—
tert-Butylbenzene	198	7	490	0.40	28	70

The ratios of the partial rate factors for toluene and tert-butylbenzene in deuterium exchange and nitration are as follows:

	f_o/f_p	f_o/f_m	f_p/f_m
Deuterium exchange	1.5	2.5	1.7
Nitration	7.1	9.5	1.2

The rate constant of hydrogen exchange in toluene depends strongly on the acidity of the medium. According to the measurements of Gold and Satchell [185], log k for the exchange of p-D-toluene is a linear function of H_0. This is shown by the following data:

H_2SO_4 (wt. %)	53.0	58.5	63.9	68.9	73.4
$-H_0$	3.57	4.16	4.76	5.39	5.96
$k \cdot 10^7$ (sec^{-1})	1.36	9.40	56.6	498	4250

Let us compare the rate constants of the exchange of hydrogen in benzene and the ortho-, meta-, and para-hydrogen atoms of toluene with some acids (Table 65). The partial rate factors for hydrogen exchange with four acids may be calculated (Table 66).

All the hydrogen atoms in the toluene molecule exchange more rapidly than those in benzene. The degree of acceleration is different in different positions and depends

---

[*] The values of H_0 in solvents differing in the dielectric constant are not fully comparable (p. 54).

very strongly on the acid used in the exchange. The latter factor has a particularly strong effect on the exchange of ortho- and para-deuterium atoms, but has little effect on the exchange of the meta-atoms. This is shown particularly clearly by a comparison of the ratios of the partial rate factors for deuterium exchange with liquid hydrogen bromide and with 68% sulfuric acid:

para-	ortho-	meta-
36	10	2

With the increase in the strength of methylbenzene as bases which occurs during progressive methylation (p. 140), there is a sharp increase in the rate of exchange of the hydrogen atoms remaining in the ring. Thus, for example, the rate of hydrogen exchange between m-xylene and DBr at 25° was so rapid that it could not be measured [95].

TABLE 65. Rate Constants of the Exchange of Deuterium and Tritium Atoms in the Aromatic Ring of Toluene at 25°

Acid	C_6H_6	ortho-	meta-	para-	Literature reference
H_2SO_4 (68%)	$4.3 \cdot 10^{-7}$	$3.56 \cdot 10^{-5}$	$8.21 \cdot 10^{-7}$	$3.56 \cdot 10^{-5}$	[182] [185]
H_2SO_4 (73.24%)	$1.2 \cdot 10^{-6}$	—	$5.9 \cdot 10^{-6}$	$3.0 \cdot 10^{-4}$	[188a]
HBr (100%)	$5 \cdot 10^{-8}$	$5 \cdot 10^{-8}$	$2 \cdot 10^{-7}$	$2 \cdot 10^{-4}$	[211]
HI (100%)	—	$1 \cdot 10^{-6}$	—	$4 \cdot 10^{-6}$	[213]

TABLE 66. Partial Rate Factors of the Exchange of Deuterium in Toluene with Acids (at 25°)

Acid	f_p	f_0	f_m	Lit. ref.
H_2SO_4 (68%)	83	83	1.9	[185]
H_2SO_4 (80.7%)	123	135	2.6	[188]
H_2SO_4 (73.24%)**	254	—	4.99	[188a]
H_2SO_4 (73.30%)**	243	—	5.02	[188a]
H_2SO_4 (77.67%)**	—	—	5.42	[188a]
H_2SO_4 (79.80%)**	—	—	5.72	[188a]
H_2SO_4 (81.14%)**	—	—	5.70	[188a]
$CF_3CO_2H + H_2SO_4$ (0.25 m)	350	245	3.2	[71]
CF_3CO_2H (96.6%)*	420	250	4	[232]
HBr (100%)	2900	830	4	[211]

* Temperature 70°.

** Experiments with tritiated toluene.

TABLE 67. Data on the Kinetics of Deuterium Exchange in Benzene and Alkylbenzenes

Hydrocarbon	log k'	H$^{\neq}$	$\Delta S_1^{\neq}$	Solvent
D-Benzene.........	− 3.1	16.6	− 9.0	$CF_3COOH + H_2SO_4$ 2 M
m-D-Toluene.......	− 2.6	−	−	$CF_3COOH + H_2SO_4$ 0.25 M
o-D-Toluene.......	− 0.73	16.3	+ 1.0	$CF_3COOH + H_2SO_4$ 0.25 M
p-D-Toluene.......	− 0.56	16.0	+ 0.8	$CF_3COOH + H_2SO_4$ 0.25 M
p-D-Ethylbenzene....	− 0.57	15.3	− 1.5	$CF_3COOH + H_2SO_4$ 0.25 M
p-D-tert-Butylbenzene.	− 0.54	15.8	+ 0.1	$CF_3COOH + H_2SO_4$ 0.25 M
D-p-Xylene........	0.00	15.0	0.0	$CF_3COOH + H_2SO_4$ 0.1 M
4-D-p-Xylene	−	16.0	−	$CF_3COOH + H_2SO_4$ 0.015 M
4-D-m-Xylene......	2.3	17.8	+ 16.3	$CF_3COOH + H_2SO_4$ 0.015 M
4-D-m-Xylene......	−	23.5	−	$CF_3COOH + H_2SO_4$ 6 M
D-Mesitylene.......	4.4	23.0	+ 24	$CF_3COOH + H_2SO_4$ 3.5 M

TABLE 68. Relative Rates of Deuteration of Polymethylbenzenes

Hydrocarbon.	Temp., °C	Relative rate of exchange	
		deter.	calc.
Benzene	70	6	6
Toluene	70	934	934
p-Xylene.........	70	$5.7 \cdot 10^3$	$3.8 \cdot 10^3$
o-Xylene.........	70	$6.2 \cdot 10^3$	$5.1 \cdot 10^3$
m-Xylene	70	$1.8 \cdot 10^5$	$2.8 \cdot 10^5$
Pseudocumene	40	$5.1 \cdot 10^5$	$6.5 \cdot 10^5$
Hemimellitene......	40	$6.6 \cdot 10^5$	$8.1 \cdot 10^5$
Durene	30	$1.7 \cdot 10^6$	$1.9 \cdot 10^6$
Mesitylene	0	$5.7 \cdot 10^7$	$8.1 \cdot 10^7$
Isodurene.........	0	$1.5 \cdot 10^8$	$2.0 \cdot 10^8$
Pentamethylbenzene ..	0	$1.6 \cdot 10^8$	$3.9 \cdot 10^8$

The rates of deuterium exchange in deuterated benzene and methylbenzenes were compared by Mackor and his co-workers [71]. Table 67 contains the following columns: log k' is the logarithm of the rate constant of the replacement of deuterium by hydrogen in the given hydrocarbon, relative to the rate constant of deuterium exchange in D-p-xylene at 25°; $H^{\neq}$ is the enthalpy of activation of the exchange, kcal/mole; $\Delta S_i^{\neq}$ is the entropy of activation (cal/deg· mole) relative to the value obtained for p-xylene. The solvent was CF_3COOH and the concentration of the sulfuric acid added is expressed in moles per 1000 g of CF_3COOH (M).

The reactivity of mesitylene was so great that the rate of deuterium exchange had to be determined relative to m-xylene in mixtures of acetic and sulfuric acids with a known value of H_0 and the rate of exchange with $CF_3COOH + H_2SO_4$, calculated, using the known value of the exchange rate constant for m-xylene in this solvent. The enthalpies of activation of hydrogen exchange of benzene and its alkyl derivatives with the same solvent (CF_3CO_2H) vary very little (15.0-17.8 kcal/mole) and the change in the exchange rates is caused by the entropy term. The entropy of activation increases by approximately 10 entropy units when an additional methyl group is introduced into the molecule in the ortho- or para-position relative to the deuterium atom. The authors rightly pointed out the great interest and fundamental importance of this unexpected fact.

Lauer and Stedman [234] compared the rates of deuteration of polymethylbenzenes in mixtures of trifluoroacetic acid and deuterium oxide. Measurements with various substances were made at temperatures of 70, 40, 30, and 0°. The rate of hydrogen exchange in m-xylene was determined at three temperatures and the activation energy calculated: E = 17.6 kcal/mole. The corresponding conversions were based on this value. Table 68 gives a summary of the results obtained on the relative rate of deuteration.

The last column of Table 68 gives the relative rate constants of hydrogen exchange calculated by Condon's method from the partial rate factors found for toluene. (The principles of the calculation are given in Section IV, p. 263). Inspite of the very great differences in the reactivities of the hydrocarbons given in Table 68 (the rate constants by up to seven orders), the calculated and observed constants agreed to within 25%. The authors also reported a linear relation between the basicity of the hydrocarbons and the deuteration rate.

d. PHENYLATED ALKANES

Deuterium exchange between phenylated alkanes and liquid DBr was studied in the Isotopic Reactions Laboratory [210]. Since the hydrogen atoms of the aliphatic CH bonds in the aliphatic-aromatic hydrocarbons do not participate in exchange with an acid (p. 174), the number of hydrogen atoms exchanging per phenyl group (n') in experiments of the same duration characterizes the reactivity of the aromatic rings in phenylated alkanes. It was established that it depends on the ratio between the number of aromatic rings and the number of aliphatic CH bonds in the α- and β-positions. These ratios are denoted by α' and β'. The values given were obtained in experiments lasting for 6 hr at 25°.

Methane derivatives:

	$C_6H_5CH_3$	$(C_6H_5)_2CH_2$	$(C_6H_5)_3CH$
α'	3	1	$\frac{1}{3}$
η'	2.3	0.9	0.2

Ethane derivatives:

	$C_6H_5CH_2CH_3$	$(C_6H_5)_2CHCH_3$	$(C_6H_5)_3CCH_3$
α'	2	$1/2$	0
β'	3	$3/2$	1
η'	1.4	$1/4$	0.5

Consequently, the hydrogen in the aromatic rings of sym-diphenylethane exchanges more rapidly than that in diphenylmethane:

	$C_6H_5CH_2CH_2C_6H_5$	$(C_6H_5)_2CH_2$
α'	2	1
β'	1.5	0.9

In symmetrical diphenylalkanes from diphenylethane to diphenylbutane, the number of α-CH bonds is fixed ($\alpha' = 2$), while the number of β-CH bonds increases and the hydrogen rate increases:

	$C_6H_5(CH_2)_2C_6H_5$	$C_6H_5(CH_2)_3C_6H_5$	$C_6H_5(CH_2)_4C_6H_5$
β'	0	1	2
η'	1.5	2.2	2.9

As is known [210a], the rate of electrophilic replacements of hydrogen in alkylbenzenes (for example, halogenation) depends on the ratio of α' and β' and this is usually explained by the hyperconjugation effect. This rule does not apply to the acid exchange of hydrogen in alkyl benzenes (p. 174), but, as we have seen, applies to exchange reactions with polyphenylated alkanes.

It was reported in [210] that the hydrogen in tetraphenylmethane ($C_6H_5)_4C$, exchanges with liquid DBr much more rapidly than that in benzene. Consequently, phenyl rings separated by a carbon atom increase each others reactivity. The formation of a bond between the aromatic rings of the diphenylmethane molecule to convert it to fluorene creates additional conditions for conjugation both between the π-electrons of the aromatic rings and between them and the σ-electrons of the aliphatic bonds of the methylene group so that the exchange rate increases sharply: 5 hydrogen atoms in the fluorene molecule exchange even after 10 min, while only 2 hydrogen atoms in diphenylmethane are replaced by deuterium after 6 hr.

C. Unsaturated Aliphatic Hydrocarbons

There is very little information on hydrogen exchange between acids and hydrocarbons of this class. This is explained by the fact that alkenes and alkynes react readily with acids chemically and the latter add to the multiple bond.

Horiuchi and Polanyi [235] established that the hydrogen in ethylene exchanges for the deuterium of deuterosulfuric acid. Phosphoric acid labeled with tritium does not produce hydrogen exchange in ethylene at 50° in 48 hr [236], while in propylene and 2-butylene, the exchange occurs at a rate which increases in the given sequence, but there is no hydrogen exchange with isobutane under the same conditions [237, 238].

A. E. Shilov, R. D. Sabirova, and V. A. Gorshkov [238a] studied the reaction between ethylene and concentrated deuterosulfuric acid at an ethylene pressure of 4 atm under conditions where there was no appreciable decomposition of the ethylsulfuric acid formed ($CH_2D - CH_2OSO_2OD$) and found that the light hydrogen from the ethylene did not pass into the sulfuric acid. The difference between this result and that published previously is explained by the fact that in the experiments described in the literature, the ethylsulfuric acid decomposed to the starting materials. Since the hydrogen atoms in the methyl group of ethylsulfuric acid are equivalent, the deuterium atoms in the methyl group of ethylsulfuric acid are equivalent, the deuterium entered the ethylene molecule. The equilibrium distribution of deuterium was reached by repeated addition and elimination of the acid. The authors interpreted their results as evidence indicating that in the reaction between ethylene and sulfuric acid, there is no reversible formation of a carbonium ion ($C_2H_4 + D^+ \rightleftharpoons DH_2C - \overset{+}{C}H_2$), as the result of this would be hydrogen exchange.

D. Saturated Hydrocarbons

Even in the first work of Ingold on hydrogen exchange between saturated hydrocarbons and deuterosulfuric acid, it was reported that exchange occurs only when the hydrocarbon contains a tertiary carbon atom. This concept has dominated the subsequent work of Burwell and Gordon, Beeck and Stevenson, and D. N. Kursanov and V. N. Setkina.

Ingold [161] explained this characteristic by the fact that tertiary carbon atoms have the highest negative charge. Burwell and Gordon [165, 166] proposed a reaction mechanism, which has been adopted as a basis by other authors studying acid exchange in saturated hydrocarbons. Results obtained for the relative rate of hydrogen exchange with 95% sulfuric acid at 60° [166] are given below.

Hydrocarbon	k_{rel}
3-Methylpentane	4200
3-Methylhexane	2100
3-Methylheptane.	1200
3-Methylheptane.	800
3-Methylundecane.	50
n-Octane.	7
2,2-Dimethylhexane . . .	1

(It is possible that the hydrogen exchange in the last two substances is explained by the presence of traces of olefins in them.)

It has been reported that the exchange rate decreases with an increase in the length of the chain of carbon atoms in the hydrocarbon molecule. The exchange is accompanied by the appearance of a color in the acid layer and the liberation of sulfur dioxide. This was also confirmed by D. N. Kursanov and V. N. Setkina [173, 174, 239, 240]. These authors also observed that all the hydrogen atoms in hydrocarbons with a tertiary carbon atom exchange, regardless of whether the hydrocarbon is aliphatic or alicyclic. A number of aliphatic (3-methylpentane, 3-methylhexane, and 2,2,4-trimethylpentane) and alicyclic compounds (methylcyclohexane and methylcyclopentane) were investigated. If the hydrocarbon molecule contains both a tertiary and a quaternary carbon atom, deuterium first replaces the hydrogen atoms in the part of the molecule right up to the quaternary carbon atom, i.e., 7 of the 16 hydrogen atoms in 2,2, 3-trimethylbutane (I) and 9 of the 18 hydrogen atoms in 2,2,4-trimethylpentane (II):

$$\underset{\text{I}}{\underset{\underset{\text{CH}_3}{|}}{\overset{\overset{\text{CH}_3}{|}}{\underset{\underset{\text{CH}_3}{|}}{\text{H}_3\text{C}-\text{C}}}}-\underset{\underset{\text{CH}_3}{|}}{\overset{\overset{\text{H}}{|}}{\text{C}}}-\text{CH}_3} \qquad \underset{\text{II}}{\underset{\underset{\text{CH}_3}{|}}{\overset{\overset{\text{CH}_3}{|}}{\text{H}_3\text{C}-\text{C}}}}-\underset{\underset{\text{H}}{|}}{\overset{\overset{\text{H}}{|}}{\text{C}}}-\underset{\underset{\text{CH}_3}{|}}{\overset{\overset{\text{H}}{|}}{\text{C}}}-\text{CH}_3 .$$

All the hydrogen atoms exchanged in long experiments with these hydrocarbons, probably as a result of isomerization, which was indicated by a change in the constants of the substances.

When no tertiary carbon atom is present, there is no exchange with sulfuric acid in the case of either aliphatic (n-heptane, n-dodecane, and 2,2-dimethylhexane) or alicyclic hydrocarbons (cyclopentane, cyclohexane, and 1,1-dimethylcyclohexane). The authors showed [175] that it is possible to use deuterium exchange with sulfuric acid to analyze mixtures of saturated hydrocarbons by determining the deuterium concentration when exchange equilibrium is reached.

Deuterium exchange in saturated hydrocarbons without a tertiary carbon atom becomes possible in the presence of aluminum and boron halides. For example, Powell and Reid [241] studied the exchange of tritium between TCl and butanes during the isomerization in the presence of $AlCl_3$. Pines and Weckher [242] carried out the isomerization and isotopic exchange in n-butane by using $AlBr_3 - DBr$ in the presence of traces of an olefin. There has been considerable study of hydrogen exchange in saturated hydrocarbons in the Isotopic Reactions Laboratory. The exchange was carried out in liquid deuterium bromide with aluminum bromide as the catalyst [190, 194, 195, 216] and in liquid deuterium fluoride with boron trifluoride as the catalyst [202].

Below we give data on the kinetics of hydrogen exchange between alicyclic hydrocarbons and DBr at 25°, in which saturated hydrocarbons are readily soluble. The reaction rate is proportional to the catalyst concentration. The rate constants (sec^{-1}) divided by the concentration C in moles of $AlBr_3$ per mole of HBr are given [216].

Hydrocarbon	$\dfrac{k \cdot 10^6}{C \cdot 10^2}$
Cyclohexane	15 ± 7
Methylcyclopentane. . . .	13 ± 3
Methylcyclohexane	20 ± 13
Cyclopentane	59 ± 17

It was shown that there is rapid isomerization of cyclohexane to methylcyclopentane in the solution ($k \cdot 10^4/C \cdot 10^2 = 12 \pm 2$).

The isomerization rate is approximately two orders greater than the deuterium exchange rate. Consequently, regardless of whether cyclohexane or methylcyclopentane is used for the experiment, the exchange occurs in the same system, which contains 12.5% of methylcyclopentane.

It is very probable that even more extensive conversions of the hydrocarbons may be produced in the presence of aluminum bromide, as is indicated by separation of the solution into layers in long experiments.

Some data on hydrogen exchange between saturated hydrocarbons and liquid deuterium fluoride containing boron trifluoride (0.4 mole/1000 g of DF) at room temperature [202] are given in Table 69.

TABLE 69. Hydrogen Exchange Between Saturated Hydrocarbons and DF + BF_3

Hydrocarbon	Catalyst	Time, hr	Degree of exchange, %
Cyclopentane.	BF_3	24	70
"	BF_3	96	~100
Cyclohexane	—	2000	0
"	BF_3	24	16
"	BF_3	208	~100
Methylcyclopentane. . .		16	10
"	—	90	16
"	—	450	60
"	BF_3	2	4
Methylcyclohexane . . .		16	40
"	—	190	70
"	—	330	80
"	BF_3	1.5	4
"	BF_3	7	40
"	BF_3	15	80
Decalin.	—	116	7
"	—	760	9
"	BF_3	7	14
3-Ethylpentane.	—	48	18
4-Methylpentane.	—	88	~100

Measurement of the rate of hydrogen exchange between saturated hydrocarbons and liquid hydrogen fluoride is hampered by their low solubility in the latter. Mixing was used in the experiments given in Table 69. Judging by the constants of the substances, cis-decalin isomerizes into the trans-form, while methylcyclopentane is converted into cyclohexane in the presence of boron trifluoride. When BF_3 was added to HF in normal aliphatic hydrocarbons (n-hexane, n-heptane, and n-octane), no hydrogen exchange was observed after 6 hr, but the substance had decomposed completely to form gaseous products after 300 hr.

Only the hydrogen in hydrocarbons containing a tertiary carbon atom (3-ethylpentane, 4-methylheptane, methylcyclopentane, methylcyclohexane, and decalin*) exchanges with liquid hydrogen fluoride without a catalyst.

In the work of Burwell and Gordon cited above, the rate of deuterium exchange between sulfuric acid and optically active 3-methylheptane, was compared with the

---

* No hydrogen exchange was observed between sulfuric acid and decalin [173].

rate of racemization of this hydrocarbon. Generalizing all their observations, the authors put forward a hypothesis on the reaction mechanism. In their opinion, it begins with the oxidation of the hydrocarbon with the tertiary carbon atom by sulfuric acid (this is indicated by the liberation of sulfur dioxide). A particle (an olefin or a carbonium ion) is formed in which hydrogen exchange occurs. For each molecule racemized, there are 10-20 molecules containing deuterium. A chain mechanism is proposed for the exchange with the transfer of hydrogen to a particle of an unsaturated nature.[*] The main concepts put forward by Burwell and Gordon were developed by D. N. Kursanov and V. N. Setkina in the USSR and Beeck and Stevenson in the USA. D. N. Kursanov and V. N. Setkina [177, 243] consider that a carbonium ion is formed by oxidation of the hydrogen of the methyne group by sulfuric acid:

$$-\overset{|}{\underset{|}{C}}-H + 2H_2SO_4 \rightarrow -\overset{|}{\underset{|}{C}}{}^{\oplus} + HSO_4^- + SO_2 + 2H_2O .$$

It is considered that the deuterium of the sulfuric acid participates in a rapid exchange with the hydrogen atoms of the carbonium ion adjacent to the "carbonium center." The latter may migrate along the chain of carbon atoms from the tertiary atom to any other carbon atom:

$$\underset{\overset{|}{H}\quad R}{R:C:C:R} \longrightarrow \underset{H\quad R}{R:\overset{\oplus}{C}:\overset{H}{C}:R} .$$

According to the scheme, the hydrogen is actually transferred with a pair of electrons in the form of the hydride ion H^-. As a result of this migration of the trivalent positively charged carbon, all the hydrogen atoms in a saturated hydrocarbon may exchange.

The hydrogen of the methyne group of another molecule may add to the carbonium ion to form a new carbonium ion and the reaction proceeds by a chain mechanism. It does not proceed in the absence of a methyne group. It is assumed that the hydrogen is transferred in the form of a hydride ion:

$$\underset{R}{R:\overset{R}{\overset{\oplus}{C}}} + \underset{R}{[H:\overset{R}{C}:R]} \longrightarrow \underset{R}{R:\overset{R}{C}:H} + \underset{R}{{}^{\oplus}\overset{R}{C}:R} .$$

Otvos, Beeck, Stevenson, and Wagner [171] consider that during the chain reaction, the hydrogen of the methyne group may be transferred only to a tertiary carbonium ion and therefore the hydrogen at the tertiary carbon atom does not participate in isotopic exchange, i.e., in a molecule containing n hydrogen atoms, only $n-1$ hydro-

---

[*] Racemization indicates the formation of a carbonium ion (p. 135). From the relation between the number of molecules racemized and the number of molecules into which deuterium had bene introduced it can be concluded that a series of molecules are successively converted into carbonium ions, i.e., the reaction proceeds by a chain mechanism.

gen atoms may exchange. In critizing this point of view, D. N. Kursanov and his co-workers used specially synthesized hydrocarbons containing deuterium at a tertiary carbon atom, for example:

$$CH_3CH_2 - \underset{\underset{D}{|}}{\overset{\overset{CH_3}{|}}{C}} - CH_2CH_2CH_3 \, ,$$

and showed [176] that in actual fact, hydrogen at a tertiary carbon atom also participates in an exchange. This followed from the fact that in a reaction with normal sulfuric acid, the latter was enriched in deuterium, while the deuterium content of the hydrocarbon decreased correspondingly.

D. N. Kursanov, V. N. Setkina, and V. F. Lavrushin [239] associate the coloration in the sulfuric acid layer in experiments with isoparaffins with the formation of carbonium ions as a result of oxidation of the methyne hydrogen by sulfuric acid. At the same time, as was established by Gordon and Burwell [166], the rate of hydrogen exchange in hydrocarbons with one side methyl group decreases as the chain of carbon atoms increases in length, especially for substances containing from 8 to 12 carbon atoms. On the other hand, the intensity of the color of the acid layer thereupon increases: color is absent in the case of 2-methylpentane, weak in experiments with 3-methylhexane, considerable in a reaction with 3-methylheptane, and very strong in a system containing 3-methylundecane. The formation of sulfur dioxide is also less clearly expressed with hydrocarbons of low molecular weight. Consequently, there is no direct relation between the color, the degree of oxidation (evolution of sulfur dioxide), and the rate of hydrogen exchange. In this connection, it can also be pointed out that in hydrogen exchange between saturated hydrocarbons and liquid deuterium bromide and fluoride, where oxidation processes are absent, a color is not observed [194, 202].

According to the work of D. N. Kursanov and his co-workers [173, 177], the hydrogen in isoparaffins does not exchange if deuteroacetic or deuterophosphoric acid is used instead of deuterosulfuric acid. The authors explain this observation by the fact that the acids named do not have an oxidizing action. It is more probable that the lower strength of these acids plays a part [194] (cf., the acidity functions H_0, p. 53). As regards hydrogen exchange with sulfuric acid, it also depends on the degree of acidity of the latter. Beeck and his co-authors [170] showed with isobutane that there is a linear relation between the exchange rate and the acidity function of sulfuric acid. In their opinion, the factor on which the exchange reaction depends may be the formation of a complex between the very weak base, namely, isobutane, and the strong acid. Ingold [161, 162] considers that hydrogen exchange between deuterosulfuric acid and a hydrocarbon (regardless of whether it is aromatic or saturated) obeys the same rules and, in particular, the parts of the molecule with a high electron density undergo attack by an acid most readily. According to Ingold, the only way in which sulfuric acid may participate in isotopic exchange is by the donation of a proton (or deuteron).

As in the exchange with deuterosulfuric acid, only hydrocarbons with a tertiary carbon atom undergo isotopic exchange with liquid deuterium fluoride without a catalyst (Table 69). At the same time, oxidation of the hydrogen of the methyne group is excluded. Consequently, the scheme examined above is inapplicable in this case. The same is true of hydrogen exchange between alicyclic hydrocarbons and a solution of aluminum bromide in liquid hydrogen bromide. It has been observed [216] that hydrogen exchange in cyclohexane and cyclopentane is accelerated by the addition of cyclohexyl bromide and cyclopentyl bromide, respectively (0.1 mole per mole of hydrocarbon) (Table 70).

TABLE 70. Effect of Additives on the Rate of Hydrogen Exchange on Alicyclic Hydrocarbons

Hydrocarbon	Additive	$\dfrac{k \cdot 10^6}{C \cdot 10^2}$
Cyclohexane	None	15 ± 7
"	$C_6H_{11}Br$	49 ± 9
Cyclopentane	None	59 ± 17
"	C_5H_9Br	410 ± 80
Methylcyclohexane	None	20 ± 13
"	$C_6H_{11}Br$	90 ± 16

This fact sheds light on the mechanism of hydrogen exchange in saturated hydrocarbons when examined in conjunction with results obtained on the isomerization of naphthalenes (cyclohexane ⇌ methylcyclopentane) [216]. It is generally considered in the literature that the isomerization proceeds through the formation of carbonium ions [245]. A careful study of the isomerization, in particular, in the work of Ipatieff and Pines, has shown that an aluminum halide alone is insufficient for isomerization to occur. It is not observed either in a system containing a hydrogen halide together with an aluminum halide. A promoter, in particular, an olefin, must be added. This reacts with the hydrogen halide to form an alkyl halide, which, in its turn, reacts with the aluminum halide to form carbonium ions (p. 132), which initiate the chain process. The same result is produced by the addition of an alkyl halide to a mixture of a hydrocarbon and an aluminum halide.

However, as was found in [216], when dissolved in liquid hydrogen bromide, methylcyclopentane is converted into cyclohexane at quite a high rate with catalysis by $AlBr_3$ or $GaBr_3$ without a promoter, while the addition of cyclohexyl bromide increases the isomerization rate (Table 70). In order to explain this difference from system in which HBr is not the solvent, but only an additive, it might be considered that a solution of $AlBr_3$ in liquid HBr, which is a very strong complex hydrogen acid (p. 42), is capable of reacting with a hydrocarbon to produce a hydride transfer (p. 134) according to the equation:

$$\gtrdot C - H + HA \rightarrow \gtrdot C^+ A + H_2.$$

In the opinion of Nenitzescu [37], the chain isomerization of methylcyclopentane is initiated in precisely this way when the catalyst is the hydrate of an aluminum halide. Nenitzescu detected the evolution of gaseous hydrogen during this reaction.

Since both isomerization and deuterium exchange with saturated hydrocarbons dissolved in liquid HBr proceed quite rapidly without a promoter when catalyzed by $AlBr_3$
or $GaBr_3$ and are accelerated by the addition of cyclohexyl bromide, the idea arises that
deuterium exchange with sufficiently strong acids is also caused by the formation of
carbonium ions by a hydride transfer scheme, whereupon free carbonium ions can hardly be formed in liquid HBr and it is more probable that a preionization state is realized.

The considerations presented, whose hypothetical character must be emphasized,
make it possible to combine the views presented above on the carbonium ion mechanism of hydrogen exchange between saturated hydrocarbons and acids and the ideas developed in the present monograph on the acid−base nature of heterolytic hydrogen exchange.

The carbonium ion scheme for hydrogen exchange in saturated hydrocarbons and
alcohols with sulfuric acid has been confirmed by experiments with tertiary alcohols
labeled with C^{14} and with hydrocarbons containing a tertiary carbon atom [245]. In the
presence of sulfuric acid (a ratio of $1:1:3$ between the components of the reaction
mixture), there is exchange even at room temperature. The tertiary alcohol is converted into a carbonium ion in the presence of sulfuric acid:

$$R_3COH + H_2SO_4 \rightarrow R_3C^+ + H_2O + HSO_4^-,$$

i.e., there is exchange between a carbonium ion and a hydrocarbon.

SUMMARY

1. Hydrocarbons can participate in equilibrium reactions with acids and under favorable conditions, a proton is added to the hydrocarbon molecule and a carbonium salt
is formed. When there is a difference in the isotopic compositions of the acid and the
hydrocarbon-base, the reaction between them is accompanied by hydrogen exchange.
Measurement of its rate makes it possible to estimate quantitatively the strengths of
hydrocarbons as bases in accordance with other methods described and also to draw conclusions on the electron-donor properties of individual parts of the molecule. Acid hydrogen exchange in combination with basic exchange reveal different aspects of the reactivity of hydrocarbons and the intereffect of atoms in their molecules.

2. Hydrocarbons, as electron-donors, participate in equilibrium reactions with
electrophilic acidlike substances. Data on the equilibrium constants of these reactions
and on the energies of formation and stability constants of molecular compounds are
in general agreement with the relative strengths of hydrocarbons as bases established
by isotopic and other methods.

3. There is a linear relation between the rate of irreversible electrophilic replacement of hydrogen in aromatic hydrocarbons and values characterizing their strengths as
bases (in particular, hydrogen exchange rates). This is explained by the fact that the
irreversible reaction proceeds through the equilibrium formation of a molecular compound of the hydrocarbon with the electrophilic reagents and the concentration of the
complex determines the over-all reaction rate.

4. Physicochemical investigations of the ternary systems ArH−HHal−MeHal substantiate the proposed interpretation of the catalytic activity of acidlike substances in
acid hydrogen exchange through their conversion into stronger hydrogen acids.

5. Hydrogen exchange between saturated hydrocarbons and sulfuric acid is normally interpreted on the basis of a carbonium ion chain mechanism in view of the oxidizing properties of sulfuric acid. However, this conception does not explain cases of exchange where oxidation of the hydrogen of the methyne group is absent.

LITERATURE CITED

1. J. E. Leffler, The Reactive Intermediates of Organic Chemistry, Ch. V–VII (New York, London, 1956).
2. H. Burton and P. F. G. Praill, Quart. Rev. 6, 302 (1952).
3. D. Bethell and V. Gold, Quart. Rev. 12, 173 (1958).
4. E. A. Alexander, Principles of Ionic Organic Reactions (New York, 1950).
5. J. Hine, Physical Organic Chemistry (New York, 1956).
6. L. P. Hammett, Physical Organic Chemistry (New York, London, 1940).
7. V. L. Tal'roze and A. K. Lyubimov, Doklady Akad. Nauk SSSR 86, 909 (1952).
8. N. E. Alekseevskii, V. L. Tal'roze, and V. L. Shelyapin, Doklady Akad. Nauk SSSR 93, 997 (1953).
9. P. Walden, Ber. 35, 2018 (1902).
10. M. Gomberg, Ber. 35, 2403 (1902).
11. W. Schlenk and A. Herzenstein, Ann. 372, 11 (1910).
12. K. Ziegler and H. Wollschitt, Ann. 479, 90 (1930).
13. K. Ziegler and W. Mathes, Ann. 479, 111 (1930).
14. N. N. Lichtin and H. Glaser, J. Am. Chem. Soc. 73, 5530, 5537 (1951).
15. A. Hantzsch, Z. Phys. Chem. 61, 257 (1908); Ber. 55, 953 (1922).
16. L. P. Hammett and A. J. Deyrup, J. Am. Chem. Soc. 55, 1900 (1933).
17. M. S. Newman and N. S. Deno, J. Am. Chem. Soc. 73, 364 (1954).
18. W. R. Ondorff, R. C. Gibbs, S. A. McNulty, and C. V. Shapiro, J. Am. Chem. Soc. 49, 1551 (1927).
19. L. C. Anderson, J. Am. Chem. Soc. 52, 4567 (1930).
20. F. Fairbrother and B. Wright, J. Chem. Soc. (1951) p. 1102.
21. V. Gold and B. W. V. Hawes, J. Chem. Soc. (1951) p. 2102.
22. V. Gold, B. W. V. Hawes, and F. L. Tye, J. Chem. Soc. (1952) p. 2167.
23. H. Hart and R. W. Fish, J. Am. Chem. Soc. 80, 5894 (1958).
24. V. A. Plotnikov, Zhur. Russ. Fiz. Khim. Obshchest. 34, 466 (1902).
25. E. Wertyporoch and T. Firla, Z. Phys. Chem. 162, 398 (1932).
26. A. Wohl and E. Wertyporoch, Ber. 64, 1357 (1931).
27. F. Fairbrother, J. Chem. Soc. (1945) p. 503.
28. N. E. Brezhneva, S. Z. Roginskii, and A. I. Shilinskii, Zhur. Fiz. Khim. 8, 849 (1936).
29. N. E. Brezhneva and S. Z. Roginskii, Uspekhi Khimii 7, 1583 (1938).
30. S. Z. Roginskii, Theoretical Principles of Isotopic Methods of Studying Chemical Reactions [in Russian] (Izd. AN SSSR, Moscow, 1956) p. 273.
31. F. Fairbrother, J. Chem. Soc. (1937) p. 503; (1941) p. 293.
32. G. Olah, S. Kuhn, and J. Olah, J. Chem. Soc. (1957) p. 2174.
33. D. N. Kursanov and M. E. Vol'pin, Doklady Akad. Nauk SSSR 113, 339 (1957); M. E. Vol'pin, Doctor's Dissertation (INÉOS, Moscow, 1959).
33a. Z. N. Parnes, M. E. Vol'pin, and D. N. Kursanov, Izvest. Akad. Nauk SSSR, Otdel. Khim. Nauk (1960) p. 763.
34. F. C. Whitmore, Chem. and Eng. News 26, 668 (1948).

35. M. E. Vol'pin, I. S. Akhrem, and D. N. Kursanov, Izvest. Akad. Nauk SSSR, Otdel. Khim. Nauk (1957) pp. 760, 1501; Zhur. Obshchei Khim. 28, 330 (1958).

36. D. N. Kursanov, M. E. Vol'pin, and Z.N.Parnes, Khim.Nauk. i Prom.2,159(1958).

37. C. D. Nenitzescu, Uspekhi Khimii 26, 398 (1957).

38. F. C. Whitmore, J. Am. Chem. Soc. 54, 3274 (1932); Ind. and Eng. Chem. 26, 94 (1934).

38a. J. A. Grace and M. C. R. Symons, J. Chem. Soc. (1959) p. 958.

39. W. von E. Doering, H. G. Boyton, H. W. Earhart, E. F. Wadeley, and W. R. Edwards, Tetrahedron 4, 178 (1958).

40. D. N. Kursanov and Z. N. Parnes, Zhur. Obshchei Khim. 27, 668 (1957).

41. D. N. Kursanov, V. N. Setkina, S. V. Vitt, and Z. N. Parnes, Collection: Problems in Kinetics and Catalysis, X. Isotopes in Catalysis [in Russian] (Izd. AN SSSR, Moscow, 1957) p. 242. Collection: Isotopes and Radiation in Chemistry [in Russian] (Izd. AN SSSR, Moscow, 1958) p. 13.

42. Z. N. Parnes, S. V. Vitt, and D. N. Kursanov, Zhur. Obshchei Khim. 28,410(1958).

43. D. N. Kursanov, E. V. Bykova, and V. N. Setkina, Izvest. Akad. Nauk SSSR, Otdel. Khim. Nauk (1958) p. 809; (1959) p. 2007.

44. V. N. Setkina, D. N. Kursanov, and E. V. Bykova, Izvest. Akad. Nauk SSSR, Otdel. Khim. Nauk (1959) p. 758.

45. A. I. Shatenshtein, Uspekhi Khimii 24, 377 (1955).

46. W. Klatt, Z. Anorg. Chem. 234, 189 (1937).

47. J. H. Simons, J. Am. Chem. Soc. 53, 83 (1931).

48. Ya. M. Varshavskii, M. G. Lozhkina, and A. I. Shatenshtein, Zhur. Fiz. Khim. 31, 1377 (1957).

49. D. A. McCaulay and A. P. Lien, J. Am. Chem. Soc. 73, 2013 (1951); Tetrahedron 5, 186 (1959).

50. E. L. Mackor and A. Hofstra, Rec. trav. chim., Pays-Bas. 75, 871 (1956).

51. A. F.Clifford,H.C.Beachell, and W.M.Jack, J.Inorg. and Nucl.Chem. 5, 57(1957).

51a. G. A. Olah, H. W. Quinn, and S. J. Kuhn, J. Am. Chem. Soc. 82, 426 (1960).

52. W. Klatt, Z. Anorg. Chem. 222, 225 (1935).

53. M. Kilpatrick and F. E. Luborsky, J. Am. Chem. Soc. 75, 577 (1953).

54. M. Kilpatrick and F. E. Luborsky, J. Am. Chem. Soc. 76, 5863 (1954).

55. G. Olah, S. Kuhn, and A. Pavlath, Nature 178, 693 (1956).

56. G. A. Olah and S. J. Kuhn, J. Am. Chem. Soc. 80, 6535 (1958).

57. G. A. Olah, A. E. Pavlath, and J. A. Olah, J. Am. Chem. Soc. 80, 6540 (1958).

57a. C. L. Bell and G. M. Barrow, J. Chem. Phys. 31, 3000 (1959).

57b. G. M. Barrow, Spectrochim. Acta 16, 799 (1960).

58. A. P. Lien and D. A. McCaulay, J. Am. Chem. Soc. 75, 2407 (1953).

59. M. Kilpatrick and F. E. Luborsky, J. Am. Chem. Soc. 76, 5865 (1954).

60. M. Kilpatrick and T. J. Lewis, J. Am. Chem. Soc. 78, 5186 (1956).

61. M. A. Reid, J. Am. Chem. Soc. 76, 3264 (1954).

62. V.Gold, B.W.Haves, and F.L.Tye, J.Chem.Soc. (1952) pp. 2167, 2172, 2181, 2184.

63. N. Muller, L. M. Pickett, and R. S. Mulliken, J. Am. Chem. Soc. 76, 4770 (1954).

63a. J. L. Down, J. Lewis, B. Moore, and G. Wildinson, J. Chem. Soc. 3767 (1959).

63b. W. Gecrard and E. D. Maclen, Chem. Rev. 59, 1105 (1959).

64. T. Morita, J. Chem. Phys. 25, 1290 (1956).

65. E. L. Mackor, G. Dallinga, J. H. Kruizinga, and A. Hofstra, Rec. trav. chim. Pays.-Bas. 75, 836 (1956).

66. A. A. Verrijn, A. A. Stuart, and E. L. Mackor, J. Chem. Phys. 27, 826 (1957).

67. G. Dallinga, E. L. Mackor, and A. A. Stuart, Molecul. Phys. 1, 123 (1958).

68. C. McLean and J. H. van der Waals, Chem. Phys. 22, 828 (1957).

69. J. Matsunaga, Bull. Chem. Soc. Japan 31, 774 (1958).

70. M. Kilpatrick and H. H. Hyman, J. Am. Chem. Soc. 80, 77 (1958).

71. E. L. Mackor, P. J. Smit, and J. H. van der Waals, Trans. Farad. Soc. 53, 1309 (1957).

72. D. A. McCaulay, B. H. Shoemaker, and A. P. Lien, Ind. and Eng. Chem. 42, 2103 (1950).

73. D. A. McCaulay, W. S. Higley, and A. P. Lien, J. Am. Chem. Soc. 78, 3009 (1956).

74. E. L. Mackor, A. Hofstra, and J. H. van der Waals, Trans. Farad. Soc. 54, 66 (1958).

75. E. L. Mackor, A. Hofstra, and J. H. van der Waals, Trans. Farad. Soc. 54, 186 (1958).

76. E. L. Mackor and J. H. van der Waals, Abstracts of papers of the XIII Internat. Congr. of Pure and Appl. Chem. (Stockholm, Uppsala, 1953) p. 41.

76a. J. A. Grace and M. C. R. Symons, J. Chem. Soc. (1959) p. 958.

76b. A. G. Evans, P. M. S. Jones, and J. H. Thomas, J. Chem. Soc. (1957) p. 104.

77. U. Pestemer, A. J. K. Schmidt, L. Schmidt-Willigut, and F. Manchen, Monatsh. 71, 432 (1937).

78. G. L. Hoijtink and W. P. Weijland, Rec. trav. chim. Pays.-Bas 76, 836 (1957).

79. P. A. Plattner, E. Heilbronner, and S. Weber, Helv. Chim. Acta 35, 1036, 1049 (1952).

80. L. H. Chopard-dit-Jean and E. Heilbronner, Helv. Chim. Acta 35, 2170 (1952).

81. H. M. Frey, J. Chem. Phys. 25, 600 (1956).

82. W. Treibs, W. Kirchhof, and W. Ziegenbein, Fortschr. Chem. Forsch. 3, 334 (1955).

83. P. Plattner, E. Heilbronner, and S. Weber, Helv. Chim. Acta. 32, 574 (1949).

84. V. Gold and F. A. Long, J. Am. Chem. Soc. 75, 4543 (1953).

85. J. Berenblum, Nature 156, 601 (1945); Chem. Abstr. 40, 1005 (1946).

86. R. J. Gillespie and J. A. Leistein, The behavior of organic compounds in sulfuric acid, Quart. Rev. 8, 40-60 (1954).

87. O. Maass and J. Russell, J. Am. Chem. Soc. 40, 1561 (1918).

88. O. Maass and J. Russell, J. Am. Chem. Soc. 43, 1227 (1921).

89. O. Maass, E. H. Boomer, and D. M. Morrison, J. Am. Chem. Soc. 45, 1433 (1923).

90. O. Maass and C. H. Wright, J. Am. Chem. Soc. 46, 2665 (1924).

91. D. Cook, Y. Lupien, and W. G. Schneider, Canad. J. Chem. 34, 957, 964 (1956).

92. D. Cook, J. Chem. Phys. 25, 788 (1956).

93. J. W. Reeves and W. G. Schneider, Canad. J. Chem. 35, 251 (1957).

94. E. Terres and M. T. Assemi, Brennst. Chem. 37, 257 (1956).

95. A. I. Shatenshtein, V. R. Kalinachenko, and Ya. M. Varshavskii, Zhur. Fiz. Khim. 30, 2098 (1956).

96. H. C. Brown and J. D. Brady, J. Am. Chem. Soc. 74, 3570 (1952).

97. H. C. Brown and W. J. Wallace, J. Am. Chem. Soc. 75, 6265, 6268 (1953).

98. H. C. Brown and H. W. Pearsall, J. Am. Chem. Soc. 74, 191 (1952).

99. R. L. Richardson and S. W. Benson, J. Am. Chem. Soc. 73, 4681 (1951).

100. H. C. Brown and H. W. Pearsall, J. Am. Chem. Soc. 73, 4681 (1951).

101. H. C. Brown, H. W. Pearsall, L. P. Eddy, W. J. Wallace, M. Grayson, and K. Le Roi Nelson, Ind. and Eng. Chem. 48, 1462 (1953).

102. C. A. Thomas, Anhydrous Aluminum Chloride in Organic Chemistry [Russian translation] (IL, Moscow, 1949).

103. P. H. Plesh (Ed.), Cationic Polymerization (Cambridge, 1953).

104. L. J. Andrews, Chem. Rev. 54, 713 (1954).

104a. G. Baddeley, G. Holt, and D. Voss, J. Chem. Soc. (1953) p. 100.

105. C. M. Fontana and R. J. Herold, J. Am. Chem. Soc. 70, 4681 (1948).

106. V. A. Plotnikov, Investigations of the Electrochemistry of Nonaqueous Solutions [in Russian] (Kiev, 1908).

107. V. A. Plotnikov, Zap. Inst. Khim. Akad. Nauk Ukr.SSR 5, 271 (1938).

108. E. Wertyporoch and B. Adamus, Z. Phys. Chem. 168, 31 (1933).

109. D. D. Eley and P. J. King, J. Chem. Soc. (1952) p. 2517.

110. J. F. Norris and J. E. Wood, J. Am. Chem. Soc. 62, 1428 (1940).

111. J. F. Norris and J. N. Ingram, J. Am. Chem. Soc. 62, 1298 (1940).

112. D. D. Eley and P. J. King, J. Chem. Soc. (1952) p. 4973.

113. H. Luther and G. Pockels, Z. Elektrochem. 59, 159 (1956).

114. A. Wassermann, J. Chem. Soc. (1955) p. 585.

115. A. Wassermann, J. Chem. Soc. (1954) p. 4329; (1959) pp. 979, 983, 986.

115a. A. Wassermann, Molecul. Phys. 2, 226 (1959).

116. A. N. Terenin, Uspekhi Khimii 24, 121 (1955).

117. A. I. Shatenshtein, Uspekhi Khimii 24, 377 (1955).

118. A. I. Shatenshtein, Zhur. Obshchei Khim. 9, 1603 (1939); Uspekhi Khimii 7, 1096 (1939).

119. A. I. Shatenshtein, Zhur. Fiz. Khim. 10, 766 (1937).

120. A. I. Shatenshtein, Zhur. Fiz. Khim. 13, 366 (1939).

121. A. I. Shatenshtein and E. A. Izrailevich, Zhur. Fiz. Khim. 13, 1791 (1939).

122. J. A. Moede and C. Curran, J. Am. Chem. Soc. 71, 852 (1949).

123. F. Carli, Gazz. chim. ital. 57, 347 (1927).

124. F. Carli, C., 1, 2904 (1926); C., 1, 1433, 1582 (1927).

125. N. N. Lichtin, R. E. Weston, and J. O. White, J. Am. Chem. Soc. 74, 4715 (1952).

126. G. H. Locket, J. Chem. Soc. (1932) p. 1501.

127. L. J. Andrews and R. M. Keefer, J. Am. Chem. Soc. 73, 4169 (1951).

128. J. Kleinberg and A. W. Davidson, Chem. Rev. 42, 601 (1948).

129. H. A. Benesi and J. H. Hildebrand, J. Am. Chem. Soc. 71, 2703 (1949).

130. T. M. Gromwell and R. L. Scott, J. Am. Chem. Soc. 72, 3825 (1950).

131. L. J. Andrews and R. M. Keefer, J. Am. Chem. Soc. 73, 462 (1951).

132. R. M. Keefer and L. J. Andrews, J. Am. Chem. Soc. 72, 4677 (1950).

133. L. J. Andrews and R. M. Keefer, J. Am. Chem. Soc. 74, 4500 (1952).

134. R. M. Keefer and L. J. Andrews, J. Am. Chem. Soc. 77, 2164 (1955).

135. R. M. Keefer and L. J. Andrews, J. Am. Chem. Soc. 72, 5170 (1952).

136. M. Tamres, D. R. Virzu, and S. Searles, J. Am. Chem. Soc. 75, 4358 (1953).

137. N. W. Blake, H. Winston, and J. A. Patterson, J. Am. Chem. Soc. 73, 4437 (1951).

138. R. L. Scott, J. Am. Chem. Soc. 75, 1551 (1953).

139. H. McConnell, J. S. Ham, and J. R. Platt, J. Chem. Phys. 21, 66 (1953).

140. G. Briegleb and C. Zekalla, Z. Elektrochem. 59, 184 (1955).

141. A. Bier, Rec. trav. chim. Pays.-Bas. 75, 866 (1956).

142. R. Foster, D. L. Hammick, and B. N. Pearson, J. Chem. Soc. (1956) p. 555.

143. G. J. Korineck and W. G. Schneider, Canad. J. Chem. 35, 1157 (1957).

144. F. Fairbrother, J. Chem. Soc. (1948) p. 1051.

145. V. M. Kazakova, Intermolecular Interaction and Dielectric Polarization of Ternary Systems of Bromine and Iodine with Various Organic Compounds in Benzene,

Dissertation [in Russian] (M. V. Lomonosov Moscow Chemicotechnological In-stitute, Moscow, 1954).

146. G. Kortüm and H. Wilski, Z. phys. Chem. 202, 35 (1953).

147. J. Kleinberg, E. Colton, and J. Sattizahn, J. Am. Chem. Soc. 75, 442 (1953).

148. F. E. Condon, J. Am. Chem. Soc. 74, 2528 (1952).

149. F. E. Condon, J. Am. Chem. Soc. 70, 1963 (1948).

150. R. M. Keefer and L. J. Andrews, J. Am. Chem. Soc. 72, 5170 (1952).

151. H. C. Brown and L. M. Stock, J. Am. Chem. Soc. 79, 5175 (1957).

152. P. B. D. de la Mare and C. A. Vernon, J. Chem. Soc. (1951) p. 1764.

153. L. J. Andrews and R. M. Keefer, J. Am. Chem. Soc. 78, 4549 (1956).

153a. L. J. Andrews and R. M. Keefer, J. Am. Chem. Soc. 81, 1063 (1959).

154. H. C. Brown and L. M. Stock, J. Am. Chem. Soc. 79, 1421 (1957).

155. R. M. Keefer and L. J. Andrews, J. Am. Chem. Soc. 78, 5623 (1956).

156. H. C. Brown and C. W. McGary, J. Am. Chem. Soc. 77, 2310 (1955).

157. H. C. Brown, B. A. Bolto, and F. R. Jensen, J. Org. Chem. 23, 417 (1958).

158. C. K. Ingold, C. G. Raisin, and C. L. Wilson, Nature 134, 734 (1934).

159. C. K. Ingold, C. G. Raisin, C. R. Bailey, and B. Topley, J. Chem. Soc. (1946) p. 915.

160. C. K. Ingold, C. G. Raisin, and C. L. Wilson, J. Chem. Soc. (1936) p. 1637.

161. C. K. Ingold, C. G. Raisin, and C. L. Wilson, J. Chem. Soc. (1936) p. 1643.

162. C. K. Ingold and C. L. Wilson, Z. Elektrochem. 44, 62 (1938).

163. S. K. Hsü, C. K. Ingold, and C. G. Raisin, J. Chim. Phys. 45, 738 (1948).

164. C. K. Ingold, Structure and Mechanism in Organic Chemistry (New York, 1953).

165. R. L. Burwell and G. S. Gordon, J. Am. Chem. Soc. 70, 3128 (1948).

166. G. S. Gordon and R. L. Burwell, J. Am. Chem. Soc. 71, 2355 (1949).

167. R. L. Burwell, R. B. Scott, L. G. Maury, and A. S. Hussey, J. Am. Chem. Soc. 76, 5822 (1954).

168. R. L. Burwell, L. G. Maury, and R. B. Scott, J. Am. Chem. Soc. 76, 5828 (1954).

169. R. L. Burwell and A. D. Shields, J. Am. Chem. Soc. 77, 2766 (1955).

170. O. Beeck, J. W. Otvos, D. P. Stevenson, and C. D. Wagner, J. Chem. Phys. 17, 418 (1949).

171. J. W. Otvos, D. P. Stevenson, C. D. Wagner, and O. Beeck, J. Am. Chem. Soc. 73, 5741 (1951).

172. D. P. Stevenson, C. D. Wagner, O. Beeck, and J. W. Otvos, J. Am. Chem. Soc. 74, 3269 (1952).

173. V. N. Setkina, D. N. Kursanov, O. D. Sterligov, and A. L. Liberman, Doklady Akad. Nauk SSSR 85, 1045 (1952).

174. D. N. Kursanov, V. N. Setkina, and O. D. Sterligov, Izvest. Akad. Nauk SSSR, Otdel. Khim. Nauk (1953) p. 1035.

175. V. N. Setkina, A. F. Plate, O. D. Sterligov, and D. N. Kursanov, Doklady Akad. Nauk SSSR 99, 1007 (1954).

176. D. N. Kursanov, V. N. Setkina, and A. Meshcheryakov, Doklady Akad. Nauk SSSR 105, 279 (1955).

177. D. N. Kursanov and V. V. Voevodskii, Uspekhi Khimii 23, 642 (1954).

178. V. N. Setkina and E. V. Bykova, Doklady Akad. Nauk SSSR 92, 341 (1953).

179. D. N. Kursanov and V. N. Setkina, Doklady Akad. Nauk SSSR 94, 69 (1954).

180. E. V. Bykova and V. N. Setkina, Doklady Akad. Nauk SSSR 103, 835 (1955).

181. V. N. Setkina, E. V. Bykova, and D. N. Kursanov, Doklady Akad. Nauk SSSR 104, 869 (1955).

182. V. Gold and D. P. N. Satchell, J. Chem. Soc. (1955) p. 3609.

183. V. Gold and D. P. N. Satchell, J. Chem. Soc. (1955) p. 3619.

184. V. Gold and D. P. N. Satchell, J. Chem. Soc. (1955) p. 3622.

185. V. Gold and D. P. N. Satchell, J. Chem. Soc. (1955) p. 2743.

186. D. P. N. Satchell, J. Chem. Soc. (1956) p. 3911.

187. S. Olsson, and L. Melander, Acta Chem. Scand. $\underline{8}$, 523 (1954).

188. L. Melander and S. Olsson, Acta Chem. Scand. $\underline{10}$, 879 (1956); Ark. kemi $\underline{14}$, 85 (1959).

188a. C. Eaborn and R. Taylor, J. Chem. Soc. (1960) p. 3301.

189. G. van Dyke Tiers, J. Am. Chem. Soc. 78, 4165 (1956).

190. A. I. Shatenshtein and Ya. M. Varshavskii, Doklady Akad. Nauk SSSR $\underline{85}$, 157 (1952).

191. A. I. Shatenshtein, Uspekhi Khimii $\underline{21}$, 914 (1952).

192. V. R. Kalinachenko, Ya. M. Varshavskii, and A. I. Shatenshtein, Doklady Akad. Nauk SSSR $\underline{91}$, 577 (1953).

193. Ya. M. Varshavskii and A. I. Shatenshtein, Doklady Akad. Nauk SSSR $\underline{95}$, 297 (1954).

194. A. I. Shatenshtein, K. I. Zhdanova, L. N. Vinogradov, and V. R. Kalinachenko, Doklady Akad. Nauk SSSR $\underline{102}$, 779 (1955).

195. Ya. M. Varshavskii, S. E. Vaisberg, and M. G. Lozhkina, Zhur. Fiz. Khim. $\underline{29}$, 750 (1955).

196. A. I. Shatenshtein, Collection: Problems in Kinetics, Catalysis, and Reactivity [in Russian] (Izd. AN SSSR, Moscow, 1955) p. 699.

197. V. R. Kalinachenko, Ya. M. Varshavskii, and A. I. Shatenshtein, Zhur. Fiz. Khim. $\underline{30}$, 1140 (1956).

198. Ya. M. Varshavskii, V. R. Kalinachenko, S. E. Vaisberg, and A. I. Shatenshtein, Zhur. Fiz. Khim. $\underline{30}$, 1647 (1956).

199. A. I. Shatenshtein, V. R. Kalinachenko, and Ya. M. Varshavskii, Zhur. Fiz. Khim. 30, 2093 (1956).

200. A. I. Korolev, A. I. Shatenshtein, S. N. Yurgina, V. R. Kalinachenko, and P. P. Alikhanov, Zhur. Obshchei Khim. $\underline{26}$, 1661 (1956).

201. Ya. M. Varshavskii and M. G. Lozhkina, Zhur. Fiz. Khim. $\underline{31}$, 911 (1957).

202. Ya. M. Varshavskii, M. G. Lozhkina, and A. I. Shatenshtein, Zhur. Fiz. Khim. $\underline{31}$, 1377 (1957).

203. A. I. Shatenshtein, E. N. Zvyagintseva, E. A. Yakovleva, E. A. Izrailevich, Ya. I. Varshavskii, M. G. Lozhkina, and A. V. Vedeneev, Collection: Isotopes in Catalysis [in Russian] (Izd. AN SSSR, Moscow, 1957) p. 218.

204. A. I. Shatenshtein, and E. A. Izrailevich, Collection: Isotopes in Catalysis [in Russian] (Izd. AN SSSR, Moscow, 1957) p. 430.

205. A. I. Shatenshtein, G. V. Peregudov, E. A. Izrailevich, and V. R. Kalinachenko, Zhur. Fiz. Khim. $\underline{32}$, 146 (1958).

206. A. I. Shatenshtein and A. V. Vedeneev, Collection: Isotopes and Radiation in Chemistry [in Russian] (Izd. AN SSSR, Moscow, 1958) p. 7.

207. A. I. Shatenshtein, A. V. Vedeneev, and P. P. Alikhanov, Zhur. Obshchei Khim. $\underline{28}$, 2638 (1958).

208. A. I. Shatenshtein, Collection: Problems in Physical Chemistry [in Russian] (Goskhimizdat, Moscow, 1958) p. 202.

209. Ya. M. Varshavskii, S. E. Vaisberg, and B. A. Trubitsyn, Doklady Akad. Nauk SSSR $\underline{122}$, 831 (1958).

210. A. I. Shatenshtein, V. R. Kalinachenko, E. N. Yurygina, and V. M. Basmanova, Zhur. Obshchei Khim. 29, 849 (1959).

210a. J. H. Baker, Hyperconjugation (Oxford, 1952) p. 46.

211. E. N. Yurygina, P. P. Alikhanov, E. A. Izrailevich, P. N. Manochkina, and A. I. Shatenshtein, Zhur. Fiz. Khim. 34, No. 3 (1960).

212. A. I. Shatenshtein, Zhur. Fiz. Khim. 34, 594 (1960).

213. A. I. Shatenshtein and P. P. Alikhanov, Zhur. Obshchei Khim. 30, 992 (1960).

214. P. P. Alikhanov and Ya. M. Varshavskii, Doklady Akad. Nauk SSSR 128, 1214 (1959).

215. K. I. Zhdanov, V. M. Basmanova, and A. I. Shatenshtein, Zhur. Obshchei Khim. (in press).

216. A. I. Shatenshtein, K. I. Zhdanova, and V. M. Basmanova, Doklady Akad. Nauk SSSR 133, 1117 (1960); Zhur. Obshchei Khim. 31, 2134 (1961).

217. G. Olah, S. Kuhn, and A. Pavlath, Nature 178, 693 (1956).

218. E. L. Mackor and J. van der Waals, Abstr. 13th Congr. Pure and Appl. Chem. (Stockholm, Uppsala, 1933) p. 150.

219. G. Dallinga, A. A. Verrijn, A. A. Stuart, P. J. Smith, and E. L. Mackor, Z. Elektrochem. 61, 1019 (1957).

220. W. G. Brown, K. E. Wilzbach, and W. H. Urry, Canad. J. Res. 276, 398 (1949).

221. W. M. Lauer and J. T. Day, J. Am. Chem. Soc. 77, 1904 (1955).

222. G. P. Miklukhin and A. F. Rekasheva, Doklady Akad. Nauk SSSR 101, 881 (1955).

223. A. I. Shatenshtein and E. N. Zvyagintseva, Zhur. Obshchei Khim. 29, 1751 (1959); 31, 1432 (1961).

224. D. P. N. Satchell, J. Chem. Soc. (1957) p. 2878.

225. D. Bethell, V. Gold, and D. P. N. Satchell, J. Chem. Soc. (1958) p. 1918.

226. D. P. N. Satchell, J. Chem. Soc. (1958) p. 3910.

227. J. Kenner, M. Polanyi, and P. Szego, Nature 135, 267 (1935).

228. A. Klit and A. Langseth, Nature 135, 956 (1935).

229. A. Klit and A. Langseth, Z. Phys. Chem. 176, 65 (1936).

230. A. Langseth and A. Klit, Kgl. danske vid. selskab. Mat.-fys. medd. 15, No. 1322 (1937); Chem. Abstr. 32, 2315 (1937).

231. M. J. S. Dewar, T. Mole, and E. W. T. Warford, J. Chem. Soc. (1956) p. 3581.

231a. C. Eaborn and R. Taylor, J. Chem. Soc. (1961) p. 1012.

231b. H. M. Frey, J. Chem. Phys. 25, 600 (1956).

231c. W. Treibs, A. Schmidt, A. Rudolph, and H. J. Schneider, Z. Phys. Chem. 214, 358 (1960).

232. W. M. Lauer, G. W. Matson, and G. Stedman, J. Am. Chem. Soc. 80, 6433 (1958).

233. W. M. Lauer, G. W. Matson, and G. Stedman, J. Am. Chem. Soc. 80, 6437 (1958).

234. W. M. Lauer, and G. Stedman, J. Am. Chem. Soc. 80, 6439 (1958).

235. J. Horiuchi and M. Polanyi, Nature 134, 897 (1934).

236. J. Turkevich and R. K. Smith, J. Chem. Phys. 16, 446 (1948).

237. A. Farkas and L. Farkas, Ind. and Eng. Chem. 34, 716 (1942).

238. T. D. Stewart and D. Harman, J. Am. Chem. Soc. 68, 1135 (1946).

238a. A. E. Shilov, R. D. Sabirova, and V. A. Gorshkov, Doklady Akad. Nauk SSSR 119, 533 (1958).

239. V. F. Lavrushin, D. N. Kursanov, and V. N. Setkina, Doklady Akad. Nauk SSSR 97, 265 (1954).

240. V. N. Setkina, D. N. Kursanov, and A. L. Liberman, Izvest. Akad. Nauk SSSR, Otdel. Khim. Nauk (1954) p. 109.

241. T. M. Powell and E. B. Reid, J. Am. Chem. Soc. 67, 1020 (1945).

242. H. Pines and R. C. Weckher, J. Am. Chem. Soc. 68, 2018 (1946).

243. D. N. Kursanov, Ukr. Khim. Zhur. 22, 34 (1956).

244. W. I. Aalbersberg, G. J. Hoijtink, E. L. Mackor, and W. P. Weijland, J. Chem. Soc. (1959) pp. 3049, 3055.

245. G. G. Pastukhov and R. F. Galiulina, Zhur. Obshchei Khim. 31, 2159 (1961).

CONCLUSION

The contents of the two chapters of this section make it possible to compare the acidic and basic properties of hydrocarbons. It is most convenient to use data on hydrogen exchange for this purpose.

Saturated hydrocarbons are the weakest acids and bases. When hydrogen atoms attached to one carbon atom in them are replaced by alkyl groups, there is an increase in the stability of the carbonium ions and a decrease in the stability of the carbanions. Therefore, branching of the carbon chain has opposite effects on the strengths of saturated hydrocarbons as bases and as acids.

Cyclopropane hydrocarbons are stronger protolytes and occupy an intermediate position between saturated and unsaturated hydrocarbons.

With a change from sp^3- to sp^2- and to sp-hybridization of the carbon valence, there is a progressive increase in the polarity of the CH bond and protonization of the hydrogen is facilitated. Therefore, alkenes and especially alkynes are stronger than alkanes. The electrons of the multiple bond fulfill a donor function and as a result, unsaturated hydrocarbons react as bases.

When a molecule contains a multiple bond, there is partial overlapping of the cloud of π- and σ-electrons (σ, π-conjugation effect). The π-electron system of an aromatic ring may participate in this conjugation instead of the π-electrons of a multiple bond. The hydrocarbon molecule becomes polar and this promotes an increase in its protolytic activity.

The presence in a hydrocarbon molecule of a large number of conjugated bonds (for example, in the carotene molecule) or many condensed rings (in the zethrene molecule) is the reason why such substances have a considerable affinity for a proton and may be converted into carbonium ions even by reaction with a carboxylic acid. The effect of conjugation on the acidic properties of hydrocarbons is shown by the example of fluoradene.

The increase in protolytic activity with an increase in the polarity of a molecule is particularly noticeable when substituents of the electron-donor and electron-acceptor type are introduced into a hydrocarbon molecule. Either the combined action of conjugation and inductive displacement of electrons is observed or one of these effects predominates. At the same time, the symmetrical and nonpolar molecules of such hydrocarbons as benzene and naphthalene have clearly expressed acidic and basic properties. The comparatively ready exchange of hydrogen in them with acids and bases can probably be ascribed to the polarizability of the molecule.

It must be considered that each of the protolytic functions is normally fulfilled by different parts of the molecule. For example, propylene is a stronger acid and, at the same time, a stronger base than ethylene, but in the first case, because of the methyl group and in the second case, because of the participation of the π-electrons of the double bond in the reaction. Toluene is a stronger acid and a stronger base than benzene, but the methyl group participates primarily in the reaction of toluene with a base, while the aromatic ring reacts exclusively with an acid. As regards the acidity of the aromatic ring of toluene, it is less than that of unsubstituted benzene, while the methyl group, in its turn, is completely inactive in an exchange reaction with an acid.

Data on hydrogen exchange are often used in the literature, but attention is not paid to the medium in which the exchange occurs and the protolytic function the substance fulfills. Thereupon, conclusions are drawn on the "protonic lability" of the hydrogen of the substance, treating it as a property of only the molecule itself. This concept of the term considered is obviously indefinite as hydrogen exchange in the same molecule may occur in completely different ways as a result of a change in the protolytic function of the substance. In hydrogen exchange with a base, protonic lability is actually shown by the hydrogen in the hydrocarbon molecule, while in acid hydrogen exchange, protonic lability is not shown by the substrate, which is a proton acceptor, but the acid reagent participating in the exchange. The term "protonic lability" should only be used in hydrogen exchange when considering exchange reactions in which there appear the acidic properties of the substances or when their potential appearance is implied. This corresponds to the established chemical tradition of considering as "labile" that hydrogen which participates in condensations with aldehydes, hydrogen found by the Chugaev-Tserevitinov method, etc., i.e., "acid hydrogen."

The material presented in Section III of the book leaves no doubt that the concepts of "acid" and "base" are completely applicable to hydrocarbons. The objectivity of the conclusions on the effect of the structure of a hydrocarbon on its relative strength as an acid or a base is confirmed by the agreement of results obtained by different methods. For example, it is sufficient to compare the numerous data on the relative strengths of methylbenzenes as bases (pp. 139, 140, 144, 145, 147, 155, 157, 162, 163, 164, 166, 178, 225, 227, 231, 263), which clearly show that with the successive intro- duction of methyl groups into the benzene ring, the hydrocarbon becomes an increas- ingly stronger base, especially if these groups are meta relative to each other.

Most of the information on the acidic and basic properties of hydrocarbons has been obtained recently, most frequently in the last few years and the accumulation contin- ues at an ever increasing rate. It may be expected, therefore, that even in the near fu- ture our knowledge will become sufficient to make a detailed comparison of these pro- perties of hydrocarbons with their structure.

MECHANISM OF ACID–BASE INTERACTION

1. BRØNSTED'S THEORY OF ACIDS AND BASES

In its time, Brønsted's theory not only brought order to the chaotic mass of isolated facts on reactions of acids and bases known when it originated in 1923, but made it possible to interpret and, moreover, predict the reactions of acids and bases in non-aqueous solution, explain phenomena of acid–base catalysis, etc.

According to Brønsted's theory, reactions between acids and bases or protolytic re-actions consist of the transfer of a proton of the acid to the base and is represented by the following equation (p. 2):

$$acid_1 + base_2 \rightleftharpoons base_1 + acid_2.$$

A detailed discussion of Brønsted's theory of protolytic equilibrium is given in [1, ch. 14]. According to this theory, the relative strength of acids and bases is independent of the solvent. To compare the strengths of protolytes, for example, the strengths of a series of acids in any solvents, it is usual to compare the equilibrium constants of pro-tolysis K_{AB_0}, involving a standard protolyte, which is marked by the subscript "zero."

$$A + B_0 \rightleftharpoons B + A_0.$$

The symbol A denotes one of the acids whose strengths are to be compared, while A_0 is some standard acid, for example, benzoic acid, relative to which the strengths of all the acids in each of the solvents is expressed. In this case, B_0 is the benzoate ion, while B is the base conjugate with the acid A (for example, an acetate ion in the case of acetic acid):

$$CH_3COOH + C_6H_5COO^- \rightleftharpoons CH_3COO^- + C_6H_5COOH.$$

It is evident that if the solvent is able to participate in protolytic reactions, the proto-lysis constant K_{AB_0} equals the ratio of the dissociation constants $K_{(A)}$ and $K_{(A_0)}$ of the acids A and A_0. In actual fact, in water, for example:

$$K_{(A)} = \frac{[H_3O^+][B]}{[A]} \; ; \; K_{(A_0)} = \frac{[H_3O^+][B_0]}{[K_0]}$$

197

and the protolysis constant

$$K_{AB_0} = \frac{[B]\,[A_0]}{[A]\,[B_0]} = \frac{K_{(A)}}{K_{(A_0)}}.$$

For determination of the protolysis constant in the case of a solvent which does not participate in the protolytic reaction (i.e., in an aprotic solvent), to the solution is added a base, for example, an indicator I_B. The addition of a proton to it forms the conjugate acid I_A, whose concentration may be measured with a spectrophotometer. The constant of the equilibrium between the indicator and the acid A

$$A + I_B \rightleftarrows B + I_A$$

is given by the equation

$$K'_{(A)} = \frac{[B]\,[I_A]}{[A]\,[I_B]}\ ,$$

while for the reaction between the indicator and the standard acid

$$A_0 + I_B \rightleftarrows B_0 + I_A$$

the equilibrium constant is given by

$$K'_{(A_0)} = \frac{[B_0]\,[I_A]}{[A_0]\,[I_B]}$$

and consequently, the protolysis constant equals the ratio $K'_{(A)}/K'_{(A_0)}$:

$$K_{AB_0} = \frac{[B]\,[A_0]}{[A]\,[B_0]} = \frac{K'_{(A)}}{K'_{(A_0)}}\ .$$

If the protolysis constants K_{AB_0} for a series of acids in two solvents are plotted along the ordinate axis, according to Brønsted's theory the points for acids with the same type of charge should lie on a straight line whose slope equals unity. With a difference in the dielectric constants of the solvents compared, parallel straight lines correspond to acids differing in the sign or magnitude of the charge.

Another consequence of the theory, which can also be checked experimentally, is as follows. If we plot along the ordinate axis the protolysis constants ($\log K_{AB_0}$) in a series of solvents, differing in dielectric constant (D) and $1/D$ along the abscissa axis, the points should lie on a straight line whose slope is given by M:

$$M = \frac{\varepsilon^2\,(2z_A - 1)}{2kT}\left[\frac{r_{A_0} - r_A}{r_A r_{A_0}}\right],$$

where ε is a single electrical charge; z_A is the charge of the acid A (it is assumed that $z_A = z_{A_0}$); r_A and r_{A_0} are the radii of the acid particles A and A_0; k is Boltzmann's constant; T is the absolute temperature. When the radii of the particles of the two acids are equal, the coefficient M = 0 and the line should lie parallel to the abscissa axis.

Brønsted's theory is based on well-known thermodynamic principles, but in the derivation of formulas which can be checked experimentally, it is assumed that acid–base reactions are described completely by the Brønsted-Lowry scheme. In addition, only electrostatic forces acting between ions of spherical form with a uniform charge distribution are considered. In actual fact, as will be shown later (p. 200), these assumptions are justified only in the very first approximation. In solutions of electrolytes, especially in solvents with a low dielectric constant, noncoulombic interaction must be considered.

2. DEVIATIONS FROM BRØNSTED'S THEORY AND THE NEED FOR DEVELOPMENT OF THE LATTER

Experimental checking confirmed the consequences of the theory on many examples [1, pp. 167-180]. However, Kilpatrick [2], Hammett [3], and Verhock [4] observed cases of serious deviation from the theory: some acids which are similar in strength in water and identically charged, but belong to different classes of chemical compounds (for example, carboxylic acids and nitrophenols) differ considerably in strength when dissolved in acetonitrile, butanol, and formamide.

Even more striking examples of similar deviations from the theory were revealed by N. A. Izmailov by a comparison of the strengths of acids in water and nonaqueous solvents. It was found that the dissociation constants of acids which are equal in water may differ by a factor of several millions in acetone. It was also found that if an acid of a different chemical nature from those measured is taken as the standard for comparison, the linear relation between the logarithm of the ratios of the dissociation constants of these acids and the reciprocal of the dielectric constant of the solvents, which is required by Brønsted's theory, is no longer obtained.

Many years' work by N. A. Izmailov and his co-workers made it possible to establish the main reason for these deviations. It was shown that the reactions between acids and bases and, in particular, their dissociation in solution, are much more complex than admitted by Brønsted's theory. The reaction by no means reduces to merely the transfer of a proton. A molecular compound of the acid with the base is formed first (one of the protolytes may be a solvent molecule). The formation of this compound is the result of the formation of a hydrogen bond between the hydrogen of the acid and the electron-donor atom of the base. Ionization occurs only as a result of subsequent interaction of the molecular compound with the solvent.

Investigations of acid–base reactions in aprotic solvents with a low dielectric constant (benzene, etc.) by a number of authors (La Mer, Davis, and others) demonstrated the predominant role of associative processes and showed from this point of view that under these conditions, reactions between acids and bases cannot be described satisfactorily by the simple Brønsted-Lowry scheme, which provides for only the transfer of the proton of the acid to the base. Davis emphasized the need for considering also the formation of a hydrogen bond by an acid–base interaction.

Numerous facts presented in the previous section of the book demonstrate quite conclusively the acid and basic properties of hydrocarbons, which Brønsted assigned to the category of aprotic substances, i.e., substances which are unable to participate in protolytic reactions.

Thus, the range of acid−base protolytic reactions is greater than that provided for by Brønsted's theory and the latter does not describe processes occurring in these reactions sufficiently accurately. Therefore, there is no doubt that further development of the theory of acids and bases is necessary, beginning with the definitions of these fundamental chemical concepts. The purpose of this section of the book is to provide a detailed substantiation of this conclusion and an explanation of the mechanism of acid−base interaction. The latter is important to us for since hydrogen exchange occurs as a result of acid−base interaction, it is impossible to obtain a true understanding of the mechanism of hydrogen exchange in solutions which is not based on a knowledge of the mechanism of reactions between acids and bases.

3. NONCOULOMBIC INTERACTION IN SOLUTIONS OF ACIDS AND SALTS

About twenty years ago, it was widely believed that in solutions of electrolytes it was necessary to consider only electrostatic interaction forces between ions. Investigators were predominantly concerned with dilute aqueous solutions. However, experiments with solutions in nonaqueous solvents with low dielectric constants altered this point of view. Thus, for example, in 1936 in a discussion of experiments on acid catalysis of ammonolysis in liquid ammonia it was noted [5] that the combination of available data indicated the presence of very considerable interionic forces in ammonia solutions of electrolytes, which evidently could not be reduced to a simple electrostatic interaction, as was clearly shown in the case of acetates, for example.

A similar conclusion was reached by V. A. Pleskov and E. N. Gur'yanova [6], who studied the electrical conductivity of solutions of acids and salts in liquid ammonia. They observed that although the ions $C_6H_5COO^-$ and CH_3COO^- have a larger radius than anions of simple inorganic acids and therefore the electrostatic interaction between them and cations should be weaker than that with other ions of the same charge but smaller radius, the electrical conductivity of solutions of organic acids and salts is considerably less than that of solutions of inorganic electrolytes with the same cation. V. A. Pleskov and E. N. Gur'yanova wrote: "This contradiction undoubtedly indicates that in addition to purely electrostatic forces, there are other forces of a chemical nature between the ions of organic acids in liquid ammonia."

This view was very clearly confirmed by measurements of the catalytic activity [7, 8], absorption spectra [9, 10], and electrical conductivity [10] of solutions of nitrophenol isomers in liquid ammonia. All the nitrophenol isomers form stable crystalline ammonium salts. Therefore, when they are dissolved in liquid ammonia, the nitrophenols are undoubtedly ionized completely, i.e., they are converted completely into the ammonium salts, as is indicated by the absorption spectra. However, despite the complete ionization in solution, the properties of solutions of the different isomers of nitrophenol are different (see Table 71).

Table 71 gives the rate constants k of the ammonolysis of santonin at a catalyst concentration of 0.05 N at 25°. The catalytic activity of ammonium p-nitrophenolate equals the catalytic activity of ammonium iodide, a strong inorganic acid, as essentially is catalyzed by the same acid, namely, the ammonium ion, in both cases. A solution of the m-nitrophenolate catalyzes ammonolysis much more weakly than a solution of the para-isomer of the same concentration. These solutions differ even more in electrical conductivity. Since the radii of the isomeric anions must be similar, the deviations can hardly be ascribed to purely electrostatic effects. It is most likely that they are explained by the slightly different nature of the bond between a solvated proton and the anions of the different isomeric forms. This bond has a more clearly ex-

TABLE 71. Properties of Liquid Ammonia Solutions of Nitrophenol Isomers

Isomer	k	Equivalent electrical conductivity at various dilutions, log v					$K_i^{H_2O}$
		1.5	2.0	3.0	3.5	4.0	
Para-	0.0331	170	216	342	395	430	$7 \cdot 10^{-8}$
Ortho-	—	47	69	162	232	308	$6 \cdot 10^{-8}$
Meta-	0.0251	36	45	102	155	236	$5 \cdot 10^{-9}$

pressed covalent character with the isomer whose ionization constant in water ($K_i^{H_2O}$) is less than that of the other isomers. Analogous observations were made in the measurement of the electrical conductivity of ammonia solutions of nitrophenol isomers[11].

The problem of noncoulombic interaction in electrolyte solutions was the subject of a paper by Kraus [12], in which, in particular, he discussed the way in which the dissociation constants K of alkylammonium salts change, depending on whether or not there is the possibility of interionic interaction resulting in the formation of a hydrogen bond, which is superposed on a purely electrostatic interaction between the ions. Experiments were carried out with nitrobenzene, a solvent with a low affinity for a proton, which therefore does not participate in the formation of a hydrogen bond. Measurements with ammonium picrates and butyl-substituted ammonium salts showed that as long as the cation contained at least one hydrogen atom, the dissociation constant was small and changed little (from $1.5 \cdot 10^{-4}$ to $1.9 \cdot 10^{-4}$). It increased by three orders when the last hydrogen atom was replaced by an alkyl group so that the formation of a hydrogen bond between the cation and the anion became impossible and only electrostatic forces acted between them.

Substance	K
Ammonium picrate	$1.5 \cdot 10^{-4}$
Butylammonium picrate	$1.5 \cdot 10^{-4}$
Dibutylammonium picrate.	$1.6 \cdot 10^{-4}$
Tirbutylammonium picrate	$1.9 \cdot 10^{-4}$
Tetrabutylammonium picrate. . . .	$> 2 \cdot 10^{-1}$

It is interesting to compare the dissociation constants of butylammonium picrate ($1.5 \cdot 10^{-4}$) and butylammonium perchlorate ($2.5 \cdot 10^{-3}$). The latter is one order higher. Kraus considered that this was due to the higher affinity of a picrate anion for a proton as compared with a perchlorate ion.

Similar results were obtained for the following salts in dichloroethane:

	K
Pyridinium picrate.	$4 \cdot 10^{-8}$
Pyridinium perchlorate.	$5 \cdot 10^{-7}$
Tetramethylammonium picrate	$3 \cdot 10^{-5}$

Kraus discussed noncoulombic interaction in ion pairs in later articles [13, 14]. He reported that there is a considerable difference in the dissociation constants of tri- and tetramethylammonium picrates in benzene ($4 \cdot 10^{-21}$ and $1.5 \cdot 10^{-17}$). In accordance with the low dielectric constant of benzene, the dissociation constants are very low and the salts are almost wholly in the form of ion pairs. The dipole moment of such an ion pair equals 15 debyes. We will limit ourselves to these examples for now as below we give additional facts that demonstrate that forces which are not of a purely electrostatic nature are of great importance in acid–base processes.

4. ACID–BASE REACTIONS IN APROTIC SOLVENTS

In Brønsted's theory, together with protophilic (basic), protogenic (acid), and amphoteric solvents, which participate in acid–base equilibria, there is also provision for aprotic solvents, which serve as an inert medium in reactions between acids and bases introduced from outside. A typical example of an aprotic solvent is benzene. As has been shown, in actual fact, benzene can participate in acid–base reactions, but is basic or acidic properties are expressed so weakly that they can be neglected when we are dealing with reactions between normal, quite strong acids and bases.

In benzene solutions, the dielectric constant of which is very low (DC = 2.2), only a very small part of a salt formed in a reaction between an acid and a base (of the order of a millionth of a percent) is in the form of free ions. Brønsted [15], La Mer [16], and V. M. Dulova [17] considered the association of ions in benzene, but normally assumed that the ions were bound by purely electrostatic forces. It has already been pointed out above that noncoulombic forces must also be considered.

A. Application of Brønsted's Theory

Even in Brønsted's first articles [15], it was stated that this theory is completely applicable to the description of acid–base reactions in benzene solutions, for example, to reactions with indicators. Subsequent work of other authors substantiated this conclusion. For example, La Mer and Downes [16], in addition to studying reactions with indicators, carried out electrometric and conductimetric titrations between acids and bases in this solvent. The titration procedure was later refined considerably [18] and despite the very low electrical conductivity of benzene solutions (a factor of 10^8 less than the electrical conductivity of corresponding aqueous solutions), completely clear and quite accurate results were obtained. Brønsted's equation relating the rate constant of a reaction catalyzed by acids in a given solvent to their dissociation constants (in water) (see p. 43) was shown to apply to benzene solutions [19].

Thus, Brønsted's theory could be used to describe acid–base processes in an aprotic solvent.

B. Inadequacy of Brønsted's Theory

Beginning in 1947, Davis and his co-workers [20] made a series of studies of indicator equilibria in benzene solutions. These studies led to a point of view on the mechanism of acid–base reactions which differed from Brønsted's opinions. The conclusions of Davis were based largely on the results of experiments with the acid indicator Bromophthalein-magenta E. With primary, secondary, and tertiary amines, this indicator gives solutions in benzene with different spectra and electrical conductivities. These characteristics are ascribed to the different nature of the hydrogen bonds in the molecular compounds with different amines. Davis formulated his ideas in the following way: "The concept of a hydrogen bond makes it necessary to modify Brønsted's hypothesis.

According to his definition, a conjugate acid and a base cannot have the same charge and at least one of the substances must be an ion. However, the proton may be only partly detached from the acid and added to the base. The formation of a hydrogen bond may be regarded as the beginning of dissociation, i.e., an intermediate stage in the complete transfer of a proton. The bond between hydrogen and the atoms between which it lies is considerably weaker than a normal ionic or covalent bond."

In the opinion of Davis, the association constants in the formation of molecular compounds with a base are a measure of the strengths of acids in their reactions in benzene solutions. These constants normally vary in parallel with the ionization constants of the same acids in aqueous solution.

In the work of N. A. Izmailov, great importance was also attached to the formation of molecular compounds of an acid and a base through a hydrogen bond. N. A. Izmailov's theory of the dissociation of acids and bases is given below.

5. N. A. IZMAILOV'S THEORY OF THE DISSOCIATION OF ACIDS AND BASES

A. Comparison of the Strengths of Acids in Different Solvents

The systematic investigations of N. A. Izmailov on the comparison of the strengths of acids in different solvents form the basis on which he developed a theory of the dissociation of electrolytes [22-40]. The role of the solvent, including the nature of its interaction with the solute, is considered much more fully than in Brønsted's theory.

In the experimental work of N. A. Izmailov and his co-workers, most attention was paid to solvents which he called, following P. I. Walden [21], differentiating solvents. In such solvents, not only acids, but any electrolytes differ in strength particularly sharply. These solvents include acetonitrile, formamide, nitrobenzene, acetone, and mixtures of it with water. Together with M. A. Bel'gova, I. F. Zabara, and others [22-26], Izmailov compared the strengths of acids in acetone and aqueous solutions by titration of acid mixtures and titration of salts by displacement and also by determination of the dissociation constants by means of cells with and without transference.

It was established that in acetone, the dissociation constants of sulfonic acids become several million and those of dinitrophenols several thousand times the ionization constants of carboxylic acids when acids of equal strength in water are compared. This makes it possible to titrate in acetone, acids which are very similar in strength in water such as a mixture of picric and trichloroacetic acids. As in the work cited above [2-4], the points for identically charged acids lie on different lines on a graph of pK in acetone against pK in water (Fig. 19). The slopes of the lines for different classes of acids in acetone and mixtures of it with water vary from 0.5 to 2.0. N. A. Izmailov also found equally great deviations from the theory for solutions of acids in dichloroethane, where the ordinates for carboxylic acids and phenols differ by more than three pK units (Fig. 20). In his investigations of tautomeric equilibria, M. I. Kabachnik [28] found new examples of deviations from Brønsted's theory like those observed by N. A. Izmailov. Thus, for example, if the ionization constants of ketones and enols as acids in water and in dioxane are plotted on a graph, the points for the different classes of compounds lie on different parallel lines.

N. A. Izmailov and I. F. Zabara also showed that the linear relation of log K_{AB_0} to $1/D$ for acetone and aqueous acetone solutions is obeyed only with acids of the same chemical nature and not with any acids with the same charge, as follows from Brønsted's theory (p. 199). If as the standard for the comparison of acid strengths A_0 = C_6H_5COOH, then the data for sulfonic acids and nitrophenols do not lie on a straight line (Fig. 21).

A linear relation is obtained when the standard acid is of the same chemical type, for example, if A_0 = 2,4-$C_6H_4(NO_2)OH$ for dinitrophenols (Fig. 22).

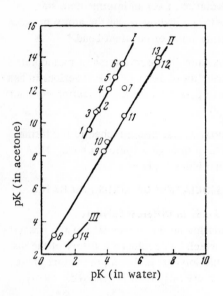

Fig. 19. Relative strengths of acids in water and in acetone: I) Aromatic carboxylic acids: 1) salicylic; 2) m-nitrobenzoic; 3) p-nitrobenzoic; 4) benzoic; 5) p-hydroxybenzoic; 6) p-aminobenzoic; 7) anthranilic. II) Nitrophenols: 8) picric acid; 9) 2,6-dinitrophenol; 10) 2,4-dinitrophenol; 11) 2,5-dinitrophenol; 12) p-nitrophenol; 13) o-nitrophenol. III) Sulfonic acids: 14) sulfosalicylic (1st constant).

B. Reasons for Deviations from Brønsted's Theory

To explain the deviations from the theory, N. A. Izmailov suggested that it is not the acids themselves that are ionized, but molecular compounds of them with the solvent formed in the solution. The different effect of the solvent on the strength of acids in chemical structure depends on the differences in the composition, polarity, and stability of these molecular compounds.

N. A. Izmailov not only put forward this quite probable hypothesis, but demonstrated with detailed investigations the existence of association products of acid and solvent molecules, determined their composition, and measured their instability constants. Cryoscopic measurements were used for this purpose [27]. The method of treating the results of the measurements made it possible to detect reliably even a weak interaction with the solvent. For example, it was found that carboxylic acids and phenols react with an equimolecular amount of acetone to give compounds, to which were assigned the structures:

$$R-C\overset{O}{\underset{OH\cdots O=C\underset{R}{\overset{R}{\big/}}}{\big\backslash}} \quad \text{and} \quad C_6H_5OH\cdots O=C\overset{R}{\underset{R}{\big\backslash}}\big/ \quad .$$

The complexes formed by acetone with carboxylic acids are much more stable than the complexes with phenols.

The products of addition of carboxylic acids and phenols to alcohols, in contrast to the associates with acetone, have a different composition (with two and with one solvent molecule):

$$O \cdots H - O - R$$
$$R - C$$
$$O - H \cdots O$$
$$R$$
$$H$$

and

$$C_6H_5OH \cdots O$$
$$R$$
$$H$$
.

As is shown by the formulas given, the formation of a molecular compound is ascribed to a hydrogen bond between the hydrogen of the acid and the oxygen mole-cule of the alcohol and acetone. The presence of hydrogen bonds of this type was confirmed by Raman spectral measurements on solutions of carboxylic acids and phenols [29, 30] in ac-cordance with the results of the work of other authors. Special cryoscopic measurements demonstrated the identity of addition products formed by solution of the substance in indivi-dual solvents and in an inert solvent with the addition of the substance studied and the sol-vent [27, 31]. In solvents of one type, acids of the same chemical group form molecular com-pounds of the same composition and structure, which are similar in polarity and stability. In the dissertation of K. P. Parkhaladze [32], which was completed under the direction of N. A. Izmailov, it was proved that in other differen-tiating solvents besides acetone, namely, ace-tonitrile and nitrobenzene, molecular compounds with carboxylic acids and phenols have the same composition as in acetone, i.e., 1:1.

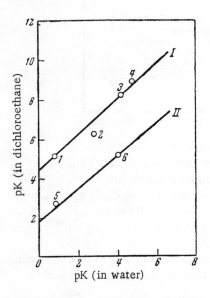

Fig. 20. Relative strengths of acids in water and dichloroethane: I) car-boxylic acids: 1) trichloroacetic; 2) monochloroacetic; 3) acetic; 4) benzoic. II) Nitrophenols: 5) picric acid; 6) 2,4-dinitrophenol.

N. A. Izmailov considers that not only the composition of the solvent addition products, but also the solvation energy of the ions and molecules of the acid depend on the structure of the latter.

N. A. Izmailov generalized the results of his experimental work in the book "The Electrochemistry of Solutions" and in several reviews [33-36]. He wrote: "Our inves-tigations have shown that Brønsted's theory is too schematic and provides for only one type of chemical interaction between dissolved acids and the solvent (only the exchange of protons), with the result that this theory does not allow the explanation of all the peculiarities in the effect of solvents and, in particular, does not explain the differen-tiating action on the strengths of acids, which is the result of more complex individual interactions in solvents of different types" ([35], p. 143). In another article [36], he stated: "A deficiency in Brønsted's theory, apart from the inaccurate scheme for the dissociation of acids, is the fact that the theory does not allow for the effect of addi-tion products on the activity of the undissociated acid, i.e., does not allow for the de-gree of its conversion into addition products, changes in its acid properties, nor the as-sociation and solvation of ions."

C. N. A. Izmailov's Theory

N. A. Izmailov put forward a theory of the dissociation of electrolytes (acids, bases, and salts). In examining the effect of the solvent on the strength of acids, he started

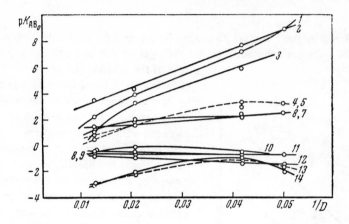

Fig. 21. Relative strengths of acids in acetone and mixtures of it with water with different dielectric constants ($A_0 = C_6H_5COOH$). 1) Sulfosalicylic; 2) picric; 3) sulfanilic; 4) 2,4-dinitrophenol; 5) 2,6-dinitrophenol; 6) monochloroacetic; 7) salicylic; 8) butyric; 9) caproic acid; 10) acetic; 11) p-hydroxybenzoic; 12) p-aminobenzoic; 13) p-nitrophenol; 14) o-nitrophenol.

from the premise that the ionization of acids in solution is effected in several stages, of which the first is the formation of the product from the addition of the solvent to the acid molecule HA:

$$HA + nM \rightleftarrows HAM_n, \tag{1}$$

where M is a solvent molecule. Ionization occurs as a result of further solvation of the addition product and the formation of solvated ions:

$$HAM_n + mM \rightleftarrows MH^+_{sol} + A^-_{sol}. \tag{2}$$

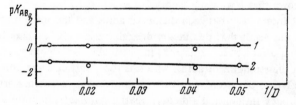

Fig. 22. Relative strengths of dinitrophenols in acetone and mixtures of it with water with different dielectric constants. $A_0 = 2,4\text{-}C_6H_4(NO_2)OH$. 1) 2,6-Dinitrophenol; 2) 2,5-dinitrophenol.

Weak acids are present in solution largely in the form of solvent addition products. The solvated ions may be associated in ion pairs:

$$MH^{+}_{sol} + A^{-}_{sol} \rightleftarrows MH^{+}_{sol} A^{-}_{sol} . \tag{3}$$

Ion pairs are not involved in the electrical conductivity of the solutions, i.e., behave like molecules in this respect. At the same time, in optical properties they normally differ from molecules are are similar to ions. Therefore dissociation constants determined electrochemically and by optical methods are often different in magnitude, especially with solvents with a low dielectric constant, where ion association processes are very marked. In addition to ion pairs, more complex aggregates of ions may be formed (cf., p. 54).

In parallel with these processes, there may also be a change in the degree of association of both acid molecules

$$(HA)_{s} \rightleftarrows s\,HA, \tag{4}$$

and solvent molecules:

$$M_{l} \rightleftarrows l M. \tag{5}$$

Thus, there are often many particles coexisting in a solution: un-ionized molecules of the acid HA (various degrees of association), solvent molecules (various degrees of association), the product (or products) of addition of solvent to the acid, ion pairs and more complex formations (various degrees of solvation), and free solvated ions. The relation between the concentrations (activities) of these particles depends on the properties of the acid and the solvent and on the concentration of the solution. In dilute solutions it is possible to ignore the existence of equilibria (4) and (5) as the former is usually displaced wholly to the right, while the degree of association of the solvent does not change appreciably.

N. A. Izmailov considers that the following equilibria exist between the addition product, ion pairs, and free ions:

$$\begin{array}{c} HAM_{n} \rightleftarrows MH^{+}_{sol} + A^{-}_{sol} \\ \searrow \qquad\qquad \nearrow \\ MH^{+}_{sol} A^{-}_{sol} \end{array} . \tag{6}$$

The ion pairs which are denoted by the formula $MH^{+}_{sol}A^{-}_{sol}$ here are not actually the same. They may differ in the degree of interaction between the ions in them. The ions are either tightly bound and surrounded by a single solvation envelope or, on the other hand, separated by a solvation envelope lying around each of them. Such an ion pair is often called "loose" (p. 287). Naturally, the degree of solvation depends on the concentration of the solution and the other substances presence in it as well as the structure of the ions and the properties of the solvent.

With a knowledge of the equilibria existing in the solution of an acid (and, correspondingly, in solutions of other electrolytes), it is possible to write out the constants of the corresponding equilibria.

The equilibrium constant of reaction (1), the addition of solvent molecules to the acid molecule, is given in the form of the instability constant:

$$K_{inst} = \frac{a^*_{HA}\, a^{*n}_{M}}{a^*_{HAM_n}}.$$ (7)

Here and below, the asterisk denotes that the activity refers to an infinitely dilute solution of the substance in the solvent M. The activity of the solvent a^{*n}_{M} is constant in dilute solutions. Therefore, a^{*n}_{M} can be combined with the constant to give a new constant, which we will denote by K^*_{inst}:

$$K'^*_{inst} = \frac{a^*_{HA}}{a^*_{HAM_n}}.$$ (7a)

The equilibrium constant of the ionization of the addition product (2)

$$K'_{ion} = \frac{a^*_{MH^+_{sol}}\, a^*_{A^-_{sol}}}{a^*_{HAM_n}}.$$ (8)

Here it is also assumed that a^*_{M} = const.

The equilibrium constant of the association of ions into ion pairs (3) is given by the equation:

$$K'^{-1}_{ass} = \frac{a^*_{MH^+_{sol}}\, a^*_{A^-_{sol}}}{a^*_{MH^+_{sol} A^-_{sol}}}.$$ (9)

Scheme (6) on p. 207 includes the equilibrium of the conversion of the un-ionized product from the addition of the solvent to the acid into an ion pair:

$$HAM_n \rightleftarrows MH^+_{sol} + A^-_{sol}.$$ (10)

The constant of the conversion, is given by the equation:

$$K_{con} = \frac{a^*_{MH^+_{sol}}\, a^*_{A^-_{sol}}}{a^*_{HAM_n}}.$$ (11)

This constant is related to K_{ion} of the product from addition of the solvent to the acid and K_{ass} of the ions by the expression:

$$K_{con} = K_{ion}\, K_{ass}$$ (12)

as is readily seen on combining Equations (8) and (9).

The effect of the equilibria (4) and (5) on the strength of acids is observed only in concentrated solutions and they are neglected in the subsequent discussion.

The over-all dissociation of an acid HA in dilute solution according to N. A. Iz-mailov is described by the following scheme:

$$HA + (n + m) M \xrightleftharpoons{K^*_{inst}} HAM_n + mM \xrightleftharpoons{K_{ion}} MH^+_{sol} + A^-_{sol}$$

$$+ \qquad\qquad\qquad \updownarrow K^{-1}_{ass} \qquad (13)$$

$$yM \xrightleftharpoons{} MH^+_{sol} A^-_{sol} + xM$$

The stoichiometric coefficients are related by the equation: $x + y = m$.

Although the scheme presented does not include all the equilibria existing in the ionization and dissociation of acids,* it still represents the phenomena occurring in electrolyte solutions quite fully.

The dissociation constant is normally understood to be the ratio of the product of the activities of the free ions to the total activity of the undissociated particles. The activity of the solvent a^*_M = const.

$$K_{norm} = \frac{a^*_{MH^+_{sol}} a^*_{A^-_{sol}}}{a^*_{undiss}} \qquad (14)$$

where K_{norm} is the normal dissociation constant.

In view of what was stated above on the fact that not only acid molecules and the products from addition of solvent to acid molecules, but also ion pairs do not conduct a current, we have:

$$a^*_{undiss} = a^*_{HA} + a^*_{HAMn} + a^*_{MH^+_{sol} A^-_{sol}} . \qquad (15)$$

If the value of a^*_{undiss} from (15) is substituted in the denominator of (14) and at the same time, a^*_{HA}, $a^*_{HAM_n}$, and $a^*_{MH^+_{sol} A^-_{sol}}$ are expressed in terms of the activities of the free ions by means of Equations (7a), (8), and (9), then after rearrangement, Equation (16) is obtained:

$$K^{-1}_{norm} = (K^*_{inst} + 1) K^{-1}_{ion} + K_{ass}. \qquad (16)$$

Considering Equation (12), according to which

$$K_{ass} = \frac{K_{con}}{K_{ion}} \qquad (12a)$$

the values of K_{ass} may be substituted in Equation (16) to give:

$$K_{norm} = \frac{K_{ion}}{1 + K^*_{inst} + K_{con}} . \qquad (17)$$

---

* The concepts of "ionization" and "dissociation" of acids should be distinguished. Ionization consists of the conversion of an electrically neutral molecule into electrically charged ions as a result of a protolytic reaction. Dissociation is the equilibrium of the decomposition of ion pairs into free ions.

This equation reveals the complex nature of the normal dissociation constant of an acid. Equation (17) assumed a simpler form in a series of particular cases. Thus, if K_{inst}^* and $K_{con} \ll 1$, then $K_{norm} \cong K_{ion}$, i.e., the solution contains practically no ion pairs and acid molecules not bound by solvent. This occurs in a solvent with a high dielectric constant, which readily forms addition products, for example, water.

If

$$K_{con} > K_{inst}^* > 1,$$

then

$$K_{norm} \cong \frac{K_{ion}}{K_{con}}$$

i.e., according to (12), $K_{norm} = K_{ass}^{-1}$. In other words, ion pairs and free ions are in equilibrium. This type of equilibrium is often encountered when an acid is dissolved in a protophilic solvent with a low dielectric constant, for example, liquid ammonia.

The way in which the form of Equation (17) changes in relation to the properties of the electrolyte and the solvent is discussed in detail in N. A. Izmailov's book "The Electrochemistry of Solutions" (p. 599).

We should note that although the formation of ion pairs in electrolyte solutions was considered previously in the theories of V. K. Semenchenko, Bjerrum, Fuoss, and Kraus, the formation of associates of the electrolyte and the solvent and their role in dissociation were not considered before the work of N. A. Izmailov and these associates appeared for the first time in the latter's theory.

N. A. Izmailov's theory gives a quantitative relation between the normal dissociation constant of an electrolyte and the properties of the solvent. The theory is based on the dissociation scheme discussed above and evaluates the work of transferring the electrolyte from the solvent M in a solution of a given concentration into vacuum. The cycle given in Table 72 is examined for this purpose.

TABLE 72. N. A. Izmailov's Cyclic Process

$$HA \underset{\rightleftarrows}{\overset{A_I}{}} H^+ + A^-$$

$$A_{mol}\Big\downarrow\Big\uparrow \qquad \Big\uparrow\Big\downarrow A_{H^+} \Big\downarrow\Big\uparrow A_{A^-} \; (\Sigma A_x)$$

$$HA\,(M) \underset{\rightleftarrows}{\overset{A_D}{}} MH^+_{sol} + A^-_{sol}$$

N. A. Izmailov uses the concept of the natural acidity of an acid in vacuum; the expression for the acidity constant is analogous to Brønsted's [1, p. 145].

$$K_{a\,(HA)} = \frac{a_{H^+}a_{A^-}}{a_{HA}}. \tag{18}$$

The absence of asterisks from the activities indicates that they refer to one standard state, which is taken as in vacuum in the following account.

The work to abstract a proton from a molecule of the acid HA in vacuum or, in other words, the work consumed in the ionization of the acid molecule to a proton and an anion (HA $\rightleftharpoons$ H$^+$ + A$^-$) is denoted by A$_I$

$$A_I = RT \ln K_a. \tag{19}$$

The magnitude of K_α depends on the affinity of the anion of the acid A$^+$ for a proton.

In the scheme in Table 72, the symbol HA(M) arbitrarily denotes different types of molecular forms in solution: unsolvated acid molecules (HA), products of the addition of solvent to it (HAM$_n$), and ionic molecules, i.e., ion pairs (MH$_{sol}^+$ A$_{sol}^-$). None of them are involved in the electrical conductivity of the solution and their total activity in solution in Equation (14) was denoted by the symbol a$^*_{undiss}$. These particles are in equilibrium with free solvated ions MH$_{sol}^+$ and A$_{sol}^-$. The work of dissociation according to the equation at the bottom of the cycle is denoted by A$_D$ and is given by the equation:

$$A_D = RT \ln K_{norm}, \tag{20}$$

where K_{norm} is the normal dissociation constant, whose relation to the properties of the solvent we wish to determine.

In addition, in the examination of the cycle, the solvation energy of the ions MH$^+$ and A$^-$, i.e., the lyonium ion and the anion of the acid HA, is considered.

The work of transferring various types of unchanged particles from the solution to vacuum is denoted by A$_{mol}$. Its value depends on the composition and stability of the addition products formed, the degree of conversion of the molecules into solvates and ion pairs, the energy of condensation of the isolated molecules into the liquid state, and the energy of solution of the liquid acid in the medium M.

For the work of transferring the ions, we adopt the notation $\Sigma A_x = A_{H^+} + A_{A^-}$ (see scheme in Table 72 on the right). The energy of solvation of the proton is composed of the energy of addition of the proton to a solvent molecule in vacuum (P, the affinity for a proton) with the formation of the lyonium ion MH$^+$ and the energy of its further solvation.

From the examination of the cycle with substitution of the values for the energy of solvation of the protons and anions (or the work of their desolvation), N. A. Izmailov obtained a final expression for the normal dissociation constant of an acid:

$$K_{norm} = \frac{K_{a(HA)} \, a_M}{K_{a(MH^+)}} \, e^{\left(\frac{\Sigma A_x - A_{mol}}{RT}\right)} \tag{21}$$

where $K_{a(HA)}$ and $K_{a(MH^+)}$ are the true acidity constants of the acid HA and the lyonium ion in the solvent M and a$_M$ is the activity of the solvent referred to vacuum as the standard state.

Equation (21) shows that the normal dissociation constant depends, firstly, on the structure of the acid molecule and the lyonium ion, which is reflected in the constants $K_{a(HA)}$ and $K_{a(MH^+)}$. In addition, the work of solvation of the acid molecules and the ions formed from them is of considerable importance. Brϕnsted's theory gives no consideration at all to the energy of interaction of the undissociated acid molecules with the dipolar solvent molecules or the association of ions and does not completely allow for the energy of solvation of the ions. Brønsted's theory considers only the change in the potential energy of an ideal spherical ion in relation to the dielectric constant of the solvent.

If the dissociation constant is expressed in the form of pK_{norm}, i.e., $-\log K_{norm}$, then from Equation (21) we obtain:

$$pK_{norm} = -\log K_{a(HA)} - \log a_M + \log K_{a(MH^+)} - \frac{\Sigma A_x}{2.3 RT} + \frac{A_{mol}}{2.3 RT}. \qquad (21a)$$

By writing the same expression for the dissociation constant of the acid in another solvent M' (for example, water) and marking the corresponding values with a dash, we can take the difference in pK_{norm} for the two solvents and obtained the following equation for ΔpK_{norm}:

$$\Delta pK_{norm} = pK_{norm} - pK'_{norm} = \log \frac{K_{a(MH^+)}}{K_{a(M'H^+)}} + \log \frac{a_M}{a_{M'}} + \frac{\Sigma A'_x - \Sigma A_x}{2.3 RT} - \frac{A'_{mol} - A_{mol}}{2.3 RT},$$

$$(22)$$

ΔpK_{norm} does not depend on the true acidity constant. The first two terms of the equation are constants for the solvents compared, regardless of the structure of the acid. The last two terms differ little for acids of similar structure. This is no longer true when acids of different structures in solvents of different types are compared.

N. A. Izmailov derived an equation analogous to (21) for the dissociation constant of a base. It differs in that it includes the true basicity constants of the substance (K_b) and the lyate ion (arbitrarily denoted by M⁻), while ΣA_x consists of the work of solvation of the ions BH⁺ and M⁻.

$$K_{norm} = \frac{K_b a_M}{K_{b(M^-)}} e^{\left(\frac{\Sigma A_x - A_{mol}}{RT}\right)}. \qquad (23)$$

A detailed derivation and analysis of the formulas is given in [39, 40] and particularly in N. A. Izmailov's book "The Electrochemistry of Solutions" (ch. 8). The latter also describes the application proposed by the author of the method of single zero activity coefficients γ_0, for which an infinitely dilute solution in water is taken as the standard state. The value of γ_0 is determined by the work for transferring the substance from an infinitely dilute nonaqueous solution to an infinitely dilute aqueous solution. By successively applying this method and expressing the work of transfer in values of γ_0, N. A. Izmailov has shown that

$$\Delta pK_{norm} = 2\log \gamma_{0(ion)} - \log \gamma_{0(mol)}. \qquad (24)$$

Substitution of the values of $\log \gamma_{0(ion)}$ and $\log \gamma_{0(mol)}$ gives equations, which, with a series of limiting conditions, are converted into the equations of Brønsted's theory. Consequently, the latter is a particular case of N. A. Izmailov's more general theory.

6. THE ROLE OF THE HYDROGEN BOND IN ACID−BASE INTERACTION

In the investigations discussed above, Davis and N. A. Izmailov clearly pointed out the important part which the formation of a hydrogen bond plays in acid−base interaction. The material gathered here provides a more complete and manifold basis for this conclusion, which is extremely important for understanding the mechanism of reactions between acids and bases and for defining these concepts. First we will consider the ideas of the nature of the hydrogen bond put forward by N. D. Sokolov, from which it follows that the formation of a hydrogen bridge is the first stage in the transfer of a proton from an acid to a base. Then we will examine the effect of the formation of a hydrogen bond on various physical parameters of a system consisting of a base and an acid. These parameters depend on the strength of the protolytes; their measurement gives a means of estimating the latter in cases where an acid—base reaction is not completed by the transfer of a proton.

A. Donor—Acceptor Interaction in the Formation of a Hydrogen Bond

There is no one point of view in the literature on the nature of a hydrogen bond, though this concept is used very often.

The actual formation of a hydrogen bond may be established quite objectively by different methods. The energy of a hydrogen bond is low; it normally varies over the range of 3-7 kcal, while the energy of a covalent chemical bond is usually of the order of 50-150 kcal/mole.

A hydrogen bond is usually formed if the hydrogen lies between the electronegative atoms such as fluorine, chlorine, oxygen, sulfur, and nitrogen. In recent years it has been established definitely that there is the possibility of the formation of a hydrogen bond with a hydrocarbon whose molecule has π-electrons and also that the hydrogen atom of a CH bond can participate in the formation of a hydrogen bond.

It is considered that the energy of a hydrogen bond is the sum of energies of different types [41, 42]. N. D. Sokolov [42-45] first pointed out the inadequacy of a purely electrostatic model of a hydrogen bond and noted that in the formation of the latter, covalent as well as electrostatic forces operate. In his opinion, a hydrogen bond is explained to a certain extent by the same donor-acceptor interaction that finally results in the transfer of a proton in a reaction between an acid and a base. N. D. Sokolov based this conclusion on a quantum mechanicam calculation. According to N. D. Sokolov, the formation of a hydrogen bond is always the first stage in the transfer of a proton in a protolytic reaction. A donor-acceptor interaction is realized at all stages in the transfer of a proton to a base, which is completed by the addition of the proton to the base under favorable conditions.

If AH is the acid and :B the base, an atom of which has a free electron pair, the formation of the hydrogen bond AH. . . :B is accompanied by stretching of the bond A—H. The hydrogen is protonized to some extent and the atom A charged negatively. This results in an additional dipole moment in the molecular compound obtained.

One of the most characteristic signs of the formation of a hydrogen bond is a change in the characteristic frequency of the bond whose hydrogen is involved in the hydrogen bridge. This is detected by a displacement of the frequency of the corresponding band in the infrared absorption spectrum (and is also shown by broadening of the band and a change in the absorption intensity). The existence of a relation between the displacement of the characteristic frequency $\Delta\nu$ and the protolytic properties of the substances involved in the formation of the hydrogen bond is in agreement with the idea of its donor-acceptor nature. A theoretical basis for this relation was provided by N. D. Sokolov. The stronger the acid AH is or the more protophilic the base B,

the stronger is the hydrogen bond, the higher its energy, and the greater the displacement of the frequency $\Delta\nu$. Below we examine appropriate experimental data agree-
ing with this conclusion.

As has already pointed out, the rate of proton transfer is normally symbatic with
the strength of the acid (p. 43). Considering what was stated above, the proportionality between the frequency displacement $\Delta\nu$ and the rate of transfer of a proton from
an acid to a base, which was pointed out by N. D. Sokolov and is shown, in particular,
in the rate constants of some reactions catalyzed by acid, is also reasonable.

The role of the hydrogen bond in the kinetics of chemical reactions was discussed
in detail in the comprehensive reviews of D. G. Knorre and N. M. Émanuél' [46, 47].
From an analysis of results obtained both by the authors of the review and by other
scientists it was concluded that the formation of hydrogen bonds between reagents and
the solvent may also have a substantial effect on reactions in solution. It may facilitate the transfer of a proton in acid—base reactions and affect the reaction rate by
changing the acid—base properties of the reagents (see p. 36 for an example of this effect on the rate of deuterium exchange). The authors hold the view that the formation
of a hydrogen bond proceeds the transfer of a proton, which is completed by ionization.
The hypothesis that the formation of a hydrogen bond (partial transfer of a proton) and
the complete transfer of a proton are closely related phenomena was emphasized.

M. I. Ryskin [Optika i Spektroskopiya, 7, 278 (1959)] also interprets the mechanism
of formation of a strong hydrogen bond on the basis of the donor-acceptor concept and
regards the formation of a hydrogen bond as a case of an acid—base interaction. He
pointed out that with a proton acceptor of a given strength, the strength of the hydrogen bond depends on conditions determining the degree of "protonization" of the hydrogen. A condition for the formation of a short hydrogen bond is fixation of the electron cloud with attraction of the proton. This situation is illustrated by infrared spectral measurements.

B. Polarity of Molecular Compounds of Acids with Bases

There is an increase in the dipole moment with the formation of a hydrogen bridge.
For example, this is the origin of the "dioxane" effect, which was first observed by
V. G. Vasil'ev and Ya. K. Syrkin [48] in comparing the dipole moments of molecules
in benzene and dioxane. It was found that the dipole moments of molecules which are
capable of forming a hydrogen bond with the oxygen atom of dioxane (for example,
aromatic amines) are greater in dioxane than in benzene: the dipole moment (μ) of
aniline in benzene equals 1.54 D, which is close to the value obtained for the vapor,
while in dioxane, $\mu = 1.77$ D.

The existence of a dioxane effect has been confirmed by other investigators [49].
When the hydrogen in the amino group is replaced completely by methyl groups, the
effect becomes small and in this case it may be caused by the formation of a weak
hydrogen bond between the hydrogen atom of the ring in the para-position and the
oxygen atom of dioxane (Table 73). In comparing measurments in dioxane and in benzene, it must be remembered that benzene itself is not a completely inert solvent.
This is shown, for example, by the fact that the dipole moment of phenol in benzene
is appreciably higher than in cyclohexane [50] as a result of the electron-donor properties of benzene.

The polarity of the products from the addition of acids to bases has been demon-strated by direct measurements of the dielectric properties of their solutions. Such

TABLE 73. Comparison of Dipole Moments in Dioxane and in Benzene (Dioxane Effect)

Substance	Dioxane, D	Benzene, D	$\Delta\mu$, D
Aniline	1.750	1.505	0.245
Methylaniline	1.833	1.643	0.190
Dimethylaniline	1.633	1.577	0.056
p-Chloroaniline	3.372	2.994	0.378
2,4,6-Tribromoaniline	1.972	1.693	0.279
2,4,6-Tribromodimethylaniline	1.029	1.048	—0.019

measurements in ternary systems consisting of an acid, a base, and a solvent (in parti-cular, benzene) have been made by several authors [14, 51, 52]. Ya. K. Syrkin and L. Sobczyk [53, 54] measured the dielectric polarization P of more than 30 systems con-sisting of solutions of stoichiometric amounts of an acid and a base in benzene and compared the values with the sums of the polarizations of the two components. The difference between the observed and calculated values ΔP characterizes the strength of the interaction between the reagents. The dipole moment of the molecular com-pounds formed in solution were also determined. The difference between the dipole moment of the molecular compound and the vector sum of the moments of the bonds of the substances forming it gives an idea of the polarity of the hydrogen bond.

With an increase in the strength of an acid, there is an increase in the polarity of its molecular compound with a base. For example, the additional polarity of the hy-drogen bond in the molecular compounds of acids with pyridine is as follows: for pro-pionic acid (1.4 D) < chloroacetic acid (2.0 D) < trichloroacetic acid (5.1 D).

Table 74 gives the values of ΔP and the dipole moments μ of the molecular com-pounds obtained for some of the systems studied, containing stoichiometric amounts of the acid and the base. The second column gives $K_i^{H_2O}$, i.e., the dissociation constants of the bases in water. Analogous data for the acids are given under their formulas.

In [54] it was concluded that: "The reaction of a strong hydrogen acid with an or-ganic base when there is no solvating solvent present leads to the formation of com-plexes with hydrogen bridges and not ionization and salt formation. In the case of strong acids AH and bases B, the bridge A . . . H . . . B is formed, where the hydrogen is strongly displaced toward B."

The authors regard the reaction between an acid and a base in a nonpolar solvent as a particular form of "preneutralization." In their opinion, the polarity of the hydro-gen bridge may serve as a measure of the strength of the acid and the base in a non-ionizing solvent. The values of ΔP are proportional to the values of pK of the acids. According to Ya. K. Syrkin [55], hydrogen is not transferred at all in the reaction of an acid and a base in benzene, but forms a hydrogen bond.

TABLE 74. Dielectric Polarizations and Dipole Moments of Molecular Compounds

Substance	$K_i^{H_2O}$	ΔP and μ	CH₃CO₂H $K_i^{H_2O} = 1.8\cdot10^{-5}$	CH₂ClCO₂H $K_i^{H_2O} = 1.6\cdot10^{-3}$	CCl₃CO₂H $K_i^{H_2O} = 2\cdot10^{-1}$	C₆H₂(NO₂)₃OH $K_i^{H_2O} = 2\cdot10^{-1}$
Pyridine	$1.7\cdot10^{-9}$	ΔP	71	297	1070	1970
		μ	2.93	4.67	7.78	10.1
Quinoline	$6.3\cdot10^{-10}$	ΔP	—	318	1250	2060
		μ	—	4.77	8.32	10.3
Acridine	$3\cdot10^{-10}$	ΔP	—	366	1340	2340
		μ	—	4.98	8.60	10.9
Triethylamine	$5.6\cdot10^{-4}$	ΔP	—	774	1520	2700
		μ	—	6.36	8.82	11.7
Piperidine	$1.2\cdot10^{-3}$	ΔP	—	76	90	3100
		μ	—	2.86	3.15	12.4

Tsubomura [56], whose investigations was similar in nature and results to the work of L. Sobczyk and Ya. K. Syrkin, measured the dipole moments of systems consisting of a base and an acid in nonpolar solvents, namely, dioxane, benzene, and carbon tetrachloride. The results he obtained are given in Table 75. The following abbreviations

TABLE 75. Dipole Moments of Complexes of Acids with Bases

No.	Acid	Base	Solvent	Dipole moment		
				μ_{obs}	μ_{calc}	$\Delta\mu$
1	$C_6H_2(NO_2)_3OH$	C_5H_5N	Dioxane	10.43	3.4	7
2	$C_6H_2(NO_2)_3OH$	$C_6H_5NH_2$	»	7.0	1.8	6
3	CCl_3CO_2H	$(C_2H_5)_3N$	»	9.6	3.0	8
4	$CHCl_2CO_2H$	$(C_2H_5)_3N$	»	7.76	3	7
5	CH_2ClCO_2H	$(C_2H_5)_3N$	»	7.15	3	6.5
6	CH_3CO_2H	$(C_2H_5)_3N$	C_6H_6	3.96	2.2	3
7	$p\text{-}C_6H_4(NO_2)OH$	$(C_2H_5)_3N$	Dioxane	5.57	5.4	0.3
8	$C_9H_{19}CO_2H$	Urotropin	CCl_4	2.05	1.55	0.5
9	C_6H_5OH	"	»	1.50	1.70	—0.2

are adopted in it: μ_{obs} is the measured dipole moment expressed in debyes; μ_{calc} is the dipole moment calculated by vector addition of the values of μ for the acid and the base; $\Delta\mu = \mu_{obs} - \mu_{calc}$. For the first five systems, $\Delta\mu$ reaches 6-8 D. The author considers that in these systems the proton of the acid is transferred to the base to form the ion pair $A^-\ldots HB^+$, in which the proton is bound covalently to the nitrogen atom of the amine. The dipole moment of the system $CH_3COOH-(C_2H_5)_3N$ in benzene is a factor of 2-3 less. It was suggested that there is partial transfer of the proton. As regards the last three systems with low dipole moments, the acid–base interaction in them is limited to the formation of a hydrogen bond.

Tsubomura attempted to relate the dipole moment of the complex to the strength of the acid and the base, expressed as their dissociation constants in aqueous solution. The pK values of the acid and the base ($pK_a^{H_2O}$ and $pK_b^{H_2O}$) are plotted along the axes of the graph in Fig. 23. The broken line roughly marks the boundary between regions of acid–base interaction leading to the formation of ion pairs and molecular complexes through a hydrogen bond. Some complexes ($CH_3CO_2H-(C_2H_5)_3N$) are of an intermediate character.

The treatment of the results given by the Japanese author differs from that proposed by Ya. K. Syrkin and L. Sobczyk [53, 54], who ascribe the excess polarization of the complex to the polarity of the hydrogen bond and say nothing of the formation of ion pairs. However, in the work of Tsubomura also there is no doubt that in nonpolar media the acid–base interaction may be limited to the formation of a hydrogen bond and may not be accompanied by complete transfer of a proton, leading to ionization of the acid and base molecules.

Similar conclusions were drawn by Barrow [57] on the basis of spectral measurements. He considered that in the reaction between a base and a strong acid in a nonpolar solvent there is formed an ion pair, joined by a hydrogen bond (as in the system $CF_3COOH-C_5H_5N$), while in the reaction with a weak acid, the hydrogen remains covalently bound to the acid molecule; the base is attached to it through a hydrogen bond (in the system $CH_3COOH - C_5H_5N$, for example).

In Barrow's opinion, with a gradual change in the strength of the acid (or the base), the proton does not occupy an intermediate position between the acid and the base,

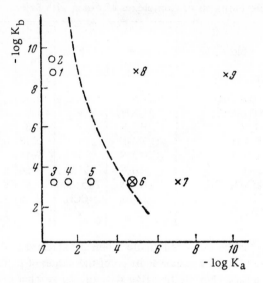

Fig. 23. Relation between the strength of an acid and a base and the structure of the complex formed by them. See Table 75 for the numeration.

which would correspond to movement of the minimum on the potential energy curve of the given reaction. It is assumed that in the case of protolytes of medium strength the potential curve has two minima,* as products of both types actually coexist and are in equilibrium:

$$R - C \underset{OH \cdots N}{\overset{O}{\diagup}} \rightleftarrows R - C \underset{\bar{O} \cdots H - \overset{+}{N}}{\overset{O}{\diagup}}$$

According to Barrow, with a change in the strength of the acid, the position of the proton in the hydrogen bridge between the acid and the base does not change, but the equilibrium is displaced toward the formation of some other form.

To support his hypothesis, the author measured infrared absorption spectra. In the spectrum of the reaction product from CF_3COOH and C_6H_5N there are bands at 6.12 and 6.77 μ, which are characteristic of the $C_5H_5NH^+$ ion and indicate complete transfer of the proton of the acid to the base molecule. These bands are absent from the spectrum of the compound of acetic acid and pyridine, indicating that the proton is not trans-

---

* They are at the distance of a covalent bond from the acid residue and from the base.

ferred in the reaction between them. The band specific to the ring of the pyridine molecule has a high intensity. It is displaced only slightly (6.24 instead of 6.33 μ) in comparison with the spectrum of a solution of pyridine in CCl_4. The direction of the displacement is the same as when pyridine is dissolved in butanol or chloroform, but the displacement is greater in magnitude. The displacement of the band is caused by the formation of a hydrogen bond (see p. 220 for more details), which is stronger in the complex with acetic acid than in complexes with the solvents mentioned.

If the absorption coefficients of the bands at 6.11, 6.77, and 6.24 μ for acids of different strengths are plotted on a graph (along the ordinate axis) against the corresponding values of $pK_i^{H_2O}$ for these acids, it is evident that as the strength of the acid increases, there is a fall in the intensity of the pyridine band and an increase in the intensities of the bands corresponding to the pyridinium ion (Fig. 24). It is possible to put forward an arbitrary scale of the degree of ionic character of the interaction product or the "degree of transfer" of the proton, expressed in percents (at the right of the graph). The fact that a change in temperature is accompanied by a change in the ratio of the intensities of the bands at 6.72 and 6.24 μ led Barrow to the idea that there is an equilibrium between the two forms. This is also indicated by the fact that an increase in the strength of the acid is not accompanied by a gradual displacement of the band. Subsequent work by Barrow [57a, 57b] has substantiated the view that there are two minima on the potential curve of systems in which a hydrogen bond is formed.

The treatment of the structure of complexes formed by an acid and a base in nonpolar, so-called nonionizing solvents is still far from clear. In the reaction between a sufficiently strong acid and a base an ion pair may be formed in which ions are firmly bound, while in the reaction between a weak acid and a weak base, a less polar complex is formed; the components in it are bound by a hydrogen bridge in which the hydrogen is slightly displaced. Some authors are inclined to think (and this is more prob-

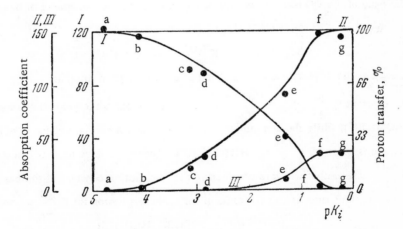

Fig. 24. Optical densities ($\log I_0/I$) of bands in the infrared spectrum of pyridine−carboxylic acid systems as functions of pK_i. I) For the band at 6.24 μ (components of the complex bound by a hydrogen bond); II) for the band at 6.72 μ; III) for the band at 6.11 μ (ion pairs are formed in the last two cases). a) Acetic; b) chloropropionic; c) iodoacetic; d) chloroacetic; e) dichloroacetic; f) trichloroacetic; g) trifluoroacetic acid.

able) that the intermediate formations are actual mixtures of the two forms in equilibrium, while other authors hold the opinion that the proton may occupy a position in the hydrogen bridge intermediate between the atoms of the acid and the base. Data on the relation of the magnitude of the band displacement in the infrared spectrum ($\Delta\nu$) to the strength of the protolyte, which we consider below, indicate that if the proton cannot occupy any position, there are in any case different positions.

C. Displacement of Bond Vibration Frequency in the Infrared Absorption Spectrum on the Formation of a Hydrogen Bond

We will not consider here the voluminous work on the spectroscopic demonstration of the formation of a hydrogen bond in various systems containing an acid and a base, but only investigations which make it possible to follow the relation between the magnitude of the band displacement in the infrared spectrum and the strength of the protolytes.

Gordy and Stanford [58-60] were the first to conclude that such a relation exists. They systematically measured the position of the band of the OD bond in the infrared spectrum of heavy methanol (CH_3OD) and heavy water when substances of the electron-donor type were dissolved in them. About 80 compounds were examined and these included amines, amides, ketoximes, nitriles, nitro compounds, ketones, aldehydes, glycols, ethers, and esters of organic and mineral acids. The position of the vibration band of the OD bond changed in the presence of these substances. This displacement is a measure of the weakening of the bond of the hydrogen atom and the oxygen in the hydroxyl group; it can therefore be used to estimate the capacity of the base to attract the proton from the molecule of the substance, i.e., gives an idea of the relative strength of bases. Thus, the use of infrared spectroscopy has opened up new possibilites for comparing the strengths of weak bases. With few exceptions, in the work of Gordy, it was established that there is a linear relation between the logarithm of the constant $K_a^{H_2O}$ of substances in aqueous solutions* and the magnitude of the displacement they produce in the band of the OD bond in the infrared spectrum of heavy methanol or heavy water:

$$\Delta\mu = 0.0147 \log K_a^{H_2O} + 0.194 \ (\text{for } CH_3OD)$$

$$\Delta\mu = 0.0175 \log K_a^{H_2O} + 0.297 \ (\text{for } D_2O),$$

where $\Delta\mu$ is the displacement of the band of the OD bond in microns.

* The constant $K_a^{H_2O} = K_W / K_b^{H_2O}$. The numerator is the ionic product of water $K_W = (H_3O^+)(OH^-)$, while the denominator is the dissociation constant of the base B in water and is given by $K_b^{H_2O} = (BH^+)(OH^-)/(B)$. Above, we have normally denoted it by $K_i^{H_2O}$, stating that we were considering the dissociation constant of the base. By substituting the values of the constants given in the expression for $K_a^{H_2O}$ we obtain

$$K_a^{H_2O} = \frac{(H_3O^+)(OH^-)(B)}{(BH^+)(OH^-)} = \frac{(H_3O^+)(B)}{(BH^+)}.$$

Consequently, $K_a^{H_2O}$ is the dissociation constant of the acid BH^+ conjugate with the base. The sum $pK_a^{H_2O} + pK_b^{H_2O} = 14$.

The linear relation is shown in Fig. 25, where $\log K_a^{H_2O}$ is plotted along the ordinate axis and $\Delta\mu$ along the abscissa axis.

Table 76 gives some data obtained with heavy methanol, which illustrate the relation between the ionization constant of the base in water and the displacement of the band in the infrared spectrum ($\Delta\mu$ in microns and $\Delta\nu$ in cm^{-1}).

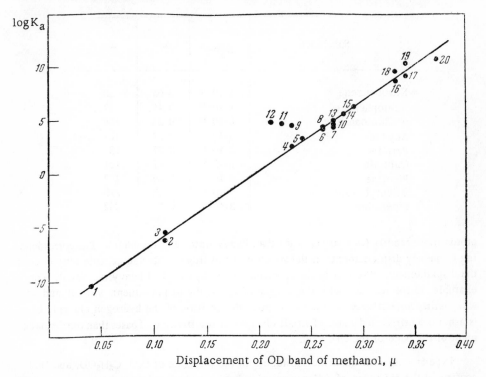

Fig. 25. Relation between basicity constants of substances and the displacement of the band of the OD bond of deuteromethanol. 1) Nitrobenzene; 2) acetophenone; 3) p-methylacetophenone; 4) o-chloroaniline; 5) m-chloroaniline; 6) o-toluidine; 7) quinoline; 8) aniline; 9) m-toluidine; 10) methylaniline; 11) N-ethyl-o-toluidine; 12) dimethylaniline; 13) pyridine; 14) quinaldine; 15) α-picoline; 16) dimethylbenzylamine; 17) benzylamine; 18) tributylamine; 19) tri-n-propylamine; 20) piperidine.

In view of the linear relation between the constant $K_a^{H_2O}$ and the frequency displacement $\Delta\nu$, Gordy and Stanford concluded that it is possible to determine the ionization constant from the magnitude of the displacement. In a brief note, Hammett [61] reported that experimental determinations have confirmed ionization constants interpolated from Gordy and Stanford's graph; the values of the latter are given in brackets: acetophenone $0.9 \cdot 10^{-20}$ ($1 \cdot 10^{-20}$), benzoic acid $6 \cdot 10^{-22}$ ($5 \cdot 10^{-22}$),* nitrobenzene $10^{-24} - 10^{-25}$ ($4 \cdot 10^{-25}$). The good agreement of the observed and calculated values indicates that the method is suitable for determining the ionization constants of extremely weak bases, which are difficult to measure by other methods.

---

*Interpolated data for methyl and benzyl benzoates. The experimental values of $K_b^{H_2O}$ were obtained by Hammett for benzoic acid as a base.

To compare the proton-acceptor power of very weak bases, Gordy [60] made use of measurements of the infrared spectra of hydrogen chloride as the latter denotes a

TABLE 76. Comparison of Dissociation Constants of Weak Bases and Displacements of the OD Band of Methanol in the Infrared Spectrum

Substance	$K_b^{H_2O}$	$\Delta\mu$	$\Delta\nu$
Nitrobenzene	$4 \cdot 10^{-25}$	0.04	28
Acetophenone	$1 \cdot 10^{-20}$	0.11	77
o-Chloroaniline	$4 \cdot 10^{-12}$	0.23	156
Aniline	$5 \cdot 10^{-10}$	0.27	181
Pyridine	$2 \cdot 10^{-9}$	0.27	181
Quinoline	$1 \cdot 10^{-9}$	0.27	181
Picoline	$4 \cdot 10^{-8}$	0.29	193
Tributylamine	$9 \cdot 10^{-5}$	0.33	218
Piperidine	$2 \cdot 10^{-3}$	0.37	242

proton more readily (is a strong acid) than heavy water and methanol. The corresponding frequency displacements in the spectrum of hydrogen chloride considerably exceed the displacements observed in the spectrum of the hydroxyl of heavy methanol. For example, in the reaction with the oxygen of ether, the displacements are 130 and 440 wave units, respectively, and consequently the position of the hydrogen chloride band is a more sensitive indicator of small changes in the strength of bases than the displacement of the band of the OD bond.

Experiments with the same substances in the presence of D_2O, C_2H_5OD, and HCl confirmed the existence of a linear relation between the band displacements in the spectra of heavy methanol and water and also in the spectra of methanol and hydrogen chloride. Thus, on the basis of measurements with one proton (deuteron) donor it is possible to predict the effect when a hydrogen bond is formed with another donor (see Fig. 26).

The frequency displacement caused by the formation of a hydrogen bridge is not at all dependent on the dipole moment of the electron-donor molecule, though this might have been expected had the deciding role been played by solely electrostatic forces. On the contrary, there is a clear connection between the magnitude of the displacement and the ionization energy of the electron-donor molecule [51]. As was mentioned on pp. 13 and 164, according to Mulliken the magnitude of the ionization potential is of great importance in a donor-acceptor interaction. These facts agree with the view that a hydrogen bond is formed as a result of a donor-acceptor interaction.

Later measurements of Tamres, Searles et al. [62] revealed exceptions from Gordy's rule. If pK_a (i.e., $-\log K_a^{H_2O}$) is plotted against $\Delta\nu$ (the displacement of the band of the OD bond of deuteromethanol), pyridine derivatives and aliphatic amines lie on two different straight lines (Fig. 27). The first line is described by the equation:

$$\Delta\mu = 0.0242 \ pK_a^{H_2O} + 0.194,$$

whose slope differs somewhat from that given by Gordy and Stanford's equation. The existence of two lines is explained by the fact that in the first case, the π-electrons of

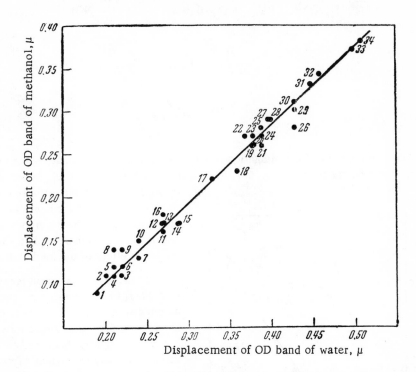

Fig. 26. Comparison of the displacements of the band of the OD bond in the spectra of CH_3OD and D_2O. 1) Acetonitrile; 2) n-butyronitrile; 3) methyl ethyl ketone; 4) acetophenone; 5) ethyl acetate; 6) acetonitrile (?) 7) diacetyl; 8) dioxane; 9) acetone; 10) 1,4-dimethoxycyclohexane;11) dimethyl ether of ethylene glycol; 12) N,N-dimethylformamide; 13) cyclohexanone; 14) n-butyl phosphate; 15) triethyl phosphate; 16) N,N-dimethylacetamide; 17) Δ^2-cyclohexenone; 18) pyrrole; 19) o-toluidine; 20) m-toluidine; 21) propylene chlorohydrin; 22) quinoline; 23) pyridine; 24) aniline; 25) quinaldine; 26) dimethylcyclohexylamine; 27) α-picoline; 28) nicotine; 29) propylene glycol; 3) ethylene glycol; 31) dimethylbenzylamine; 32) benzylamine; 33) piperidine; 34) ethyl methyl ketoxime.

the heterocycle participate in the formation of the hydrogen bond, while the electron pair of the aliphatic amine participate in the reaction between the latter and an alcohol molecule. There is an analogy with the characteristics of the interaction in the ionization of acids and bases, which was examined above (p. 203) and as a result of which points corresponding to different groups of compounds fall on different lines.

In another article of these authors [63], additional examples are given of deviations from Gordy's rule, which are not always explainable. Points corresponding to pK_i and $\Delta\nu$ of cyclic imines with four-, five-, and six-membered rings (azetidine II, pyr-

rolidine III, and piperidine IV) lie on the line for acylic aliphatic amines, while eth-
ylenimine (I) lies close to the line for nitrogen heterocycles (pyridine and its derivatives). This is

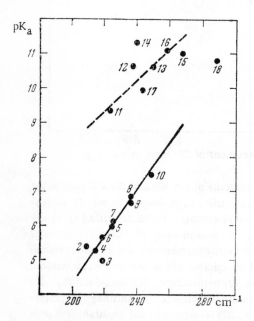

ascribed to the presence of conjugation of the ring with the electron pair of the nitrogen in the three-membered ring. It is less comprehensible why the replacement of a
hydrogen bound to the nitrogen in cyclic imines by a methyl group displaces the points toward the line for pyridine derivatives. The authors explain this by the steric effect of the methyl group.

We should also mention recent work [63a] in which data were presented on the displacement of the band of the OD bond in the infrared spectrum of heavy methanol by many ethers of mono-, di-, and triethylene glycols, etc. A comparison of the results obtained on the basic function of oxygen in organic compounds by optical and other methods is given in [63b].

In recent years it has also been confirmed that Gordy's rule also applies to an acid−base interaction in which the base is the molecule of an aromatic hydrocarbon, in particular, an alkylbenzene, whose strength as a base increases with the successive replacement of hydrogen in the benzene ring by alkyl groups (benzene < toluene < m-xylene < < mesitylene < durene < pentamethylbenzene < hexamethylbenzene) (see p. 196 for more details). Josien and his co-workers [64-66] give the results of measuring the infrared frequencies of the N−H bond in pyrrole, O−H bond in phenol, and Cl−H bond in hydrogen

Fig. 27. Relation between the displacement of the band of the OD bond in the CH_3OD spectrum and pK_a of nitrogen bases. 1) Pyrrole; 2) isoquinoline; 3) quinoline; 4) pyridine; 5) α-picoline; 6) β-picoline; 7) γ-picoline; 8) 2,4-lutidine; 9) 2,6-lutidine; 10) 2,4,6-collidine; 11) benzylamine; 12) n-butylamine; 13) n-propylamine; 14) di-n-butylamine; 15) diethylamine; 16) diisopropylamine; 17) tri-n-butylamine; 18) triethylamine.

chloride dissolved in aromatic hydrocarbons, whose strength as bases was successively increased. Table 77 gives the values of $\left(\frac{\Delta\nu}{\nu_g}\right)_{X-H}\cdot 10^2$, i.e., the displacement of the X−H bond frequency (relative to the bond frequency in the gas state ν_g).

TABLE 77. Displacement of Frequencies in Infrared Spectra

Substance	Pyrrole	Phenol	HCl
C_6H_5Cl	—	21.2	37.1
C_6H_6	20.4	24.9	47.1
$C_6H_5CH_3$	23.4	29.0	49.2
$o\text{-}C_6H_4(CH_3)_2$	—	31.6	56.5
$m\text{-}C_6H_4(CH_3)_2$	24.1	—	56.5
$1,2,4\text{-}C_6H_3(CH_3)_3$	—	33.6	58.1
$1,3,5\text{-}C_6H_3(CH_3)_3$	25.8	33.7	60.2

On making analogous measurements with solutions of three hydrogen halides (HCl, HBr, and HI) in aromatic hydrocarbons, Josien noted the proportionality of the displacements $\Delta\nu$ for solutions of each of the pairs of hydrogen halides in the corresponding hydrocarbon and also proportionality between the magnitude of the displacement and the deviation from Henry's law when the hydrogen halide was dissolved in n-heptane with the addition of the aromatic hydrocarbon, which is also a function of its strength as a base (see p. 154). The results are illustrated in Fig. 28.

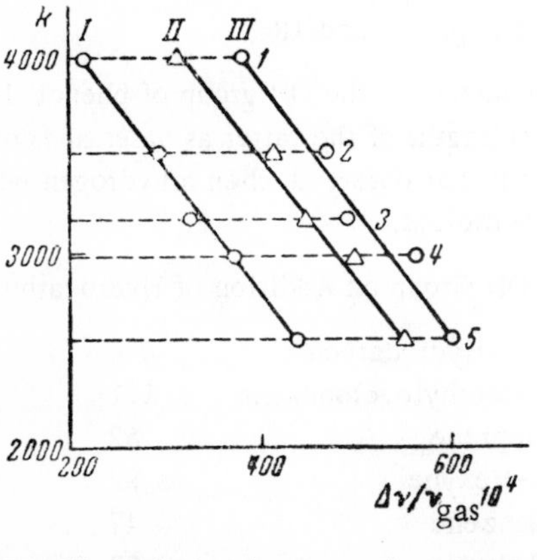

The association constants of complexes were determined [66a] from the intensity of the infrared absorption bands of the valence vibrations of the OH group of phenol or the NH group of pyrrole and of the bands of the same groups in substances in a complex with a proton acceptor (i.e., a base). The proton acceptors were pyridine (55; 2.7) and a group of aromatic compounds: chlorobenzene (0.30; 0.23), benzene (0.37; 0.30), mesitylene (0.58; 0.40), and hexamethylbenzene (0.85; 0.72). The values in brackets are the association constants $K_{ass} = a_c/a_d \times a_a$, where a_c, a_d, and a_a are the activities of the complex and the proton donor and acceptor. The first of the constants given was obtained for the complex of the base with phenol and the second for the complex with pyrrole. The solvent was carbon tetrachloride. There is a linear relation between K_{ass} and $\Delta\nu/\nu_{gas}$ and either of them may be used as an approximate measure of the relative strength of bases. The article gives

Fig. 28. Relation between the constant of Henry's law k and the displacement of the valence vibration band of hydrogen halides. I) HI; II) HBr; III) HCl. 1) Chlorobenzene; 2) benzene; 3) toluene; 4) m-xylene; 5) mesitylene.

literature data on the association constants of complexes of phenols with stronger bases than hydrocarbons: acetone (8.5), acetonitrile (9.3), ethyl ether (10), triethylamine (83), etc. The association constants are given in brackets.

Cook [67] measured the displacement of the characteristic frequency in the infrared spectrum of hydrogen chloride (cm^{-1}) as a result of the formation of complexes with alkylbenzenes, dissolved in carbon tetrachloride (see Table 78).

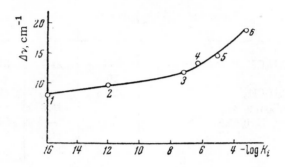

Fig. 29. Relation of the displacement of the C $\equiv$ N valence vibration band to the logarithm of the ionization constant of the solvent. 1) Methanol; 2) phenol; 3) acetic acid; 4) formic acid; 5) chloroacetic acid; 6) trichloroacetic acid.

There is a linear relation between the values of ν_{HCl} and the ionization potentials of alkylbenzenes, which were taken from [68] (cf., pp. 13 and 164).

West [68b] used the displacement of the frequency of the OH group of phenol in CCl$_4$ in the presence of olefins to compare the strengths of the latter as bases and compared the frequency displacement $\Delta\nu$ (cm^{-1}) with that observed when a hydrogen bond is formed between phenol and benzene and its homologs.

Displacement of Infrared Frequency of Phenol OH Group on Addition of Hydrocarbons

Hydrocarbon	$\Delta\nu$	Hydrocarbon	$\Delta\nu$
1-Octene	62	1-Methylcyclohexene	111
1-Hexene	60	Isoprene	52
2-Hexene	85	1-Hexyne	92
2-Pentene	86	Benzene	47
Cyclohexene	95	Toluene	58
2-Methyl-1-pentene	100	p-Xylene	66
2-Methyl-1-butene	104	Mesitylene	73
2-Methyl-2-butene	108		

Among the olefins, the weakest bases are the hydrocarbons with the double bond in position 1. However, they are superior to benzene and toluene in strength and approach p-xylene. The electron-donor power of olefins, like aromatic hydrocarbons, is increased by methylation (p. 156). Measurements with isoprene showed that the basicity of the π-electrons of a double bond is decreased by conjugation with another double bond. The electron-donor power of a triple bond is shown more clearly than that of a double bond, judging by the fact that 1-hexyne produces a greater displacement of the frequency of the OH group of phenol than 1-hexene.

TABLE 78. Displacement of HCl Frequency on Reaction with Alkylbenzenes

Substance	v	Δv
C_6H_6	2765	—
$C_6H_5CH_3$	2735	30
$1,3\text{-}C_6H_4(CH_3)_2$	2733	32
$1,3,5\text{-}C_6H_3(CH_3)_3$	2718	47
$1,2,4,5\text{-}C_6H_2(CH_3)_4$	2725	60
$C_6H(CH_3)_5$	2692	73
$C_6(CH_3)_6$	2683	82

The hydrogen of a CH bond of a hydrocarbon can also participate in the formation of a hydrogen bridge.* Thus, according to the communication of D. N. Shirogin, N.M. Shemyakin, and M. N. Kolosova [Izvest. Akad. Nauk SSSR, Otdel. Khim. Nauk·(1958) p. 1133], this type of intermolecular reaction is observed when a sufficiently acid hydrocarbon, in particular, phenylacetylene, is dissolved in electron-donor solvents. The formation of a hydrogen bond is accompanied by a fall in the frequency of the CH bond valence vibrations of the $\equiv C-H$ group, which equals 3315 cm^{-1} for the substance as the vapor and as a solution in carbon tetrachloride. The relative band displacement reached 3%.

Solvent	Δv, cm^{-1}
Acetone	60
Dioxane	65
Ether	67
Pyridine	95

Measurements were made with mixtures consisting of one part of hydrocarbon to ten parts of solvent.

More detailed information on the acidic and basic properties of many acetylene derivatives is given in [68c, d].

A regular displacement of the characteristic frequency of a bond vibration in the infrared spectrum is also observed when the strength of the acids reacting with the same

---

* This hypothesis was put forward first to explain deviations in the solubility from the "ideal" in electron-donor solvents and was confirmed by measurements of the infrared spectra of systems containing chloroform, in which the hydrogen of the CH bond is labile due to the inductive displacement of the electrons toward the chlorine atoms (p. 229). It has been suggested [68a] that the formation of a hydrogen bridge through the hydrogen of a CH bond of hydrocarbons of comparatively low acidity may be observed if a very strong base is used (for example, a solution of potassamide in liquid ammonia).

base is varied. This phenomenon was described by V. M. Chulanovskii and his co-workers [69]. The frequency of the $C \equiv N$ bond in the spectrum of acetonitrile was displaced regularly when the substance was dissolved in methanol (10^{-16}), phenol(10^{-10}), acetic acid (10^{-5}), formic acid (10^{-4}), monochloroacetic acid (10^{-3}), and trichloroacetic acid (10^{-1}) (the ionization constants of the acids in water are given in brackets). As the ionization constant of the acid increased, the band was displaced toward higher frequencies (Fig. 29). V. M. Chulanovskii explains this regularity by the fact that with an increase in the ionization constant, there is an increase in what he calls the "exposure" of the proton in the hydrogen bridge. In other words, in the reaction of an acid molecule AH with a base molecule B, the electron of the hydrogen atom enters a single electron shell with the electrons of the atom A, which is attached more strongly to the latter than to the proton. The formation of a hydrogen bond is accompanied by partial protonization of the hydrogen atom of the hydroxyl group of the acids and the degree of protonization (or "exposure" of the proton) is greater, the stronger the acid.

Barrow [70] measured the displacement of the OH bond frequency of alcohols and the change in infrared absorption intensity caused by a hydrogen bond between the hydroxyl group of the alcohols and two bases of different strengths, namely, ethyl ether and trimethylamine. In Table 79, the frequency displacement $\Delta \nu$ is given relative to the frequency ν (cm^{-1}) of the OH band for alcohol in an inert solvent, namely, carbon tetrachloride. The magnitude of the frequency displacement and the intensity change correspond to the relative strength of the solvent as a base and the alcohol as an acid.

TABLE 79. Displacement of the Infrared Frequencies of Alcohol OH Groups

Alcohol	Frequency displacement, cm^{-1}		Intensity change, liter·mole^{-1} × × cm$^{-1} \cdot 10^{-4}$		K_i	$k_{120°} \cdot 10^7$, sec^{-1}
	$(C_2H_5)_2O$	$(CH_3)_3N$	$(C_2H_5)_2O$	$(CH_3)_3N$		
tert-Butanol	116	337	2.6	4.6	—	—
Isopropanol	122	374	3.0	4.8	1	7
sec-Butanol	123	375	3.2	5.3	—	—
Ethanol	133	393	3.8	6.1	12	20
n-Butanol	135	400	3.8	6.7	—	—
Methanol	136	401	3.9	6.2	50	30
2-Ethoxyethanol	143	422	3.5	7.1	100	—
2-Chloroethanol	177	479	4.6	8.2	—	—
Ethylene glycol	—	—	—	—	547	90
Phenol	276	—	8.5	—	—	—

The relative acidity of alcohols of different structures was determined by Hine and Hine [71] by the indicator method. The ionization constant of isopropanol was arbitrarily taken as unity. These relative ionization constants of alcohols K_i are given in the last column but one in the table, while the last column gives the rate constants of deuterium exchange ($k_{120°}$, sec^{-1}) between the hydrogen of the methyl group of quinaldine and deuteroalcohol [71a]. The acidity of some substituted methanols was measured by Ballinger and Long [71b].

The hydrogen bond becomes more polar with an increase in the strength of the base and the acid. This is demonstrated directly by the dipole moments of alcohols in the solvents named [70] (Table 80).

TABLE 80. Dipole Moments of Alcohols in Various Solvents

Alcohols	Solvent		
	CCl_4	$(C_2H_5)_2O$	$(CH_3)_3N$
Tertiary and secondary	1.0	2.5	3.2
Primary	1.2	2.9	4.4
Phenol	1.5	3.9	—

D. Other Manifestations of a Hydrogen Bond

The above discussion of the formation of a hydrogen bond in reactions between acids and bases is completed by an account of the results of some work in which the solubility, freezing points, thermal effects, and proton magnetic resonance spectra of systems containing an acid and a base were measured. It can be shown that there is a linear relation between values obtained by different methods.

a. SOLUBILITY

By measuring the solubility of halogen derivatives of methane and ethane in many solvents, Zellhoefer, Copley, and Marvel [72-76] found that in solvents containing an electron-donor nitrogen or oxygen atom (amines, amides, ethers, esters, etc.), the solubility is higher than the "ideal," i.e., that calculated from Raoult's law. The solubility increases as the hydrogen in methane or ethane is replaced by a halogen, for example, from CH_3X to CH_2X_2 and then to CHX_3, but then falls sharply again when the last hydrogen atom in the molecule is replaced by a halogen (CX_4 or C_2X_6) [72].

The authors ascribed the increase in the solubility to an interaction between the solution components, which appears as the formation of a hydrogen bond between the electron-donor atom and the hydrogen atom of the halogen derivative ($R_3N...HCHal_3$). With an increase in the number of halogen atoms, there is an increase in the lability of the hydrogen atoms remaining in the molecule (see p. 96). When the last of them is replaced by a halogen, there is no possibility of the formation of a hydrogen bond. It is understandable that an increase in solubility is also observed when organic substances of the electron-donor type are dissolved in halogen derivatives of methane [73].

Like the solubility of other weak bases, the solubility of aromatic hydrocarbons (anthracene, phenanthrene, naphthalene, and mesitylene) in chloroform is greater than in carbon tetrachloride [72, 73].

The solubility of acetylene in solvents of the electron-donor type deviates from the "ideal" [74] because acetylene shows acidic properties (see p. 84).

The formation of a hydrogen bond in systems consisting of chloroform and an electron-donor substance was demonstrated by infrared spectral measurements. Analogous investigations were subsequently carried out with deuterochloroform [79, 80]. It has the advantage over normal chloroform that the overlapping of the infrared frequencies of some substances and the CH bond frequency of chloroform can be avoided. The latter frequency is displaced to different extents in the presence of electron donors of different strengths. The absorption intensity of the CH bond of deuterochloroform is considerably increased in various electron-donor solvents, including benzene and mesitylene, which also indicates the formation of a hydrogen bond [81].

The change in the solubility of HCl in ethers, alcohols, and carboxylic acids corresponds to the different electron densities at the oxygen atoms of these compounds [81-83]. There is good correlation between Gordy's data on the displacement of the band frequency of the OD bond in the infrared spectrum of deuteromethanol and the values of Henry's constant, as was found for the solubility of HCl in alkylbenzenes [66, 67] (pp. 155 and 224).

b. HEAT OF MIXING

One of the effects of the interaction of substances which appears as the formation of a hydrogen bond is the liberation of heat when they are mixed. Plotting the mole fraction of the electron-donor substance along the abscissa axis and the heat effect when it is mixed with chloroform, for example, along the ordinate axis gives a curve with a maximum, whose position corresponds to an equimolecular ratio between the mixture components [75, 76]. The same type of curve is obtained on measuring the heats of mixing of phenylacetylene with ether, acetone, cyclohexylamine, and methyl acetate [74].

Since the effects examined above (increased solubility, evolution of heat on mixing, and displacement of a band in the infrared spectrum) are the consequences of the same phenomenon, namely, the formation of a molecular compound through a more or less strong hydrogen bond, there may be a linear relation between the results of independent measurements.

Marvel and Copley [76] showed that there is quite a good linear relation between the amount of heat evolved when equimolecular amounts of chloroform and an electron-donor substance (solvent) are mixed and a value expressing the ratio of the solubility of dichlorofluoromethane, $CHCl_2F$, in the same solvent to the ideal solubility. This ratio also serves as a measure of the intermolecular reaction.

It was found that there is a linear relation between the solubility of CHF_2Cl and $\Delta\nu$, the displacements of the band of the OD bond in the infrared spectrum of C_2H_5OD and also between the heats of mixing of electron-donor liquids with chloroform ΔH and $\Delta\nu$. The existence of the latter relation was also confirmed later by Searles, Tamres, and Barrow [84-86] with simple and cyclic ethers, lactones, ketones, sulfoxides, and nitrogen bases. We should note that points for substances belonging to different classes of nitrogen bases lie on a common line, which is expressed by the equation:

$$\Delta H = 8.33\,\Delta\nu - 1240,$$

where ΔH is the heat of mixing in cal/mole (with equimolecular amounts of chloroform and electron donor), though the points in the coordinates of $\log K_a^{H_2O}$ and $\Delta\nu$ for aliphatic amines and heterocycles lie on different lines (p. 222). The line corresponding to the above equation is given in Fig. 30.

It should be emphasized particularly that Tamres [87] showed that there is a linear relation between the heats of mixing of aromatic hydrocarbons with $CHCl_3$ and the displacement of the OD bond frequency of heavy methanol produced by interaction with the same hydrocarbons. No evolution of heat is observed when aromatic hydrocarbons are mixed with CCl_4, with which the formation of a hydrogen bond is impossible as a result of the absence of hydrogen from its molecule (Table 81).

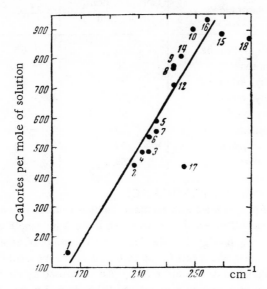

Fig. 30. Relation between the displacement of the band of the OD bond in deuteromethanol and the heat of mixing of nitrogen bases with chloroform at 25°. The numeration is the same as in Fig. 27.

c. FREEZING POINT

A direct demonstration of the formation of molecular compounds of a definite stoichiometric composition in systems consisting of chloroform and an electron-donor solvent is provided by the presence of maxima on the freezing point curves [88, 88a]. In most cases their position corresponds to a 1:1 composition. Measurements in Schneider's laboratory [89] with systems consisting of chloroform and an aromatic or ethylenic hydrocarbon also revealed maxima on the composition-freezing curves. Consequently, unsaturated hydrocarbons react with chloroform like other electron donors-bases (Table 82).

TABLE 81. Heat of Mixing and Infrared Frequency Displacement

Substance	Heat of mixing, cal/mole		Frequency displacement, $\Delta\nu$, cm^{-1}
	$CHCl_3$	CCl_4	
1,2-Dichlorobenzene . .	0	− 69	13
Chlorobenzene	29	− 34	21
Benzene	90	− 38	24
m-Xylene	202	− 3	20
Mesitylene.	216	− 35	34

In contrast to HCl, which is a stronger acid than chloroform, the latter does not give molecular compounds with benzene and cyclohexene, which are weaker bases than those listed in Table 82.

TABLE 82. Freezing Points of $CHCl_3$ -Electron-Donor Systems

Substance	Complex composition, mol. %	M.p., °C
Triethylamine	50	−85
Acetone	50	−109
Ethyl ether	33	−93
Ethyl ether	50	−95
Propionitrile	50	−90.5
Toluene	50	−106
Mesitylene	50	−49
Hexene	66.6	−83

d. PROTON MAGNETIC RESONANCE

Measurements of proton magnetic resonance make it possible to detect the formation of a hydrogen bond. This was first discovered in 1951 by Arnold and Packard [90]. They observed that while the signal of the CH_2 and CH_3 groups of ethanol are independent of temperature, a rise and a fall in the latter displaces the signal of the proton of the OH group of alcohol in opposite directions. This is caused by the fact that at low temperature there is most complete association of the alcohol molecules, while at a high temperature, the complex formed through a hydrogen bond is almost completely dissociated.

In the opinion of I. V. Aleksandrov [91], the sign and magnitude of the displacement occurring, for example, when the hydrogen bond O−H... O is formed may be explained by assuming that the O−H bond is polarized to a considerable extent as a result of a donor-acceptor interaction of the hydrogen atom with the atom of the adjacent molecule. The electron density at the hydrogen atom is thereupon reduced to approximately 15% (see also [92]). The formation of a complex between chloroform and triethylamine through a hydrogen bond was confirmed by proton magnetic resonance [92a]. Korinek and Schneider [88] compared the displacements of the signal of the chloroform proton in the formation of a hydrogen bond with a series of substances, differing in electron-donor properties. Because of the self-association of chloroform molecules, the measurements were made with different concentrations of it in a mixture with the base and extrapolated to zero concentration. For comparison, analogous measurements were made with solutions of chloroform in saturated hydrocarbons (cyclopentane, cyclohexane, and n-hexane). The displacements of the chloroform proton signal are given below. The magnitude of the signal depends on the strength of the hydrogen bond and is greater, the stronger the bond.

Electron donor	Displacement, cycles/sec
Triethylamine	48.0
Acetone	32.4
Diethyl ether	30.5
Propionitrile.	22.0
Propyl fluoride	13.0

Together with a hydrogen bond whose formation involves the free pair of electrons of an N, O, Cl, or F atom, the association between chloroform and an aromatic or olefinic hydrocarbon through the π-electrons of the hydrocarbon has been studied by the same method: the π-electrons of the double bond in cyclopentadiene, cyclohexene, and hexene react with the hydrogen of chloroform somewhat more weakly than the electron pair of a chlorine or fluorine atom. The formation of a weak hydrogen bond produces a small displacement of the proton signal (6-8 cps) in the same direction as the displacement caused by reaction with the bases examined.

The interaction of chloroform with aromatic hydrocarbons (benzene, toluene, and mesitylene) produces a displacement of the signal in the opposite direction by 54-57 cps. Judging by the freezing curves, alkylbenzenes form molecular compounds (1:1) with chloroform, while benzene, as a weaker base, forms an analogous complex only with hydrogen chloride.

There is a correspondence between the magnitude of the displacement of the proton magnetic resonance signal of chloroform and the ionization potential of the electron donors [93-95] and also the intensity of the band of the OD bond in the infrared spectrum of deuterated chloroform [81] (cf., p. 227).

7. RANGE OF PROTOLYTIC REACTIONS

Thus, it is evident from what has been presented above that with the present level of our knowledge of reactions between acids and bases, Brønsted's theory is already inadequate to describe them; further development of the theory of acids and bases is therefore required.

Even when the acid−base reaction is completed by ionization of electrically neutral molecules, its mechanism is considerably more complex than that presented in the Brønsted-Lowry scheme. Consideration of the equilibria existing in solutions of acids and bases enabled N. A. Izmailov to give a more complete and more accurate picture of the dissociation of protolytes and to refine the quantitative side of the theory. It is not difficult to demonstrate that even if a reaction between an acid and a base leads to the formation of a salt which is readily isolated in a crystalline state, there is not always complete transfer of a proton. In this connection we should mention the measurements of vibration spectra by P. G. Yaroslavskii [96] (in A. N. Terenin's laboratory) and P. P. Shorygin [97]. It was shown that compounds of aniline with not only weak acids (phenol and acetic acid), but even hydrochloric acid are not true salts in which the proton of the acid is transferred completely to the nitrogen atom of the base, but are molecular compounds. No less clear are the results of experiments on deuterium exchange between aromatic amines and carboxylic acids [98], which have been mentioned already on p. 37.

Aromatic amines have dual reactivity, which appears in their reaction with acids through not only the nitrogen atom, but also the ortho- and para-carbon atoms. The first reaction results in the formation of an amine salt, while the second reaction leads to isotopic exchange when there is a difference in the isotopic compositions of the amine and the acid. The addition of a proton to the nitrogen atom saturated the free electron pair of the latter and disrupts the conjugation of the free pair of electrons of the nitrogen atom with the π-electron system of the aromatic ring. Had the valence state of the nitrogen atom changed identically in the formation of a salt with any acid (as follows from the Brønsted-Lowry scheme), the retarding effect would have been the same. This is not confirmed by experiment. Measurement of the rate of deuterium exchange between anhydrous carboxylic acids (acetic, formic, chloroacetic, trichloroacetic, and fluoroacetic acids) and p-deuterodimethylaniline shows that in actual fact, the degree of retardation of the exchange depends on the strength of the acid and its dielectric constant. The rate of the reaction with triphenylamine, which does not give salts with carboxylic acids, increases as the strength of the acid increases, while in experiments with p-deuterodiphenylmethylamine, which forms salts only with the strongest acids, the exchange rate constant passes through a maximum.

The sequence of the rate constants of hydrogen exchange of dimethylaniline and triphenylamine with trifuloroacetic acid is opposite to that found for deuterium exchange with acetic acid. The retardation of hydrogen exchange in dimethylaniline with an increase in the strength of the acid and an increase in the dielectric constant is caused by the increase in the activation energy of the exchange reaction.

Reactions consisting of the formation of molecular compounds and associates of different degrees of strength in which the molecules are connected by a hydrogen bond were examined in detail above. Freezing curves of the systems give information on the composition of the molecular compounds formed. Dipole moment measurements indicate the polarity of the complexes obtained. The degree of reaction between substances can be assessed from the solubility, heat of solution, displacement of infrared absorption frequency, displacement of the proton signal in proton magnetic resonance measurements, association constants of complexes between proton donors and acceptors, and rate constants of hydrogen exchange. All the values quantitatively express different aspects of the same phenomenon. There is therefore proportionality between them. There is also proportionality between these values and the ionization potentials of the bases participating in the reactions. This agrees with the idea of the donor-acceptor interaction between the reagents. The proportionality between the given values and the dissociation constants of the substances in aqueous solution is the result of the fact that the formation of a hydrogen bond between them and their ionization as a result of proton transfer are essentially different stages of an acid—base interaction, which begins with the formation of a molecular compound between the acid and the base and ends with their ionization. Only the last stage is represented by the Brønsted-Lowry scheme for an acid—base reaction. Consequently, this scheme does not cover all stages and cases of protolytic processes.

The potential curves of donor-acceptor interactions, which were examined in Section I (p. 14), also apply in general to protolytic reactions. In the absence of a polar medium (or with reagents of very low "protolytic activity"), the molecular compounds obtained are of the type which Mulliken called "outer" complexes. Without a polar medium, the potential barrier is too high for the change to an "inner" complex, i.e., for strong polarization of the bond right up to its ionization. In a polar medium (especially with reagents of high "protolytic activity"), energetically favorable conditions are created for the conversion of the "outer" complex into an "inner" complex, i.e., an ion pair is formed, which decomposes to free ions as a result of a secondary process (for example, solvation). In the work of A. N. Terenin [99], it was shown by direct experiment that the formation of a hydrogen bond proceeds ionization in an acid—base interaction and that the two stages are distinguished energetically. Mixing solutions of succinic acid and a nitrogen base, acridine, forms a molecular compound:

Substance	P, kcal/mole	Substance	P, kcal/mole
NH_2^-	419	C_6H_6	150
OH^-	383	CH_4	122 ± 8
NH_3	214	H_2	70 ± 9
H_2O	182	C_2H_6	61; 110 ± 10
C_2H_4	174	C_3H_8	61

The partial protonization of the hydrogen atom, which occurs in the formation of the hydrogen bridge, is completed by the complete transfer of the proton, i.e., ionization of the acid and the base, when the solution is irradiated with light. The molecular compound is converted into a salt through the energy supplied by the light quanta:[*]

This is an example illustrating the connection between "outer" and "inner" complexes in acid−base interaction. A. N. Terenin recognizes the possiblity of incomplete transfer of a proton and writes of it in the work. In the adsorption of bases (ammonia, amines, and aromatic hydrocarbons) on silicate and aluminosilicate catalysts, the OH groups of the latter fulfill a proton-donor function [100, 100a]. As emphasized by A. N. Terenin [101], a detailed spectroscopic study of the behavior of the N−H bond frequencies of adsorbed ammonia molecules "give no grounds for assuming that there is complete transfer of the proton of the OH surface group to the nitrogen atom of the adsorbed base. The acid−base interaction is not completed by salt formation here."

The increasing number of examples of the effects of acid−base interaction provided by future investigations will make it possible to trace more fully the relation between the formation of unstable associates and ionization, to characterize quantitatively all the stages of protolytic reactions, and to describe clearly their relation to the structure of the reagents. However, even now it is possible to use the differences between their initial and final stages to classify protolytic reactions.

According to Brønsted, bases are substances with an affinity for a proton. The affinity for a proton of some ions, for example, carbanions, an amide ion, and others may be very high and reach 400 and more kcal/mole. The whole range of affinities of ions and molecules for a proton has the lower limit of zero.

For example, the affinity of a molecule of the hydrocarbon ethylene for a proton reaches 174 kcal/mole [102]. This value differs from zero by approximately the same amount as the affinity of a hydroxyl ion for a proton is greater than the affinity of the ethylene molecule for a proton. Undoubtedly, the range down to zero may be filled with many substances with the properties of weak bases. On the other hand, in principle, any substance containing hydrogen and hydrogen itself under suitable conditions may be acids.

---

[*] It has been reported [99a] that on irradiation, acridine becomes a stronger base (pK = = 10.65) than normal acridine (pK = 5.45) as a result of a change in its electron structure under the action of light quanta.

This statement, which was made several years ago in connection with a detailed examination of the acid – base properties of hydrocarbons [103], is undoubtedly correct and the gaps in the scale of acids and bases are being filled gradually with the increase in the sensitivity of investigation methods playing an important part in this.

V. L. Tal'roze and E. L. Frankevich [104] determined the affinity for a proton of molecules of some saturated hydrocarbons (methane, ethane, and propane) and the hydrogen molecule, using the ionic collision method. The values of P, the affinity for a proton, for a series of saturated hydrocarbons and the hydrogen molecule were given in their article. In 1959, American authors [104a] used a mass-spectrometric method to determine the affinity for a proton of the benzene molecule and considered that it equaled 145-150 kcal/mole.

The values of the affinity for a proton of ions and molecules given below are taken from several sources [104-106] and were determined and calculated in different ways. Their reliability and comparableness are not examined here. It should be noted that the values obtained for the same substance sometimes differ considerably. For example, Lampe and Field [104b] consider that the affinity for a proton of an ethane molecule is 102-121 kcal/mole, while according to Tal'roze and Frankevich, it equals 61 kcal per mole. According to the calculations of Hartmann and Grein [104c], the affinity for a proton of the CH_4 molecule is also greater (161 kcal/mole):

$$
\begin{array}{c}
O \\
\diagdown \\
C - CH_2 - CH_2 - C \\
\diagup \\
HO
\end{array}
\qquad
\begin{array}{c}
O \\
\diagup\diagup \\
\\
\diagdown \\
O - H \cdots N
\end{array}
$$

The width of the acidity scale is shown by the dissociation constants of **hydrides**, which were taken from Bell's book [107]. Naturally, some of the values given are quite provisional, but they still give an idea of the parameter interesting us.

CH_4 (10^{-58})	NH_3 (10^{-35})	OH_2 (10^{-16})	FH (10^{-3})
	PH_3 (10^{-27})	SH_2 (10^{-7})	ClH (10^{7})
		SeH_2 (10^{-4})	BrH (10^{9})
		TeH_2 (10^{-3})	IH (10^{10})

Both the proton-donor activity of substances containing hydrogen and the affinity for a proton are functions of the composition and structure of substances. It is obvious that a hydrogen-containing substance may have a definite affinity for a proton and act as a base. Both parameters vary strongly. Therefore, it is not only the strength of the bond between the product of a donor-acceptor interaction, but also its character that must vary. Ionization and even more so, the formation of free ions is only the limiting case of these observed in reactions between acids and bases.

8. DEFINITIONS OF ACIDS AND BASES

If it is shown that the scheme of the acid – base process postulated by Brønsted and Lowry is inaccurate and inadequate, then it is obviously necessary to change the definitions of an acid and a base derived from this scheme. This idea arose long ago. Thus, in 1939 Gordy [58] wrote: "The contemporary definition of a base as a proton acceptor means that the basicity of a substance is caused by its tendency to add a proton. Why should not the contemporary definition of an acid as a proton donor and a base as a proton acceptor include all substances which participate in the formation of an inter-molecular hydrogen bond? Substances whose hydrogen participates in the formation of a hydrogen bond should be regarded as acids."

Similar views were put forward in 1947 and later by Davis [20] (p. 203). In 1949 Kolthoff [108] observed: "The last word has certainly not been said on the mechanism of an acid—base reaction. It is at least debatable whether the initial reaction between bases and Bronsted acids is the direct transfer of a proton. The initial reaction between an acid and a base may consist of the formation of a hydrogen bond. In basic solvents with a high dielectric constant, the acid reacts with the solvent-base with the initial formation of a hydrogen bond. After this there is dissociation with the formation of a solvated proton and the base conjugate with the acid. However, in solvents with a low dielectric constant, an acid and a base may react to form a stable addition product through a hydrogen bond. Even in an aqueous medium there are often reactions associ-ated with the formation of a hydrogen bond. The reaction between water and ammonia is an example." Kolthoff suggested that substances which donate a proton to a base or combine with a base with the formation of a hydrogen bond should be called acids.

In view of these opinions and the multiplicity of new facts which became known in subsequent years, in 1955 definitions of acids and bases were suggested [103], which were genetically related to Brønsted's definitions, but represented the next step toward extending the range of protolytic reactions in accordance with new information on the reactions between acids and bases available at the present time. In addition, the posi-tion of acids and bases in Ingold's more general classification of reagents (Section I) was indicated in these definitions. In a somewhat modified form, the definitions are as follows: a b a s e is an electron-donor reagent with an affinity for a proton; an a c i d is an electron-acceptor reagent whose hydrogen participates in an equilibrium reaction with a base. An acid combines with a base through a hydrogen bond or donates a proton to it.

The initial and final stages of the acid—base interaction are mentioned. In con-trast to the definitions published previously [103, 109], the present definitions refer to "reagents" and not "substances." This emphasizes the conditional nature of the concepts "acid" and "base" as applied to an individual substance as its protolytic function appears only during a reaction, i.e., when the substance becomes a reagent. This eliminates the shortcoming of the previous definitions, which was pointed out by N. A. Izmailov [110, p. 568].

In contrast to acids, electron-acceptor reagents whose equilibrium reactions with bases do not involve hydrogen are called a c i d l i k e, s u b s t a n c e s [111].

9. SUMMARY

1. Reactions between acids and bases do not reduce to merely the transfer of a proton from the acid to the base in accordance with the Brønsted-Lowry scheme. In actual fact, the mechanism of acid—base interaction is more complex. During a pro-tolytic reaction, molecular compounds are formed between and acid and the base through a hydrogen bond and the reaction may be limited to this initial stage. Systems which include ions have specific characteristics and rules. The substantial difference between the initial and final stages of a protolytic reaction may be used for the classi-fication of protolytic reactions, which will be worked out in detail in the future.

2. As the Brønsted-Lowry scheme is inadequate to describe the mechanism of an acid—base process, the definitions of an "acid" and a "base" derived from it do not cover the whole range of protolytic reactions. It follows logically from this that the definitions must be changed. The new definitions also indicate the position of acids and bases in the more general classification of reagents proposed by Ingold. A b a s e is an electron-donor reagent with an affinity for a proton; an a c i d is an electron-acceptor reagent whose hydrogen participates in an equilibrium reaction with a base. An acid combines with a base through a hydrogen bond or donates a proton to it. An electron-acceptor reagent whose equilibrium reaction with a base does not involve hydrogen is called an a c i d l i k e s u b s t a n c e.

3. The inferences of Brønsted's quantitative theory of protolytic equilibrium usually apply only to protolytes which are similar in structure in solvents of the same chemical type. N. A. Izmailov developed a theory of the dissociation of acids, bases, and salts, which gives a fuller description of the phenomena occuring in solutions of electrolytes. In particular cases, its equation reduces to the equation of Brønsted's theory, which has played a dominant part in the development of chemistry in the last twenty-five years.

4. At the present time, methods are being developed for the quantitative assessment of the protolytic activity of substances under conditions where they do not react according to the Brønsted-Lowry scheme.

LITERATURE CITED

1.   A. I. Shatenshtein, Theories of Acids and Bases [in Russian] (Goskhimizdat, Moscow, 1949).
2.   M. Kilpatrick and M. Kilpatrick, Chem. Rev. 13, 131 (1933).
3.   L. A. Wooten and L. P. Hammett, J. Am. Chem. Soc. 57, 2289 (1935).
4.   F. N. Verhock, J. Am. Chem. Soc. 58, 2577 (1936).
5.   A. I Shatenshtein, Zhur. Fiz. Khim. 6, 33 (1936).
6.   V. A. Pleskov and E. N. Gur'yanova, Zhur. Fiz. Khim. 8, 346 (1936).
7.   G. S. Markova and A. I. Shatenshtein, Zhur. Fiz. Khim. 13, 1367 (1939).
8.   E. A. Izrailevich and A. I. Shatenshtein, Zhur. Fiz. Khim. 17, 24 (1943).
9.   A. I. Shatenshtein and E. A. Izrailevich, Zhur. Fiz. Khim. 13, 1791 (1939).
10. N. M. Dukhno and A. I. Shatenshtein, Zhur. Fiz. Khim. 22, 461 (1948).
11. N. M. Dukhno and A. I. Shatenshtein, Zhur. Fiz. Khim. 25, 670 (1951).
12. C. A. Kraus, J. Phys. Chem. 43, 231 (1939).
13. C. A. Kraus, J. Phys. Chem. 60, 129 (1956).
14. C. A. Kraus, J. Chem. Educ. 35, 324 (1958).
15. J. N. Brønsted, Ber. 61, 2049 (1928).
16. V. K. La Mer and H. C. Downes, J. Am. Chem. Soc. 53, 888 (1931); Chem. Rev. 13, 47 (1933).
17. V. I. Dulova and N. V. Vostrilova, Doklady Akad. Nauk Uzb.SSR No. 12, 14 (1948); No. 12, 20 (1951).
18. A. A. Mariott, J. Res. Nat. Bur. Stand. 38, 527 (1947); 41, 1, 7 (1948).
19. R. P. Bell, Acid-Base Catalysis (Oxford, 1941).
20. M. M. Davis, P. J. Schuman, H. B. Hetzer, and E. A. McDonald, J. Res. Nat. Bur. Stand. 39, 21 (1947); 41, 27 (1948); 42, 595 (1949); 46, 496 (1951); 48, 381 (1952); 60, 569 (1958); J. Am. Chem. Soc. 71, 3544 (1949); 76, 4247 (1954).
21. P. I. Walden, Transactions of the Jubilee Mendeleev Conference [in Russian] (Izd. AN SSSR, Moscow, 1938) Vol. I, p. 513.

22. N. A. Izmailov and M. A. Bel'gova, Zhur. Obshchei Khim. 9, 453 (1939).
23. N. A. Izmailov, M. V. Shustova, and N. Vodorez, Zhur. Obshchei Khim. 9, 598 (1939).
24. N. A. Izmailov and I. F. Zabara, Zhur. Fiz. Khim. 20, 165 (1946).
25. N. A. Izmailov and M. A. Bel'gova, Collection of Work on Physical Chemistry [in Russian] (Izd. AN SSSR, Moscow, 1947) Vol. I, p. 301.
26. N. A. Izmailov and I. F. Zabara, Collection of Work on Physical Chemistry [in Russian] (Izd. AN SSSR, Moscow, 1947) Vol. I, p. 310.
27. N. A. Izmailov, L. L. Spivak, and V. Levchenkova, Uch. Zap. Khar'kov. Gos. Univ. 30, 123, 135, 153 (1950).
28. M. I. Kabachnik and S. T. Ioffe, Doklady Akad. Nauk SSSR 91, 833 (1953); M. I. Kabachnik, Uspekhi Khimii 25, 137 (1956).
29. N. A. Izmailov and L. M. Kutsyna, Izvest. Akad. Nauk SSSR, Ser. Fiz. 17, 740 (1953).
30. E. V. Titov and N. A. Izmailov, Trudy Khim. Fakul'teta i NII Khim. Khar'kov. Gos. Univ. 16, 139 (1957).
31. N. A. Izmailov and V. A. Kremer, Izvest. Akad. Nauk SSSR, Ser. Fiz. 15, 565 (1951).
32. K. P. Parkhaladze, Reaction of Acids with Differentiating Solvents. Dissertation [in Russian] (Khar'kov Gos. Univ., 1953).
33. N. A. Izmailov, Zhur. Anal. Khim. 4, 267 (1947).
34. N. A. Izmailov, Zhur. Fiz. Khim. 24, 321 (1950).
35. N. A. Izmailov, Trudy NII Khim. Khar'kov. Gos. Univ. 9, 139 (1951).
36. N. A. Izmailov, Trudy NII Khim. Khar'kov. Gos. Univ. 10, 5 (1953).
37. N. A. Izmailov, Trudy NII Khim. Khar'kov. Gos. Univ. 10, 49 (1953).
38. N. A. Izmailov, Zhur. Fiz. Khim. 28, 2047 (1954).
39. N. A. Izmailov, Trudy Khimfaka i NII Khim. Khar'kov. Gos. Univ. 14, 5(1956).
40. N. A. Izmailov, Zhur. Fiz. Khim. 30, 2164 (1956).
41. C. A. Coulson, Research 10, 149 (1957).
42. N. D. Sokolov, Uspekhi Fiz. Nauk 57, 205 (1955).
43. N. D. Sokolov, Doklady Akad. Nauk SSSR 58, 611 (1947).
44. N. D. Sokolov, Zhur. Eksp. Teor. Fiz. 23, 315 (1952).
45. N. D. Sokolov, Ukr. Khim. Zhur. 22, 19 (1946).
46. D. G. Knorre and N. M. Émanuél', Collection: Problems in Catalysis, Chemical Kinetics, and Reactivity [in Russian] (Izd. AN SSSR, Moscow, 1955) p. 106.
47. D. G. Knorre and N. M. Émanuél', Uspekhi Khimii 24, 275 (1955).
48. V. G. Vasil'ev and Ya. K. Syrkin, Acta Physicochim. URSS 14, 414 (1941).
49. A. V. Few and J. W. Smith, J. Chem. Soc. (1949) pp. 753, 2663.
50. R. L. Schupp and R. Mecke, Z. Elektrochem. 52, 54 (1948).
51. L. Hofacker, Z. Elektrochem. 61, 1048 (1957).
52. C. H. Giles, T. J. Rose, and D. G. M. Vallance, J. Chem. Soc. (1952) p. 3799.
53. L. Sobczyk and Ya. K. Syrkin, Roczniki chemii 30, 881, 893 (1956); 31, 197, 1245 (1957); see also [112] p. 323.
54. L. Sobczyk, Dielectric Polarization of Systems with a Hydrogen Bond. Dissertation [in Russian] (M. V. Lomonosov Moscow Chemicotechnological Institute, Moscow, 1954).
55. Ya. K. Syrkin, Ukr. Khim. Zhur. 22, 23 (1956).
56. H. Tsubomura, Bull. Chem. Soc. Japan 31, 435 (1958).
57. G. M. Barrow, J. Am. Chem. Soc. 78, 5802 (1956).

57a. C. L. Bell and G. M. Barrow, J. Chem. Phys. 31, 3000 (1959).

57b. G. M. Barrow, Spectrochim. Acta 16, 799 (1960).

58. W. Gordy, J. Chem. Phys. 7, 93 (1939).

59. W. Gordy and S. C. Stanford, J. Chem. Phys. 8, 170 (1940); 9, 204 (1941).

60. W. Gordy, J. Chem. Phys. 9, 215 (1941).

61. L. P. Hammett, J. Chem. Phys. 8, 644 (1940).

62. M. Tamres, S. Searles, E. M. Leighly, and D. W. Mohrman, J. Amer. Chem. Soc. 76, 3983 (1954).

63. S. Searles, M. Tamres, J. Block, and L. A. Quarterman, J. Am. Chem. Soc. 78, 4917 (1956).

63a. J. L. Down, J. Lewis, B. Moore, and G. Wilkinson, J. Chem. Soc. (1959) p. 3767.

63b. W. Gerrard and E. D. Maclen, Chem. Rev. 59, 1105 (1959).

64. M. L. Josien, Compt. rend. 237, 175 (1953).

65. M. L. Josien, G. Sourisseau, and C. Castinel, Bull. Soc. Chim. France 178, 1539 (1955).

66. M. L. Josien, J. Lacombe, and J. P. Leicknam, Compt. rend. 246, 1419 (1958).

66a. N. Fuson, P. Pinau, and M. L. Josien, J. Chem. Phys. 55, 454 (1958); see also [112] pp. 129 and 169.

67. D. Cook, J. Chem. Phys. 25, 787 (1956).

68. J. L. Franklin, J. Chem. Phys. 22, 1304 (1954).

68a. A. I. Shatenshtein, Ukr. Khim. Zhur. 32, 3 (1956).

68b. R. West, J. Am. Chem. Soc. 81, 1614 (1959).

68c. R. West and C. S. Krainzel, J. Am. Chem. Soc. 63, 765 (1961).

68d. J. C. D. Brand, G. Eglington, and J. F. Morman, J. Chem. Soc. (1960) p. 2526.

69. V. M. Chulanovskii, M. P. Bugrova, G. S. Denisov, and E. L. Zhukova, Proceedings of the Tenth All-Union Conference on Spectroscopy [in Russian] [L'vov, 1957] Vol. 1, p. 42.

70. G. M. Barrow, J. Phys. Chem. 59, 1129 (1955).

71. J. Hine and M. Hine, J. Am. Chem. Soc. 74, 5266 (1952).

71a. A. I. Shatenshtein and E. N. Zvyagintseva, Doklady Akad. Nauk SSSR 117, 852 (1957).

71b. P. Ballinger and F. A. Long, J. Am. Chem. Soc. 82, 795 (1960).

72. G. F. Zellhoefer, M. J. Copley, and C. S. Marvel, J. Am. Chem. Soc. 60, 1337, 2666 (1938); 61, 3550 (1939).

73. C. L. Marvel, F. C. Dietz, and M. J. Copley, J. Am. Chem. Soc. 62, 2273 (1940).

74. M. J. Copley and C. E. Holley, J. Am. Chem. Soc. 61, 1599 (1939).

75. G. F. Zellhoefer and M. J. Copley, J. Am. Chem. Soc. 60, 1345 (1938).

76. C. S. Marvel, M. J. Copley, and E. Ginsberg, J. Am. Chem. Soc. 62, 3109 (1940).

77. W. Gordy, J. Am. Chem. Soc. 60, 605 (1938); J. Chem. Phys. 7, 163 (1938).

78. A. M. Buswell, W. H. Rodebush, and M. F. Roy, J. Am. Chem. Soc. 60, 2528 (1938).

79. G. M. Barrow and E. A. Yerger, J. Am. Chem. Soc. 76, 5247 (1954).

80. R. C. Lord, B. Nolin, and H. D. Stidham, J. Am. Chem. Soc. 77, 1365 (1955).

81. C. H. Higgins and G. C. Pimental, J. Chem. Phys. 23, 896 (1955).

82. W. Gerrard and E. D. Maclen, Proc. Chem. Soc. (1958) p. 200.

83. W. Strohmejer and A. Echte, Z. Elektrochem. 61, 549 (1957).

84. S. Searles, and M. Tamres, J. Am. Chem. Soc. 73, 3704 (1951).

85. S. Searles, M. Tamres, and G. M. Barrow, J. Am. Chem. Soc. 75, 71 (1953).

86. M. Tamres, S. Searles, E. M. Leighly, and D. W. Mohrman, J. Am. Chem. Soc. 76, 3983 (1954); 81, 2100 (1959).

87. M. Tamres, J. Am. Chem. Soc. 73, 3704 (1951).

88. G. J. Korinek and W. G. Schneider, Canad. J. Chem. 35, 1157 (1957).

88a. A. N. Campbell and E. M. Kartzmark, Canad. J. Chem. 38, 652 (1960).

89. D. Cook, J. Lupion, and W. G. Schneider, Canad. J. Chem. 34, 957, 964 (1956).

90. J. T. Arnold and M. G. Packard, J. Chem. Phys. 19, 1608 (1951).

91. I. V. Aleksandrov, Some Problems in the Theory of Nuclear Magnetic Resonance, Dissertation [in Russian] (Institute of Chemical Physics, Acad. Sci. USSR, Moscow, 1958).

92. I. V. Aleksandrov and N. D. Sokolov, Doklady Akad. Nauk SSSR 124, 115 (1959).

92a. J. D. Morrison and A. J. C. Nicolson, J. Chem. Phys. 20, 1021 (1952).

93. C. A. Huggins, G. S. Pimental, and J. N. Schoolery, J. Chem. Phys. 23, 1244 (1955).

94. W. C. Price, Chem. Rev. 47, 257 (1947).

95. K. Watanabe, J. Chem. Phys. 26, 542 (1957).

96. P. G. Yaroslavskii, Zhur. Fiz. Khim. 22, 265 (1948).

97. P. P. Shorygin, A. Kh. Khalilov, and Z. B. Alaune, Zhur. Fiz. Khim. 25, 1475 (1951); 33, 717 (1959).

98. A. I. Shatenshtein and E. N. Zvyagintseva, Zhur. Obshchei Khim. 29, 1751 (1959).

99. A. N. Terenin, Izvest. Akad. Nauk SSSR, Ser. Biol. (1947) p. 369.

99a. H. Leonhardt and A. Weller, Naturwiss. 47, 58 (1960); Z. Elektrochem. 61, 956 (1957).

100. A. V. Kiselev, Uspekhi Khimii 25, 705 (1956).

100a. A. Terenin and V. Filimonov in the book [112] p. 545.

101. A. N. Terenin, Collection: Surface Chemical Compounds and Their Role in Adsorption Phenomena [in Russian] (Izd. MGU, Moscow, 1957) p. 260.

102. E. C. Bauham, M. G. Evans, and M. Polanyi, Trans. Farad. Soc. 37, 377 (1941).

103. A. I. Shatenshtein, Uspekhi Khimii 24, 377 (1955).

104. V. L. Tal'roze and E. L. Frankevich, Doklady Akad. Nauk SSSR 111, 376 (1956); 119, 1174 (1958).

104a. J. L. Franklin, F. W. Lampe, and H. E. Lumpkin, J. Am. Chem. Soc. 81, 3153 (1959).

104b. F. W. Lampe and F. H. Field, J. Am. Chem. Soc. 81, 3242 (1959).

104c. H. Hartmann and F. Grein, Z. Phys. Chem. (Frankf. Ausg.) 22, 305 (1959).

105. H. Briegleb, Naturwiss. 30, 469 (1942).

106. K. B. Yatsimirskii, Thermochemistry of Complex Compounds [in Russian] (Izd. AN SSSR, Moscow, 1951) p. 41.

107. R. P. Bell, Proton in Chemistry (Cornell University Press, New York, 1959).

108. I. Kolthoff, Ind. and Eng. Chem. (News Ed.) 27, 835 (1949).

109. A. I. Shatenshtein, Uspekhi Khimii 29, 3 (1959).

110. N. A. Izmailov, Electrochemistry of Solutions [in Russian] (Khar'kov, 1959) p. 568.

111. A. I. Shatenshtein, Zhur. Obshchei Khim. 9, 1603 (1939).

112. Hadzi, H. W. Thompson (Ed), Hydrogen Bonding (London, New York, Pergamon Press, 1959).

MECHANISMS OF HYDROGEN REPLACEMENT REACTIONS

INTRODUCTION

Ingold was the first to note the great similarity between the exchange of hydrogen in aromatic compounds with deuterosulfuric acid and electrophilic replacements of hydrogen in the same compounds [1]. This similarity extends completely to hydrogen exchange with other acids (HBr [2], HF [3], CF_3COOH [4]).

It has already been pointed out above (p. 15) that these reactions evidently proceed through the formation of a π- or σ-complex. The intermediate complex can be isolated sometimes in an individual state. Moreover, the strength of the hydrocarbon as a base (electron donor) and the electron density distribution at individual carbon atoms in its molecule are of great importance.

According to Ingold's classification, nucleophilic replacement of hydrogen is opposite in nature to electrophilic replacement. In this case, the aromatic compound is an electrophilic (electron acceptor), while the reagent acts as a nucleophile (base) (cf., p. 7).

The two types of reaction are illustrated schematically by the following equations:

$$\text{type } S_E \quad \overset{\delta-}{RC} - H + E^+ \rightarrow \left[RC \overset{\displaystyle E}{\underset{\displaystyle H}{\diagup\!\!\!\diagdown}} \right]^+ \rightarrow \overset{\delta-}{RC} - E + H^+$$

$$\text{type } S_N \quad \overset{\delta+}{RC} - H + N^- \rightarrow \left[RC \overset{\displaystyle N}{\underset{\displaystyle H}{\diagup\!\!\!\diagdown}} \right]^- \rightarrow \overset{\delta+}{RC} - N + H^-$$

where E^+ is an electrophilic reagent (NO_2^+, H^+, D^+); and N^- is a nucleophilic reagent (NH_2^-, OH^-).

Nucleophilic replacement of hydrogen has been studied in less detail than electrophilic replacement. Deuterium exchange in aromatic hydrocarbons with a nucleophilic reagent (a solution of KND_2 in ND_3) was first investigated in the Isotopic Reactions Laboratory. In the first stage of this work it was impossible to exclude a priori the possibility of the addition of an ND_2^- ion to the most positively charged carbon atom of the aromatic ring or heterocycle, though this direction of the reaction seemed less probable than the direct attack of the nucleophilic reagent (base) on the hydrogen atom [5, p. 705] [6, p. 8].

The great similarity between deuterium exchange with strong bases and metallation, which was observed even in the first work on deuterium exchange with liquid ammonia [6a], made it necessary to adopt similar mechanisms for these reactions, whose characteristic trait is the protonization of the hydrogen atom of the CH bond in its reaction with the base. At that time, this hypothesis was contradictory to the generally accepted views of Morton on the mechanism of metallation. This difficulty disappeared when Bryce-Smith and other authors established that in the metallation of a hydrocarbon by an organoalkali compound, the carbanion of the latter protonizes the hydrogen of the CH bond, i.e., there is reaction between a hydrocarbon-acid and carbanion-base:

$$\overset{\delta+}{R}CH + B \rightarrow RC^- + BH^+.$$

Symmetrical reactions of the type

$$\overset{\delta-}{R}CH + H^+ \rightarrow RC^+ + H_2,$$

in which the hydrocarbon reacts with an acid with the elimination of a hydride ion instead of a proton and the conversion of the hydrocarbon to a carbonium ion, have been discussed in the literature. It is possible that the hydrogen in saturated hydrocarbons exchanges with strong acids by this scheme (pp. 133 and 186).

Thus, if we limit ourselves to heterolytic rupture of a carbon-hydrogen bond, it is necessary to distinguish between the following four cases:

1) attack of a nucleophilic reagent on the hydrogen atom;

2) attack of a nucleophilic reagent on the carbon atom;

3) attack of an electrophilic reagent on the hydrogen atom;

4) attack of an electrophilic reagent on the carbon atom.

This classification is undoubtedly simplified as the reaction may actually proceed by a more complicated route, for example, by a termolecular mechanism with the formation of cyclic activated complexes. In addition, when an aromatic molecule reacts, its π-electron system may participate in the reaction as a whole.

Reactions in which the rupture of a C−H bond is accompanied by the transfer of hydrogen within the molecule are discussed in [7].

The mechanisms of heterolytic replacements of hydrogen are examined in this section of the book. The simplest of these reactions is isotopic exchange of hydrogen in solutions. There is no thermal effect in these reactions if we neglect the slight changes in free energies as a result of differences in the zero point energies of C−H,

C−D, and C−T bonds (because of the different masses of the hydrogen isotopes). Consequently, the course of the reaction is determined solely by kinetic parameters. In addition, when a hydrogen atom is replaced by one of its isotopes, there is little change in the structure of the molecule of the organic substance. Therefore, when a molecule contains several hydrogen atoms, the replacement of one of them by an isotope has hardly any effect on the rate of replacement of the other atoms. When a hydrogen atom is replaced by a halogen, metal, or polar group, new substances are obtained in which the normal reactivity of the remaining hydrogen atoms is changed.

Another important circumstance in the study of the reactivity of organic substances and the intereffect of atoms in their molecules is the fact that it is often possible to replace the same hydrogen atoms by isotopes by means of reagents of different or even opposite natures (acid−base). Results obtained in the replacement of hydrogen by one of its isotopes are most readily comparable and therefore make it possible to determine particularly clearly the effect of the reagent and the medium on the reactivity of a substance.

In other respects, isotopic and chemical reactions* have much in common. This has already been demonstrated, for example, by the comparison of hydrogen exchange with bases and the chemical reactions of metallation, isomerization, and alkylation (see Section III). Therefore, a knowledge of existing ideas on the mechanisms of chemical reactions involving irreversible replacement of hydrogen is essential to the discussion of the mechanisms of hydrogen exchange at the end of this book.

The first chapter covers the mechanisms of chemical reactions involving the replacement of hydrogen (in atomatic compounds) induced by nucleophilic reagents.

The important question of the effect of the reagent on the reactivity of a substance in chemical reactions and on the mechanism of the latter has probably been analyzed most thoroughly for the case of changes in orientation rules in the electrophilic replacement of hydrogen in aromatic compounds and the second chapter is devoted to this.

In the third chapter of this final section of the book, the rules of chemical reactions involving the replacement of hydrogen and isotopic exchanges of hydrogen are compared in order to substantiate the mechanisms of the latter.

CHAPTER 1. REPLACEMENT OF HYDROGEN BY REACTION WITH A NUCLEOPHILIC REAGENT

1. ATTACK OF THE REAGENT ON THE HYDROGEN ATOM OF A CH BOND

Let us consider the mechanism of metallation by means of organoalkali compounds − a reaction that has been described already (Shorygin's reaction, see p. 86).

---

* This distinction of the reactions is naturally quite arbitrary as isotopic reactions are, at the same time, chemical reactions.

A. Electrophilic Substitution Hypothesis and Criticism of It

The mechanism of metallation by organoalkali compounds has been disputed in the literature until recently. Morton [8-10] defended the view that metallation proceeds as a result of electrophilic attack of the cation of the organometallic compound on a carbon atom with a high electron density. He based this conclusion, in particular, on the fact that in aromatic compounds with the groups NH_2, $N(CH_3)_2$, and OCH_3, which are ortho- and para-directing in electrophilic substitution, metallation generally occurs in the ortho-position. However, a check showed [11] that the ortho-isomer is also obtained in good yield in the metallation of benzotrifluoride although the CF_3 group is meta-directing in electrophilic substitution. In reactions with n-butyllithium, the meta-isomer was obtained in a yield equivalent to only 20% of the yield of the ortho-isomer while no para-isomer was produced.

The hypothesis on the electrophilic nature of replacement of hydrogen by a metal was not confirmed either in experiments with 2-methoxynaphthalene. It was metallated in position 3, although the α-position is normally involved in electrophilic subsitution [12]. Similar deviations were observed in the metallation of methoxyphenanthrene and 2,2'-dimethoxynaphthalene [13].

Morton's point of view [8-10, 14] was also contradicted by the results of the experiments of Gronowitz [15] with thiophene compounds. While electrophilic reagents replace the hydrogen atom in position 3 in thionaphthene (I), the hydrogen atom in position 2 is replaced in the reaction with n-butyllithium. Electrophilic substitution occurs at position 2 in 3-methylthiophene (II), while the hydrogen atom in position 5 is replaced in metallation in the same way as the position most remote from the methyl group, i.e., the para-position is metallated predominantly in toluene [16]:

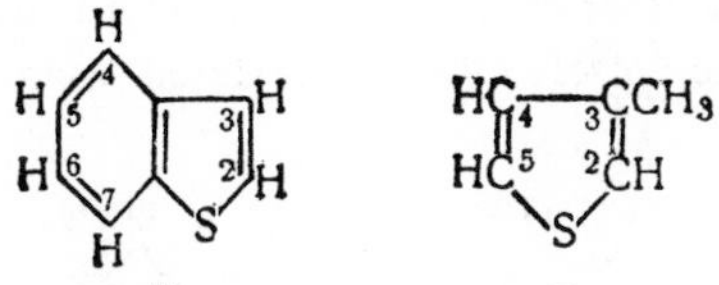

B. Metallation — A Protophilic Substitution Reaction

a. ATTACK OF A CARBANION ON THE HYDROGEN OF A CH BOND

In the opinion of Roberts and Curtin [11], metallation essentially consists of protonization of the hydrogen of a CH bond of the organic substance by the anion of the organoalkali compound (carbanion), which is a very strong base. The predominant metallation in the ortho-position relative to substituents of the first type (NH_2, OCH_3, etc.) is explained by coordination of the cation of the organoalkali compound with the electron pair of the nitrogen or oxygen atom. This results in weakening of the bond between the metal ion and the carbanion with the substituent. The adjacent CH bond is polarized most strongly, the hydrogen is charged positively, and its elimination as a proton is facilitated. In other words, protonization is promoted by the formation of a cyclic complex of the type:

$$R'-O \cdots Me$$
$$-H \cdots R \; .$$
III

Gilman [12, 17] also considers this explanation probable.

b. INDUCTIVE EFFECT OF THE SUBSTITUENT

Roberts and Curtin [11] consider that the increase in the lability of the hydrogen atoms in the ring of benzotrifluoride in comparison with benzene and the increase in their acidity is caused by the inductive effect of this substituent. Wittig [18, 19] also considers the influence of the inductive effect in metallation. For example, he reported that in the reaction with phenyllithium, the yields of the ortho-isomer decrease in the sequence: fluorobenzene > anisole > dimethylaniline. According to Wittig, the reason for this is the progressive decrease in the degree of electronegativity of the atom attached to the benzene ring, namely, F > O > N. The atom whose affinity for an electron is greater polarizes the adjacent CH bond more strongly and the corresponding hydrogen atom becomes "more acid" and is transferred to a base more readily as a proton. The inductive effect falls off rapidly as the distance of the CH bond from the substituent increases. The reaction therefore occurs predominantly in ortho-position. This rule is observed almost always when the ring is adjacent to or includes an atom of oxygen (anisole [20-24], diphenyl ether [25], and furan [26]), nitrogen (dimethylaniline [27] and diphenylamine [10]), or sulfur (thiophene [27]) (see also [28, 29]).

If two substituents with an electronegative atom lie meta to each other in the ring, the hydrogen atom between them is replaced by a metal particularly readily. For example, the dimethyl ether of resorcinol (IV) is metallated by phenyllithium at room temperature, while anisole (V) is metallated at an appreciable rate only at 100°:

OCH_3 ... OCH_3

IV   V

Methane is evolved readily when m-difluorobenzene is treated with methyllithium [23, 30]. When a methoxyl group and a halogen atom lie meta to each other, the yield of the organoalkali compound increases with an increase in the electronegativity of the halogen: m-iodoanisole (40%) < m-bromoanisole (60%) < m-chloroanisole (75%) < m-fluoroanisole (80%) [31].

Wittig did not support Roberts' view that the ortho-hydrogen atom is activated as a result of coordination of the cation of the organometallic compound with the heteroatom. This hypothesis is completely eliminated in the case of metallation of hydrocarbons.

If the inductive effect of the substituent, for example, an alkyl group, has the opposite sign, i.e., it does not attract, but repels electrons, then the acidity of the CH bonds will obviously be reduced and especially so in the ortho-position. According to Bryce-Smith [16], the isopropyl group in isopropylbenzene deactivates all the positions in the aromatic ring with respect to metallation. On an average, the reaction rate is a factor of four to five less than that for unsubstituted benzene. The rates of replacement of different hydrogen atoms relative to the rates of metallation of benzene are as follows: para (0.43) > meta (0.38) > ortho (0.10). The author considers that these figures may serve as a measure of the relative acidities of the corresponding hydrogen atoms.* tert-Butylbenzene is metallated predominantly in the para-position [14].

Shirley and Lehto [33] confirmed that a methyl group deactivates a benzene ring with respect to metallation. In the reaction of 4-tert-butyldiphenyl sulfone (VI)

$$\langle\ \rangle - SO_2 - \langle\ \rangle - C(CH_3)_3$$

VI

with n-butyllithium, they obtained 69% of a product in which metallation had occurred in the ortho-position relative to the sulfone group in the ring without a substituent and only 31% of the metallation occurred in the ring containing the tert-butyl group.

By treating a mixture of thiophene and 3-methylthiophene (VII) with an insufficient amount of reagent (competing metallation), Gronowitz [15] showed that the former is metallated three times faster than the latter. Position 5 is least deactivated as it is most remote from the methyl group and the substance is metallated precisely here. On the other hand, with compounds (VIII) and (IX) the lithium enters position 2 in the same way as anisole is metallated in the ortho-position:

HC — CCH₃	HC — COCH₃	HC — CSCH₃
‖ ‖	‖ ‖	‖ ‖
LiC CH	HC CLi	HC CLi
\S/	\S/	\S/ .
VII	VIII	IX

c. RATE DETERMINED BY PROTON ELIMINATION

According to Bryce-Smith [16], the transition complex whose structure is represented by formula X is formed during metallation:

$$>C\cdots\overset{\delta-}{}\cdots\overset{\delta+}{H}\cdots\overset{\delta-}{B} .$$
$$Me^+$$

X

Elimination of the proton is promoted by the strong field of the cation. The reaction rate is determined by the rupture of the carbon-hydrogen bond. This was demon-

---

*In disputing with Bryce-Smith, Morton [32] pointed out that with a suitable reagent, isopropylbenzene is metallated almost exclusively in the α-position and consequently it is the reagent rather than the acidity of the hydrocarbon which determines the position of attack.

strated conclusively by Bryce-Smith, Gold, and Satchell [34] by comparing the rates of metallation of a deuterated and a normal hydrocarbon. (Deuterium was introduced into the methyl group in toluene.) A considerable isotope effect was found: benzene and toluene react with ethylpotassium two and five times faster,[*] respectively, than their deuterated derivatives. Consequently, the reaction rate is limited by rupture of the CH (or CD) bond regardless of whether it is in the aromatic ring or the methyl group. The same conclusions were drawn by Gronowitz and Halvarson [35], who studied the isotope effect in the metallation of thiophene: protium was replaced at least six times faster than tritium.

The facts presented substantiate the mechanism for the replacement of hydrogen in CH bonds by a metal in which the hydrogen is protonized by a base (for example, the carbanion of an organoalkali compound). Bryce-Smith [16] suggested that this should be called protophilic replacement in order to show the acid−base nature of the reaction.

d. NUCLEOPHILIC REPLACEMENT OF HYDROGEN IN METALLA-TION

It was found [36] that the most reactive organoalkali compounds can react with benzene in two ways:

$$RK + C_6H_6 \rightarrow RH + C_6H_5K,$$

i.e., by a protophilic replacement mechanism, and

$$RK + C_6H_6 \rightarrow C_6H_5R + KH,$$

i.e., by a nucleophilic hydrogen replacement mechanism in which a hydride ion is eliminated. Reactions with nitrogen heterocycles generally proceed in the second direction (p. 249).

2. ATTACK OF THE REAGENT ON THE CARBON ATOM OF A CH BOND

A. General Principles

The displacement of hydrogen in the form of a hydride ion (H^-) by a nucleophilic reagent (N^-),

$$\equiv C-H + :N^- \rightarrow \equiv C-N + :H^-,$$

is much less favored energetically than its elimination as a proton

$$\equiv C-H + E^+ \rightarrow \equiv C-E + H^+;$$

therefore, nucleophilic replacement of hydrogen in a ring is encountered less frequently than electrophilic replacement. Normally, for nucleophilic replacement to occur it is necessary not only to use a sufficiently strong reagent, but also to activate the aromatic ring by an electronegative substituent. Substituents fall into roughly the following order with respect to activating power [37, 38]:

---

[*] The reaction with benzene was at a temperature 50° higher than that with toluene. The smaller isotope effect may be connected with this.

$$F \rangle NO_2 \rangle RSO_3 \rangle NR_3^+ \rangle CF_3 \rangle RCO \rangle CN \rangle Cl_2;\ Br_2;\ I_2 \rangle$$

$$\rangle CO_2^- \rangle C_6H_5 \rangle OAr \rangle OR \rangle SR \rangle NR_2^* .$$

An electron-acceptor substituent attracts electrons from the carbon atoms of the ring and thus facilitates the addition of an electron donor. Naturally, groups which are meta-directing in electrophilic replacement are ortho- and para-directing in nucleophilic replacement. In the presence of groups which are normally ortho- and para-directing in electrophilic replacement, nucleophilic replacement is generally hampered, but if at all, it occurs in the meta-position.†

Nitrogen heterocycles are activated with respect to nucleophilic replacement [39]. Heterocyclic nitrogen affects the reactivity like the nitro group in nitrobenzene. For example, pyridine is nitrated and sulfonated only at temperatures above 300° and then in the β-position, but it is aminated by alkali metal amides under quite mild conditions in the α- and γ-positions. This results from the fact that the electrons density at the carbon atoms in these positions is low because of the electronegativity of the nitrogen.

Some reactions involving nucleophilic replacement of hydrogen are facilitated by the addition of an oxidant (KNO_3, $K_4FE(CN)_6$, etc.), to which the electron of the hydride ion is transferred. Sometimes, another molecule of the substrate (for example, a nitro compound) acts as the oxidant.

In a number of cases it has been possible to detect the formation of an alkali metal hydride in the nucleophilic replacement of hydrogen in an aromatic or heterocyclic ring. The hydride ion is a very strong base. Molecular hydrogen is the acid corresponding to it. If the system contains substances with labile hydrogen atoms, molecular compounds may be liberated during the reaction. It is considered that the first stage of a nucleophilic replacement consists of the formation of an intermediate product as a result of the addition of the nucleophilic reagent. This point of view is confirmed by the examples given below. The mechanism of nucleophilic replacement has been substantiated most thoroughly by experiments with nitrogen heterocycles [39-42].

B. Chichibabin's Reaction

The direct replacement of hydrogen in nitrogen heterocycles by means of strong bases (amides and alkalis) was discovered by A. E. Chichibabin. The first paper, which he published together with O. A. Zeide [43] in 1914, was devoted to the preparation of aminopyridine. During the next few years, A. E. Chichibabin and his co-workers developed a new and important field in organic synthesis. The Chichibabin reaction may be applied to quinoline, isoquinoline, nicotine, etc., in addition to pyridine. The work of the first years was gathered together in A. E. Chichibabin's monograph [44]. The detailed determination of the chemical behavior or substances in the Chichibabin reaction made it possible to establish its mechanism. This was formulated by Ziegler and Zeisel [45] in analogy with the mechanism they proposed for the

---

*The approximate relative activity of nucleophilic reagents is as follows: $Alk^- \rangle$ $\rangle NH_2^- \rangle (Ar)_3C^- \rangle ArNH^- \rangle OH^- \rangle SO_3^{--} \rangle Cl^-$.

† In the second chapter of this section, it is shown that the division of substituents into ortho- and para-, and meta-directing groups is to some extent conditional.

reaction of alkyllithiums with pyridine. The authors found that equimolecular amounts of an organolithium compound (C_4H_9Li) and pyridine give an addition product, which is quite stable in the cold. Heating (70-100°) gives a colorless precipitate of lithium hydride, which reacts vigorously with water to liberate hydrogen. If the heating is sufficiently prolonged, the stoichiometric amount of lithium hydride is formed. At the same time, an alkylpyridine with the alkyl group in the α-position is obtained:

$$\text{(pyridine)} + \text{LiR} \xrightarrow{25°} \left[\text{(ring with } R, H) \right]^- \text{Li}^+ \xrightarrow{100°} \text{(ring with } R) + \text{Li}^+\text{H}^- .$$

In the article cited it was noted that this alkylation by an organoalkali compound is a very close analog of the Chichibabin reaction, which evidently also proceeds through the formation of an intermediate addition product:

$$\text{(pyridine)} + \text{NaNH}_2 \rightarrow \left[\text{(ring with } NH_2, H) \right]^- \text{Na}^+$$

which then decomposes according to the equation:

$$\left[\text{(ring with } NH_2, H) \right]^- \text{Na}^+ \rightarrow \text{(ring with } NH_2) + \text{Na}^+ \text{H}^- .$$

Finally, the sodium hydride and aminopyridine react with the evolution of hydrogen:

$$\text{(ring with } NH_2) + \text{NaH} \rightarrow \text{(ring with } NHNa) + \text{H}_2,$$

i.e., the hydride ion (base) adds a proton of the acid amino group. This scheme for the Chichibabin reaction was confirmed by M. I. Kabachnik [46], A. V. Kirsanov [47, 48], and others [42]. Bergstrom [40, 49-54], who investigated the direct introduction of an amino group into a heterocyclic and an aromatic nucleus in liquid ammonia, was able to detect the intermediate addition products of the amide ion and the heterocyclic compounds. This was first achieved with isoquinoline [50]:

$$\text{(isoquinoline structure)} \ .$$

Isoquinoline and potassamide react in liquid ammonia at $-33°$ to form a red-brown solution. If the solution concentration is low no hydrogen is liberated over a period of several hours, but acidification with ammonium bromide liberates isoquinoline and the amount of it is greater the shorter the time of the reaction between the amide and the heterocycle. If the amide concentration or the temperature is raised, evolution of hydrogen begins immediately, the solution becomes green, and the potassium salt of aminoisoquinoline is obtained. Quinoline is aminated in the presence of KNO_3, which is thereupon reduced to KNO_2. Bergstrom regarded the results of his experiments as confirmation of the mechanism for the Chichibabin reaction presented above. The author considered that when the product from the addition of the amide ion to the carbon of the heterocycle had been formed, the excess amide converted it into the potassium salt:

$$\left[\underset{N}{\overset{H}{\bigvee\bigvee}}NH_2\right]^- + NH\bar{2} \rightleftarrows \left[\underset{N}{\overset{H}{\bigvee\bigvee}}NH\right]^{--} + NH_3$$

The third stage of the reaction consisted of the elimination of a hydride ion and the substance with which it could react was an ammonia molecule:

$$\left[\underset{N}{\overset{H}{\bigvee\bigvee}}NH\right]^{--} + NH_3 \rightleftarrows \underset{N}{\bigvee\bigvee}\overset{NH^-}{} + NH\bar{2} + H_2$$

since $H^- + NH_3 \rightarrow H_2 + NH\bar{2}$.

As was pointed out, the presence of an oxidant (for example, a nitrate) facilitates amination. The yields of the amine are considerably increased if an electronegative group (carboxyl or phenyl) is introduced into the quinoline, while electropositive groups (hydroxyl and amino) completely prevent the amination of quinaldine. In addition to amide ions and carbanions, other strong bases react with heterocycles, analogously. For example, A. E. Chichibabin [44] showed that pyridine can be hydroxylated directly by heating with caustic alkali:

$$\underset{N}{\bigvee} + K^+OH^- \rightarrow \underset{N}{\bigvee}O^-K^+ + H_2.$$

C. Replacement of Hydrogen in Nitro Compounds

A nitro group is one of the strongest activators of the aromatic nucleus with respect to nucleophilic replacement of hydrogen. Wohl [55] discovered long ago that moderate heating (60-70°) of nitrobenzene with potassium hydroxide forms o-nitrophenol. With three nitro groups arranged symmetrically, the aromatic ring is strongly activated; trinitrobenzene is readily converted into picric acid even by the action of an aqueous solution of alkali in the presence of potassium ferricyanide [56] and even such a weak base as the CN^- ion replaces hydrogen by a CN group in trinitrobenzene [57]. Treatment of m-dinitrobenzene with alkali yields 2,4- and 2,6-dinitrophenol [56].

The reaction between nitrobenzene and an alkali metal amide in liquid ammonia is complex [58], but nitrobenzene reacts quite smoothly with sodium diphenylamide in this solvent to form p-nitrotriphenylamine in 45% yield [59].

Intermediate complexes of nitro compounds with bases are often stable and usually intensely colored. The literature on them is extremely voluminous (see, for example [60, 61]) and we cannot dwell on it here. As in reactions with heterocycles, bases add to one of the most positively charged carbon atoms, i.e., to ortho- or para-atoms.

A convincing demonstration of this mechanism for the formation of the complex was provided by Meisenheimer [62], who prepared complexes of the same composition from trinitrophenetole and potassium methoxide and from trinitroanisole and potassium ethoxide:

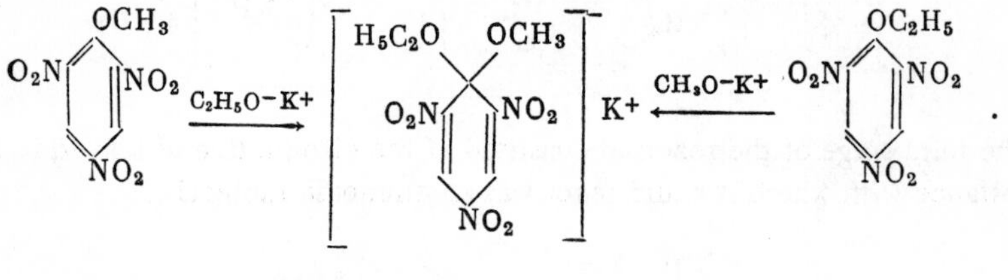

Acidification of the complex yielded a mixture of trinitrophenetole and trinitroanisole. This would not have occurred had the base added to the nitro group (as was considered by some authors). Acidification would then have regenerated the starting nitro compounds. In work with another pair of substances (amyl and isobutyl picrates), Meisenheimer obtained an analogous result. The accuracy of Meisenheimer's conclusions was confirmed by the work of Jackson [63] and S. S. Gitis [64]. The mechanism described above has been supported by many investigators [65-70]. In particular, data on the behavior of solutions of nitro compounds in liquid ammonia are in complete agreement with it.

Ammonia solutions of di- and trinitro compounds are intensely colored. This indicates the formation of molecular compounds. It is evident that the nitro derivatives is connected to an amide ion (NH_2^-), obtained by dissociation of liquid ammonia. Therefore, the addition of an acid (NH_4^+) displaces the equilibrium to the left:

$$2NH_3 \rightleftarrows NH_2^- + NH_4^+$$

$$Ar(NO_2)_n + NH_2^- \rightleftarrows Ar(NO_2)_n NH_2^-$$

$$Ar(NO_2)_n NH_2^- + NH_4^+ \rightleftarrows Ar(NO_2)_n + 2NH_3$$

and the solution is decolorized [61] as a nitro compound absorbs in the ultraviolet. Solutions of polynitro compounds in liquid ammonia conduct an electric current and the electrical conductivity increases with a decrease in temperature because of the increase in the concentration of the complex and at the same time, the intensity of the solution color increases [71]. On electrolysis, hydrogen is liberated at the cathode and nitrogen at the anode in complete accordance with the presence of the ions NH_4^+ and $Ar(NO_2)n \cdot NH_2^-$ [72].

D. Replacement of Hydrogen in Hydrocarbons

Nucleophilic replacement of hydrogen in hydrocarbons has been scantily studied. Only the strongest bases can induce such reactions without activators such as a nitro group or a heteroatom. Thus, by fusing naphthalene with sodamide, Sachs [73] obtained naphthylamine and 1,5-naphthylenediamine. There was vigorous evolution of hydrogen when the melt was poured into water. This resulted from the presence of sodium hydride in the melt. We should also mention that benzene is alkylated by some organoalkali compounds [36] by a nucleophilic replacement mechanism (p. 248).

3. SUMMARY

1. The metallation of organic substances by organoalkali compounds is an acid-base reaction in which the carbanion attacks the hydrogen atom of a CH bond and protonizes it. The metallation of some substances may also be achieved by means of other bases such as KNH_2, KOH, and $KOCH_3$. The reaction rate is limited by the rupture of the CH bond. This was demonstrated by a study of the isotope effect in metallation. It has been suggested that metallation is described as protophilic replacement of hydrogen.

2. The inductive effect of a substituent in the aromatic ring is of decisive importance in the metallation of aromatic compounds. If the substituent is an electronegative group (for example, CF_3) or an electronegative atom (F, Cl, Br, I, and even O, S, and N atoms in groups such as OR, SR, and NR_2), the displacement of electrons produces polarization of the CH bonds of the ring, in which there is an increase in the lability of the hydrogen atoms in comparison with unsubstituted benzene. The acidity of the CH bonds in an aromatic ring decreases with the distance from the substituent in the sequence: ortho > meta > para. Normally, the most acid, i.e., the ortho-hydrogen atom, is replaced predominantly by a metal. The inductive effect of an electron-donor substituent acts in the opposite way and there is the greatest hindrance to metallation in the ortho-position.

3. Under certain favorable conditions, metallation may be accompanied by nucleophilic replacement of hydrogen with liberation of a hydride ion.

4. Nucleophilic replacement of hydrogen in an aromatic CH bond normally requires activation by an electronegative substituent. The reaction is facilitated by the presence of an oxidant. The replacement proceeds through the formation of an intermediate product from the addition of the base to the most positive carbon atom of the ring. These intermediate compounds are often stable. Their relative stability is explained by the fact that the elimination of hydrogen as a hydride ion requires a considerable activation energy.

5. The conversion of the intermediate complex into the final product is accompanied by the formation of a hydride, which can be detected some times. More often there are secondary reactions such as the liberation of hydrogen or the reduction of an oxidant added specially or another substance present in the reaction system.

LITERATURE CITED

1. C. K. Ingold, C. G. Raisin, and C. L. Wilson, J. Chem. Soc. (1936) pp. 915,1637, 1643.
2. A. I. Shatenshtein, V. R. Kalinachenko, and Ya. M. Varshavskii, Zhur. Fiz. Khim. 30, 2093, 2098 (1956).

3. Ya. M. Varshavskii, M. G. Lozhkina, and A. I. Shatenshtein, Zhur. Fiz. Khim. 31, 1377 (1957).

4. E. L. Mackor, P. J. Smith, and J. H. van der Waals, Trans. Farad. Soc. 53, 1309 (1957).

5. A. I. Shatenshtein, Collection: Problems in Chemical Kinetics, Catalysis, and Reactivity [in Russian] (Izd. AN SSSR, Moscow, 1955) p. 699.

6. A. I. Shatenshtein, Ukr. Khim. Zhur. 22, 3 (1956).

6a. A. I. Shatenshtein and Yu. P. Vyrskii, Doklady Akad. Nauk SSSR 70, 1029 (1950).

7. E. S. Lewis and M. C. R. Symons, Quart. Rev. 12, 230 (1958).

8. A. Morton, E. L. Little, and W. O. Strong, J. Am. Chem. Soc. 65, 1339 (1943).

9. A. Morton, J. Am. Chem. Soc. 69, 969 (1947).

10. A. Morton, Chem. Rev. 35, 1 (1944).

11. J. D. Roberts and D. J. Curtin, J. Am. Chem. Soc. 68, 1658 (1946).

12. S. V. Sunthankar and H. Gilman, J. Org. Chem. 16, 8 (1951).

13. H. Gilman and T. H. Cook, J. Am. Chem. Soc. 62, 2813 (1940).

14. A. A. Morton and E. L. Little, J. Am. Chem. Soc. 71, 487 (1949).

15. S. Gronowitz, Ark. kemi 7, 361 (1954); 12, 239 (1958); 13, 269, 295 (1958).

16. D. Bryce-Smith, J. Chem. Soc. (1954) p. 1079.

17. H. Gilman, Collection: Organic Reactions, Edited by R. Adams [Russian translation] (IL, Moscow, 1956) No. 8, p. 333.

18. G. Wittig, Naturwiss. 30, 698 (1942).

19. G. Wittig, Angew. Chem. 66, 10 (1954).

20. H. Gilman and R. L. Bebb, J. Am. Chem. Soc. 61, 109 (1939).

21. A. A. Morton and A. F. Brachtman, J. Am. Chem. Soc. 76, 2973 (1954).

22. A. A. Morton and J. Hechenbleiner, J. Am. Chem. Soc. 58, 2599 (1936).

23. G. Wittig, U. Pockels, and H. Dröge, Ber. 71, 1903 (1938).

24. L. A. Wiles, Chem. Rev. 56, 329 (1956).

25. A. Luttringhaus and G. Saaf, Z. angew. Chem. 52, 578 (1939); Ann. 542, 241 (1939).

26. H. Gilman and F. Breuer, J. Am. Chem. Soc. 56, 1123 (1934).

27. J. Blanschette and E. V. Brown, J. Am. Chem. Soc. 74, 1848 (1952).

28. K. A. Kocheshkov and T. V. Talalaeva, Synthetic Methods in the Field of Organometallic Compounds of Lithium, Sodium, Potassium, Rubidium, and Cesium [in Russian] (Izd. AN SSSR, Moscow, 1949).

29. R. A. Benseker, D. J. Foster, D. M. Jauve, and J. F. Nobis, Chem. Rev. 57, 867 (1957).

30. G. Wittig, and W. Markle, Ber. 75, 1495 (1942).

31. G. Wittig, G. Pieper, and G. Fuhrman, Ber. 73, 1193, 1197 (1940).

32. A. A. Morton and E. J. Lanpher, J. Org. Chem. 23, 1636 (1958).

33. D. A. Shirley and E. A. Lehto, J. Am. Chem. Soc. 79, 3481 (1957).

34. D. Bryce-Smith, V. Gold, and D. P. N. Satchell, J. Chem. Soc. (1954) p. 2743.

35. S. Gronowitz and K. Halvarson, Ark. kemi 8, 343 (1955).

36. D. Bryce-Smith and E. E. Truner, J. Chem. Soc. (1953) p. 861.

37. J. F. Bunnett and R. E. Zahler, Chem. Rev. 49, 273-412 (1951).

38. J. F. Bunnett, Quart. Rev. 12, 1-16 (1958).

39. R. Elderfield, Heterocyclic Compounds [Russian translation] (IL, Moscow, 1953) Vol. I; (1955) Vol. 4.

40. F. W. Bergstrom and W. C. Fernelius, Chem. Rev. 12, 43-179 (1933); 20, 413-481 (1937); 54, 467 (1954).

41. M. T. Leffler, Collection: Organic Reactions, Edited by R. Adams [Russian translation] (IL, Moscow) Vol. 1, p. 115.

42. C. L. Deasy, "Mechanism of amination of heterocyclic bases by metal amides," J. Org. Chem. 10, 141 (1944).

43. A. E. Chichibabin and O. A. Zeide, Zhur. Russ. Fiz. Khim. Obshchest. 46, 1216 (1914).

44. A. E. Chichibabin, Investigations in the Field of Pyridine Bases [in Russian] (Moscow, 1918) No. 1.

45. K. Ziegler and H. Zeisel, Ber. 63, 1847, 1851 (1930).

46. M. I. Kabachnik, Izvest. Akad. Nauk SSSR, Otdel, Khim. Nauk (1935) p. 971.

47. A. V. Kirsanov and Ya. N. Ivashchenko, Zhur. Obshchei Khim. 5, 1494 (1935).

48. A. V. Kirsanov and I. M. Polyakova, Zhur. Obshchei Khim. 6, 1715 (1936).

49. F. W. Bergstrom, J. Am. Chem. Soc. 56, 1748 (1934).

50. F. W. Bergstrom, Ann. 515, 34 (1934).

51. F. W. Bergstrom, J. Org. Chem. 2, 411 (1937).

52. F. W. Bergstrom, J. Org. Chem. 3, 233 (1938).

53. F. W. Bergstrom, J. Org. Chem. 3, 424 (1938).

54. F. W. Bergstrom, Chem. Rev. 35, 77 (1944).

55. A. Wohl, Ber. 32, 3486 (1899).

56. Hepp, Ann. 215, 352 (1882).

57. A. Hantzsch and H. Kissel, Ber. 32, 3137 (1899).

58. W. Bradley and R. Robertson, J. Chem. Soc. (1932) p. 1254.

59. F. W. Bergstrom, J. M. Granara, and V. Erixon, J. Org. Chem. 7, 98 (1942).

60. P. Pfeiffer, Organische Molekulverbindungen (Stuttgart F. Enke, 1927).

61. A. I. Shatenshtein and E. A. Izrailevich, Zhur. Fiz. Khim. 16, 73 (1942).

62. J. Meisenheimer, Ann. 323, 219 (1902).

63. C. L. Jackson, Am. Chem. J. 29, 89 (1903).

64. S. S. Gitis, Author's abstract of dissertation [in Russian] (Dnepropetrovsk, Universitet, 1953).

65. G. M. Bennett and R. L. Wain, J. Chem. Soc. (1936) p. 1108.

66. G. N. Lewis and G. Seaborg, J. Am. Chem. Soc. 62, 2122, 3529 (1940).

67. H. Brockmann and E. Meyer, Ber. 87, 86 (1954).

68. J. F. Bunnett and R. E. Zahler, Chem. Rev. 49, 273 (1951).

69. T. Canbäck, Acta Chem. Scand. 3, 946 (1949).

70. A. I. Shatenshtein and E. A. Izrailevich, Zhur. Fiz. Khim. 26, 377 (1952).

71. A. I. Shatenshtein and N. M. Dykhno, Theories of Acids and Bases [in Russian] (Goskhimizdat, Moscow, 1949) p. 290.

72. J. D. Farr, C. C. Bard, and G. W. Wheland, J. Am. Chem. Soc. 74, 1574 (1952).

73. F. Sachs, Ber. 39, 3006 (1906).

CHAPTER 2. REPLACEMENT OF HYDROGEN BY REACTION WITH AN ELECTROPHILIC REAGENT

1. ORIENTATION RULES

Not long ago it was considered that some substituents in the aromatic ring are always ortho-para-directing (substituents of the first type: OH, OCH$_3$, NH$_2$, N(CH$_3$)$_2$, CH$_3$, C(CH$_3$)$_3$), while others are meta-directing (substituents of the second type:

$N(CH_3)_3^+$, NO_2, $COOH$, CN, etc.). To assign substituents to one or the other group, similar electrophilic replacement reactions were studied and usually the yields of mononitro derivatives were determined as recommended by Holleman [1]. Table 83 gives an idea of the relative yields of isomers in mononitration according to contemporary data [2]. The first line gives the starting material, namely, unsubstituted benzene. The figures in brackets are the conditional yields corresponding to the statistically probable distribution through all the positions with allowance for the fact that the molecule contains two ortho-positions, two meta-positions, and one para-position.

In actual fact, the directing effect of a substituent depends not only on whether it belongs to the electron-donor or electron-acceptor group, but also on whether the aromatic substance is reacting with an electrophile or a nucleophile (see [3] for more details). Groups which direct electrophilic replacement of hydrogen into the meta-position (fof example, a nitro group) direct nucleophilic replacement into the ortho- and para-positions and vice versa. Moreover, the relative yields of the isomers with the same substituent may vary considerably, even if similar reagents are used. For example, when the benzene nucleus contains a methyl group, i.e., a typical ortho-para-directing substituent in electrophilic replacement, the introduction of an iso-propyl group by the Friedel-Crafts reaction gives a high yield of the meta-isomer, although alkylation catalyzed by an aluminum halide is a clear example of electrophilic replacement. For a long time, this type of replacement was regarded as a pe-

TABLE 83. Yields of Mononitro Compounds

Substituent	Distribution of isomers		
	ortho-	para-	meta-
H	(40)	(40)	(20)
OH	40	—	60
Cl	29.6	0.9	69.5
CH_3	58.5	4.4	37.1
$C(CH_3)_3$	15.8	11.5	72.7
CH_2Cl	32.0	15.5	52.5
$CHCl_2$	23.3	33.8	42.9
$Si(CH_3)_3$	26.6	41.7	31.6
CCl_3	6.8	64.5	28.7
$COOC_2H_5$	24.0	72.0	4.0
$COOH$	18.5	80.2	1.3
NO_2	6.4	98.3	0.3
$N(CH_3)_3^+$	—	100	—

culiar anomaly. Brown and his co-workers demonstrated that the ratio of the yields of isomers in electrophilic replacement change regularly in relation to the "activity" of the reagent. There is no doubt that the same rule should also apply to nucleophilic replacement.

2. RELATION OF THE REACTIVITY OF A SUBSTANCE TO THE "ACTIVITY" OF THE REAGENT

As is well known, in acid−base reactions, the stronger the acid and the higher the dielectric constant of the solvent, the smaller are the differences in strength of bases reacting with it. An analogous phenomenon is observed in reactions with bases. For example, in an acid−base reaction with liquid ammonia and especially with solutions of amides in it, there is a sharp reduction in the differences in the strengths of acids (p. 22). The same occurs in the replacement of hydrogen in an aromatic ring. Acidlike substances [for example, Br_2, Br^+, NO_2^+, and $(CH_3)_2CH^+$], participating in bromination, nitration, isopropylation, etc., like hydrogen acids, are electrophilic reagents (p. 7), while the aromatic ring in these reactions is an electron-donor, i.e., a base. The equilibrium stage of the formation of an intermediate complex between hydrocarbons and acidlike substances has been demonstrated for some reactions. The stronger the acid, the smaller are the differences in the strengths of bases reacting with it and the more "active" the electrophilic reagent, the closer will be the rate constants of the reactions of benzene and its derivatives. Thereupon, the distribution of isomers approaches the statistically probable distribution, i.e., 40% of ortho- and meta-isomers and 20% of the para-isomer as there is a reduction in the differences in degree of "basicity" of the nonequivalent carbon atoms in the aromatic nucleus.

TABLE 84. Relative Rates of Electrophilic Replacement

Substance	Bromination with Br_2 [4,5,6,11]	Chlorination with Cl_2 [7,13]	Bromination with Br^+ [8,11]	Nitration with NO_2^+ [9,10,14]	Alkylation with C_6H_{11} [15]
C_6H_6	1	1	1	1	1
$C_6H_5CH_3$	$6 \cdot 10^2$	$3.4 \cdot 10^2$	$3.6 \cdot 10^1$	$4.2 \cdot 10^1$	2.2
$C_6H_5C_6H_5$	—	$1.2 \cdot 10^{3*}$	$1.3 \cdot 10^2$	$1.6 \cdot 10^{1**}$	1.6
$C_{10}H_8$	—	$1.0 \cdot 10^5$	—	$4 \cdot 10^2$	3.1
$C_6H_5OC_6H_5$	$1 \cdot 10^7$	—	—	$8 \cdot 10^7$	—
$C_6H_5OCH_3$	$1 \cdot 10^9$	—	—	—	1.3

* $6 \cdot 10^2$ according to [6].
** $4 \cdot 10^1$ according to [16].

Table 84 gives the ratios of the rate constants of electrophilic replacement of hydrogen (bromination by molecular bromine, chlorination by molecular chlorine, bromination by Br^+ ions, nitration, and alkylation by the Friedel-Crafts method with cyclohexyl bromide + $AlCl_3$) of benzene, toluene, biphenyl, naphthalene, diphenyl ether, and anisole. For example, toluene is brominated 600 times as fast as benzene and alkylated twice as fast. The rate of bromination of anisole is ten orders higher than that of benzene, though anisole is alkylated only 1.3 times as fast as benzene.*

---

*It is probable that this low reactivity of anisole is also caused by deactivation of the aromatic ring by coordination saturation of the oxygen atom of the methoxyl group, which disrupts the p, π-conjugation (p. 63). However, if we consider reactions with naphthalene where this effect is absent, the chlorination rate is still five orders higher than that of benzene, while the rates of alkylation of these hydrocarbons differ by a factor of only three.

The reduction in the differences in the reactivity of a substance is this great when it reacts with the most "active" reagents.

The effect of this rule on the distribution of isomers in the electrophilic replacement of hydrogen in toluene, biphenyl, and naphthalene is considered below.

3. PARTIAL RATE FACTORS OF THE REPLACEMENT OF HYDROGEN ATOMS IN AN AROMATIC RING

A. Definition

For the quantitative treatment of the results of studying electrophilic replacements, Brown and Nelson [17, 18] used the concept of partial rate factors in the replacement of aromatic hydrogen, which was introduced by Ingold [19]. They characterize the reactivity of each of the nonequivalent positions in the aromatic ring of the benzene derivative relative to the reactivity of each of the six equivalent positions in the benzene molecule. The partial rate factor of replacement is expressed by the equation

$$f_i = \frac{n_i \, k_{C_6H_5X} \, \alpha_i}{k_{C_6H_6}},$$

where $k_{C_6H_5X}/k_{C_6H_6}$ is the ratio of the rate constants for the replacement of hydrogen in the benzene derivative with the substituent X and in benzene itself; α_i is the fraction of replacement in the given position i; n_i is a statistical correction, which allows for the fact that in monosubstitution in a monosubstituted benzene there are two equivalent ortho-positions, two equivalent meta-positions, and one para-position, while in benzene all six positions are equivalent. Therefore, for the ortho-isomer the partial rate factor of replacement is expressed by the formula

$$f_o = \frac{6k_{C_6H_5X} \, \alpha_o}{2k_{C_6H_6}} = \frac{3k_{C_6H_5X} \, \alpha_o}{k_{C_6H_6}},$$

for the meta-isomer

$$f_m = \frac{3 \, k_{C_6H_5X} \alpha_m}{k_{C_6H_6}}$$

and for the para-isomer

$$f_p = \frac{6 \, k_{C_6H_5X} \alpha_p}{k_{C_6H_6}}.$$

If the yield of the isomer is expressed in percents P, then:

$$f_o = \frac{0.03 \, k_{C_6H_5X}}{k_{C_6H_6}} P_o$$

$$f_m = \frac{0.03 \, k_{C_6H_5X}}{k_{C_6H_6}} P_m \text{ and } f_p = \frac{0.06 \, k_{C_6H_5X}}{k_{C_6H_6}} P_p$$

$$F_S = \log \frac{f_p}{f_m}.$$

B. Relation Between "Activity" of the Reagent and the "Selectivity" of the Replacement

Brown and Nelson [17] defined the "activity" of a reagent as the ratio of the rate constants of replacement reactions $k_{C_6H_5X}/k_{C_6H_6}$. Brown and McGary [20] found the value f_p more convenient. In contrast to replacement in the ortho-position, replacement in the para-position is not subject to steric hindrance. The conjugation effect on which the interaction of the aromatic compound with the electrophilic reagent depends does not extend mainly to meta-replacement. Therefore, it was suggested in [21] that the selectivity of replacement should be expressed in the form of the logarithm of the ratio f_p/f_m, which is called the replacement selectivity factor: $F_S = \log f_p/f_m$. The ratio f_o/f_m depends on steric hindrance and therefore a comparison of f_o/f_m and f_p/f_m helps to differentiate between the role of conjugation and the steric factor in electrophilic replacement.

Brown used the most reliable available literature data on partial rate factors in electrophilic replacement reactions and supplemented them considerably with careful measurements of kinetics and isomer yields. The results obtained in this way are compared in Table 85, which gives the following reactions with toluene and benzene: bromination [4, 8], chlorination [22, 23, 24], chloromethylation [17], nitration [10], mercuration [25, 26], benzoylation [27-29], and alkylation by the Friedel-Crafts method [21].

The ratio of the reaction rate constants for toluene and benzene $k_{C_6H_5CH_3}/k_{C_6H_6}$ or, more briefly, k_T/k_B varies over a very wide range. The rate constants of the bromination of toluene and benzene by molecular bromine differ by a factor of 600, while the rate of isopropylation of toluene is only twice the rate of the reaction with benzene. Correspondingly, the yield of the meta-isomer in bromination (as in chlorination by molecular chlorine) is close to zero, while in isopropylation, it reaches

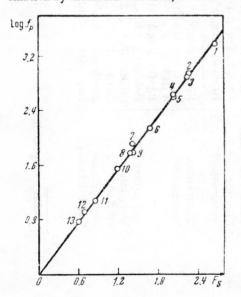

Fig. 31. Relation between the partial rate factor f_p and the selectivity factor F_S in electrophilic replacement of hydrogen in toluene. 1) Bromination (Br_2, 25°); 2) chlorination (24°); 3) benzoylation (nitrobenzene, 25°); 4) chloromethylation (60°); 5) basicity (HF); 6) basicity ($HF-BF_3$); 7) brominolysis of $ArB(OH)_2$ (25°); 8) nitration (45°); 9) bromination (Br^+, 25°); 10) mercuration (25°); 11) methylation ($GaBr_3$, 25°); 12) sulfonation; 13) ethylation ($GaBr_3$, 25°).

TABLE 85. Electrophilic Replacement of Hydrogen in Toluene

Reaction	Reaction conditions (reagents, solvent, temperature)	$\dfrac{k_T}{k_B}$	Distribution of isomers, %			Partial rate factor			F_S	Literature reference
			ortho	meta	para	f_o	f_m	f_p		
Bromination	Br₂ in 85% AcOH*, 25°	605	32.9	0.3	66.8	600	5.5	2420	2.648	[5]
Chlorination	Cl₂ in AcOH, 24°	353	57.7	0.5	41.8	611	5.3	887	2.227	[23]
"	Cl₂ in AcOH, 25°	344	59.78	0.48	39.78	617	4.95	820	2.219	[22]
Benzoylation	C₆H₅COCl, 25°	110	9.3	1.45	89.3	30.7	4.8	589	2.089	[29]
Chloromethylation	CH₂O + HCl + ZnCl₂ in AcOH	112	34.7	1.3	64.0	117	4.4	430	1.993	[17]
Nitration	HNO₃ in 90% AcOH, 45°	24.5	56.5	3.5	40.0	42	2.5	58	1.366	[10]
Bromination	HBrO + HClO₄ in dioxane + water 25°	36.2	70.3	2.3	27.4	76	2.5	59	1.373	[8]
Chlorination	HClO + HClO₄	60	74.6	2.2	23.2	134	4	82	1.312	[24]
Mercuration	Hg(OAc)₂ in AcOH + HClO₄, 50°	7.0	20.0	11.5	68.5	4.2	2.4	28.8	1.089	[25]
"	Hg(OAc)₂ in AcOH, 50°	5.0	30.7	13.7	56.1	4.6	2.0	17.0	0.850	[26]
Sulfonation	MeSO₂Cl + AlCl₃, 100°	3.7	49	15	36	5.4	1.7	8.0	0.680	—
Methylation	CH₃Br + Ga₂Br₆, 25°	5.7	55.7	9.9	34.4	9.5	1.7	11.8	0.841	[21]
Ethylation	C₂H₅Br + Ga₂Br₆, 25°	2.5	38.3	21.1	40.6	2.8	1.6	6.0	0.586	[21]
Isopropylation	iso-C₃H₇Br + Ga₂Br₆, 25°	2.2	28.4	33.4	38.2	1.5	1.5	5.2	0.548	[21]
tert-Butylation	tert-C₄H₉Br + Ga₂Br₆, 25°	1.0	0.0	32.1	67.9	0.0	1.6	6.6	0.624	[21]

*AcOH = CH₃COOH.

33%. For these reactions, f_p decreases from 2420 to 5 and the selectivity factor F_S changes from 2.6 to 0.5 (cf., Table 85).

If the values of log f_p are plotted against F_S, then, as is shown by Fig. 31, which was taken from [27], the points for all the electrophilic replacement reactions lie on one straight line expressed by the equation:

$$\log f_p = 1.310 \, F_S.$$

There is also a linear relation between log f_m and F_S:

$$\log f_m = 0.309 \, F_S.$$

Points corresponding to ortho-replacement lie on one straight line if the reactions are not complicated by steric hindrance:

$$\log f_0 = 1.215 \, F_S.$$

Having found a quantitative relation on the reaction rate to the relative yields of different isomers, Brown was able to predict parameters which were not available for some reactions and also to examine critically data available in the literature[21]. For example, if the ratio of the yields of para- and meta-isomers, i.e., F_S is known, it is not difficult to calculate the ratio of the rate constants from the equation:

$$\log k_T/k_B = 1.310 \, F_S - \log 6 - \log \alpha_p.$$

Brown and Young [27] calculated the partial rate factors of benzoylation and the values obtained were in good agreement with experiment [30].

	k_T/k_B	f_0	f_m	f_p
Calculated.........	149	32.0	4.9	817
Experimental data....	151	32.6	5.0	831

When this book had been sent to press there appeared an article by Stock and Brown [33a], which contains an exhaustive summary of data on the selectivity of electrophilic replacement of hydrogen in toluene. Data for 47 reactions were used and it was shown that they all satisfy the linear relation between the logarithm of the partial rate factor f_p and the selectivity factor F_S. For a series of critically selected reactions, the ratio log f_p/f_m is constant:

$$\log f_p/\log f_m = 4.04 \pm 0.37.$$

Having demonstrated that the selectivity relation applies to many reactions involving electrophilic replacement of hydrogen in toluene, Brown continued his investigations and extended the generalizations made to reactions with tert-butylbenzene or anisole [30a].

A comparison of the results obtained both by Brown and his co-workers and by other authors for 11 reactions of tert-butylbenzene shows that the ratio log $f_p/\log f_m$ is constant for them as for reactions with toluene. This ratio equals 2.89 $\pm$ 0.49. A linear relation between f_p and F_S was found for reactions with anisole. In addition, the interesting observation was made that there is a linear relation between the values of f_p for reactions of toluene and tert-butylbenzene and also for toluene and anisole.

Table 86 summarizes some results available in the literature on replacement reactions with biphenyl and naphthalene. It is very clear that there is a reduction in the differences in the reactivities of the α- and β-positions in the naphthalene molecule with a change from chlorination to nitration.

TABLE 86. Partial Rate Factors for the Replacement of Hydrogen Atoms in Biphenyl and Naphthalene

Substance	Reagent	$k/k_{C_6H_6}$	f_o	f_m	f_p	F_S	Literature reference
Biphenyl	Br^+	126	10.7	0.28	15.6	1.746	[12]
"	NO_2^+	40	41	<0.6	38	>1.799	[16]
"	NO_2^+	16	19	—	11	—	[14]*
Naphthalene	Cl_2	$1.05 \cdot 10^5$	$1.4 \cdot 10^8$	$1.4 \cdot 10^6$		—	[13]
"	NO_2^+	400	470	50		—	[14]
Substance	Reagent	$k/k_{C_6H_6}$	f_α	f_β		—	

*The values of the partial rate factors given in the original were calculated incorrectly, as was pointed out in [12].

C. Change in the Reactivity of a Substance in Catalysis

Brown and Nelson [17] put forward the hypothesis that in parallel with the acceleration of electrophilic replacement, the introduction of a catalyst should change the ratio of the isomers. In actual fact, this hypothesis was confirmed in the later work of Brown and McGary [20, 25]. When mercuration was catalyzed by perchloric acid, with the change in the ratio of the rates k_T/k_B there was a corresponding change in the relative yields of the isomers so that the linear relation between $\log f_p$ and F_S was retained (Fig. 32).

A vigorous electrophilic catalyst reduces the selectivity of replacement, i.e., decreases the ratio of the rates of bromination of toluene and benzene and increases the yield of the meta-isomer of the substituted toluene. For bromination in the absence of a catalyst, $k_T/k_B = 605$, while if this reaction is catalyzed by iodine [31], this ratio equals 465; it falls to 150 when a more active catalyst, namely, $ZnCl_2$, is used [32]. The yield of m-bromotoluene increases more, the more electrophilic the catalyst: $I_2 < SnBr_4 < FeBr_3 < AlBr_3$ (see unpublished data in the review [2]). In bromination catalyzed by $AlBr_3$, the yield of m-bromotoluene is 100 times that obtained in the absence of a catalyst [4, 20].

	ortho-	meta-	para-
Without catalyst ...	32.9	0.3	66.8%
$AlBr_3$ added	20	30	50%

The distribution of the isomers is nearer to the statistically probable distribution, the more active the catalyst.

This is illustrated by a comparison of the results of experiments on the methylation of toluene in the presence of $GaBr_3$ and $AlBr_3$ [21, 33].

Catalyst	k_T/k_B	ortho-	meta-	para-	f_o	f_m	f_p	F_S
$GaBr_3$ ...	5.7	55.7	9.9	34.4	9.51	1.70	11.80	0.847
$AlBr_3$ ...	2.5	49.8	20.9	29.3	3.73	1.56	4.39	0.449

TABLE 87. Calculated and Experimental Realtive Rates of Bromination of Methylbenzenes

Hydrocarbon	Calculated	Found
Benzene	1	1
Toluene	605	605
o-Xylene	$5.5 \cdot 10^3$	$5.3 \cdot 10^3$
m-Xylene	$5.4 \cdot 10^5$	$5.1 \cdot 10^5$
p-Xylene	$2.2 \cdot 10^3$	$2.5 \cdot 10^3$
Hemimellitene	$2.7 \cdot 10^6$	$1.7 \cdot 10^6$
Pseudocumene	$1.7 \cdot 10^6$	$1.5 \cdot 10^6$
Mesitylene	$4.4 \cdot 10^8$	$1.9 \cdot 10^8$
Prehnitene	$1.5 \cdot 10^7$	$1.1 \cdot 10^7$
Isodurene	$1.6 \cdot 10^9$	$0.4 \cdot 10^9$
Durene	$3.6 \cdot 10^6$	$2.8 \cdot 10^6$
Pentamethylbenzene	$4.4 \cdot 10^9$	$0.8 \cdot 10^9$

D. Calculation of Relative Reactivity of Substances from Partial Rate Factors

According to Condon [23], if the partial rate factors for the replacement of hydrogen in toluene are known, it is possible to calculate the relative rate at which any polymethylbenzene reacts. From the definition, the rate of the reaction of toluene relative to that of benzene:

$$\frac{k_T}{k_B} = \frac{2f_o + 2f_m + f_p}{6}.$$

If it is assumed that the reactivity of a substance is made up additively of the partial rate factors of each of the separate positions, then for the isomers of xylene we obtain:

$$\frac{k_{o\text{-xylene}}}{k_{benzene}} = \frac{2f_o f_m + 2f_m f_p}{6}; \quad \frac{k_{m\text{-xylene}}}{k_{benzene}} = \frac{f_o^2 + 2f_o f_p + f_m^2}{6}; \quad \frac{k_{p\text{-xylene}}}{k_{benzene}} = \frac{4f_o f_m}{6}.$$

To explain this, let us consider, for example, the formula of m-xylene in which the carbon atoms are numbered. Atom 2 is in the ortho-position relative to each of the methyl groups, i.e., its relative reactivity equals the product $f_o f_o = f_o^2$. Atom 4 is simultaneously in ortho- and para-positions, i.e., for it the component equals $f_o f_p$. The same is true of atom 6. Atom 5 is in the meta-position relative to both methyl groups, i.e., the term f_m^2 is added. Summing, we obtain $f_o^2 + 2f_o f_p + f_m^2$.

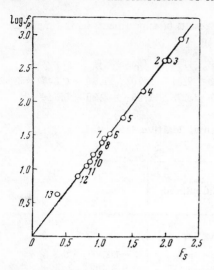

Fig. 32. Relation between the partial rate factor f_p and the selectivity factor F_S during electrophilic replacement of hydrogen in toluene (catalysis of mercuration by perchloric acid). 1) Chlorination (24°); 2) chloromethylation (60°); 3) basicity (HF); 4) basicity (HF − BF$_3$); 5) nitration (45°); 6) mercuration (HClO$_4$, 25°); 7) the same, 50°; 8) the same, 75°; 9) mercuration (without catalyst, 50°); 10) the same, 70°; 11) the same, 90°; 12) sulfonation; 13) isopropylation (40°).

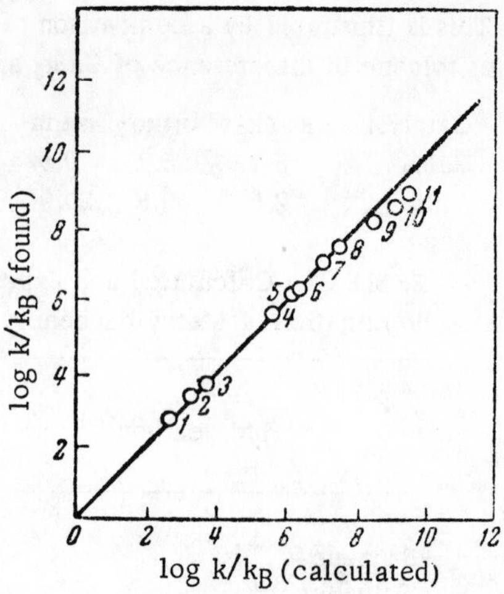

Fig. 33. Calculated and observed rates of bromination of methylbenzenes. 1) Toluene; 2) p-xylene; 3) o-xylene; 4) 5) hemimellitene; 6) pseudocumene; 7) durene; 8) prehnitene; 9) mesitylene; 10) isodurene; 11) pentamethylbenzene.

Condon [23] based his opinion on the data on the chlorination of methylbenzenes available at the time. Brown and his co-workers, using the reliable values of the partial rate factors obtained, demonstrated that it is possible to calculate similarly the relative rates of mercuration [26], bromination [4], chlorination [22], and benzoylation [29] of polymethylbenzenes in good agreement with the experimental values, which were summarized in Table 58 (p. 166).

As an example, the calculated and experimental relative rates of bromination at 25° are compared in Table 87. The mean discrepancy between the calculated and experimental values is 28%. This result is excellent in view of the fact that the extreme members of the series react at rates which differ by nine orders.

For mercuration, the mean discrepancy between the calculated and observed values is 12%. The relative rate constants may be calculated with even higher accuracy (the mean discrepancy does not exceed 6%) if we take into consideration and introduce appropriate correction factors for steric hindrance to replacement in the ortho-position.

Figures 33 and 34 give a graphical comparison of the observed and calculated relative rates of bromination and mercuration of methylbenzenes.

4. REASON FOR SELECTIVITY OF REPLACEMENT

In Brown's opinion [20, 21, 26], the degree of selectivity of replacement depends on the degree of participation of the electrons of the nucleophilic aromatic ring in the formation of the bond in the transition state. When the reaction involves sub-stances with a low affinity for electrons, the differences in electron density distribu-tion at the carbon atoms of the aromatic compound, caused by the presence of a polar substituent in the ring, are particularly marked.

As Nelson [18] commented, it is evident that the free energies of activation for replacement of ortho-, meta-, and para-hydrogen atoms are not quite the same, though they do not differ very much. In a reaction with an "active" reagent, the free energy of activation is comparatively low and a considerable number of the attacking particles have sufficient energy to produce the transition state in the re-action with any carbon atom in the ring. The rates of replacement differ little and the distribution of isomers approaches the statistically probable distribution and the selectivity of this reaction is low. On the other hand, when the free energy of acti-

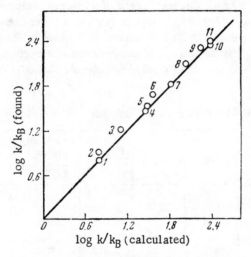

Fig. 34. Calculated and observed rates of mercuration of alkylbenzenes. 1) Toluene; 2) p-xylene; 3) o-xylene; 4) durene; 5) m-xylene; 6) pseudocumene; 7) hemimellit-ene; 8) prehnitene; 9) mesitylene; 10) pen-tamethylbenzene; 11) isodurene.

vation of the reaction is relatively high, only a few particles have sufficient energy to activate any of the atoms. The bulk of them will react at the position where the activation energy is minimal.* The highest selectivity of replacement corresponds to these conditions.

It is considered that the carbonium ion $CH(CH_3)_2^+$ and the nitronium ion NO_2^+ participate in isopropylation and nitration, while chlorination and bromination in acetic acid solution involved halogen molecules.

---

* There are indications that the relative rates of replacement in an aromatic ring in deuterium exchange, for example, may be determined by the magnitudes of the ac-tivation entropy [34] (see p. 178).

A comparison [18] of the energies of formation of $C-C$ (80 kcal/mole) and $C-N$ bonds (68 kcal/mole) shows that the activation of isopropylation in the transition state requires a considerably lower activation energy than nitration. As has been mentioned (p. 260), the distribution of isomers in isopropylation is quite close to the statistical distribution, while nitration yields largely ortho- and para-isomers and little of the meta-isomer. The formation of a $C-Cl$ bond in chlorination (formation energy 78 kcal/mole) requires an additional consumption of energy for rupture of the $Cl-Cl$ bond in the chlorine molecule (57.1 kcal/mole). It is clear therefore that chlorination requires a higher activation energy than nitration. Reactions with molecular halogens are some of the most selective (see Table 85).

In catalytic halogenation, the strongly polarized complex $Br - Br + AlBr_3 \rightleftharpoons$
$\rightleftharpoons Br - \overset{\delta^+}{Br} \cdot \overset{\delta^-}{AlBr_3}$ (it might be more accurate to write this equilibrium as follows:
$Br - Br + Al_2Br_6 \rightleftharpoons Br - \overset{\delta^+}{Br} \cdot \overset{\delta^-}{Al_2Br_6}$) reacts instead of a halogen molecule. Because of the polarity of the bond, its heterolytic rupture requires a lower activation energy and the rate of the process increases with a simultaneous increase in the yield of m-bromotoluene (p. 262). The selectivity of halogenation is strongly reduced when a Hal^+ or $HalOH_2^+$ ion reacts instead of a halogen molecule [35, 36] (Table 88).

TABLE 88. Partial Rate Factors in the Halogenation of Toluene

Reagent	f_o	f_m	f_p	F_S	Ratio of yields, para/ortho	Literature reference
Br_2	600	5.5	2420	2.648	2.0	[5]
Br^+	76	2.5	59	1.373	0.3	[8]
Cl_2	611	5.3	887	2.227	0.7	[23]
Cl^+	134	4.0	82	1.312	0.3	[24]

Thus, the relative rate of the reactions and the distribution of isomers depend to a large extent on the polarity of the reagent. Ions react least selectively, and intermediate position is occupied by polar particles, and, finally, a particularly high selectivity of replacement is shown by reagent molecules. In other words, the position of the points for the corresponding reaction on the line of log f_p against F_S gives some idea of the degree of polarity of the reagents and the bond formed in the transition state. We will return to this later in the discussion of conclusions on reaction mechanism (pp. 269 and 280).

5. ROLE OF THE STERIC FACTOR IN THE REPLACEMENT OF HYDROGEN
Steric hindrance is an important factor which affects the distribution of isomers in the replacement of hydrogen in aromatic compounds as in other reactions [37].

Deviations from linearity when F_S was plotted against log f_0 rather than log f_p were mentioned above. This is because of steric hindrance in the formation of the activated complex.

Holleman [1] mentioned the effect of the dimensions of the attacking particle in replacement in the ortho-position. For example, he considered that the volume of the reagent particle is associated with the fact that in chlorination, nitration, bromination, and sulfonation of chlorobenzene, the yields of the ortho-isomers successively decrease (45, 30, 3, and 0%). Brown and Smooth [21] pointed out that the relation between log f_0 and F_S should take into account the difference in the activation energy of replacements in the ortho- and para-positions ΔE_{act}, which depends on steric hindrance. The following equation was proposed for the replacement of one hydrogen atom in toluene:

$$\log f_o = 1.215 F_S - \frac{\Delta E_{act}}{2.3 RT}.$$

Returning to halogenation, it is noteworthy that when a halogen ion participates in the reaction, the yield of the ortho-isomer is greater than in the reaction with a molecular halogen (p. 260). In the opinion of de la Mare, in principle, the ortho-position is more reactive than the para-position, but this advantage does not normally show as a result of steric hindrance. In the bromination of toluene by Br^+ ions and in its nitration by NO_2^+ ions, the partial rate factors f_p are almost identical (59 and 58) and likewise for f_m (2.5 and 2.5). At the same time, for the first reaction $f_0 = 76$ and for the second reaction $f_0 = 42$ [4, 24]. According to de la Mare, this is connected with the fact that the effective radius of the NO_2^+ ion is greater than that of the Br^+ ion. He confirmed this idea [24] by comparing the yields of the para-isomers in the electrophilic replacement of hydrogen in toluene and tert-butylbenzene (Table 89).

TABLE 89. Yields of Para-Isomers (%) in the Halogenation of Toluene and tert-Butylbenzene

Reagent	Toluene	tert-Butylbenzene
Cl^+	23	42
Br^+	27.4	53.2
NO_2^+	40	79.5
Cl_2	40	76
Br_2	67	92

The yield of the para-isomer increases with an increase in the radius of the reagent particle particularly noticeably in the reaction with tert-butylbenzene as the tert-butyl group occupies a larger volume than the methyl group.

In examining the simplest steric models which de la Mare used, it must be remembered that some reactions may involve ion pairs as well as free ions, that they are solvated, etc., and therefore the approach proposed should be regarded as approximate although the explanation is probable in principle. As just mentioned, with branching of the alkyl radical, there is an increase in the steric hindrance to the replacement of a hydrogen atom in the ortho-position. The isomer yields obtained in the nitration of various monoalkylbenzenes are in agreement with this [38] (Table 90).

TABLE 90. Distribution of Isomers (%) in the Nitration of Alkylbenzenes

Alkyl group	ortho-	meta-	para-	Ratio of yields		
				o/p	o/m	p/m
CH_3	58.5	4.4	37.2	1.57	13.3	8.5
CH_2CH_3	45.0	6.5	48.5	0.93	6.9	7.5
$CH(CH_3)_2$	30.0	7.7	62.3	0.48	3.9	8.1
$C(CH_3)_3$	15.8	11.5	72.5	0.22	1.4	6.3

In the series examined, the ratio of the yields of para- and meta-isomers remains almost constant (8.5-6.3), while the ratio of the yields of ortho- and meta-isomers decreases by a factor of more than 9 (from 13.3 to 1.4). The change in the ratio of ortho- and para-isomers is explained by the steric effect for had the conjugation effect been of decisive importance, this would have been reflected in the ratio of para- and meta-isomers as conjugation does not affect replacement in the meta-position.

The dimensions of the alkyl group of the alkylbenzene has an equally clear effect on the partial rate factor of isopropylation in the ortho-position without affecting the values of f_m and f_p (Table 91).

The role of the steric factor also becomes increasingly clear in the alkylation of toluene as the degree of branching of the alkyl group of the reagent increases (Table 92).

TABLE 91. Partial Rate Factors in the Isopropylation of Alkyl-benzenes

Alkyl group	f_o	f_m	f_p	f_p/f_m	f_o/f_m
CH_3	2.4	1.8	4.3	2.4	1.3
$CH(CH_3)_2$	0.4	2.2	5.1	2.3	0.2
$C(CH_3)_3$	(0)	2.0	4.0	2.0	—

The yield of the ortho-isomer falls from 54% to zero when the $(CH_3)_3C^+$ ion participates in the reaction instead of the CH_3^+ ion. The yield of the para-isomer increases correspondingly from 29 to 93%.

There recently appeared [38a] additional confirmation of the effect of the steric factor in electrophilic replacement reactions (benzoylation and acetylation of monoalkylbenzenes in dichloroethane at 25°). As a result of an increase in the volume of the alkyl group, the partial rate factor f_0 successively decreased with a change from toluene (29.5; 4.5) to ethylbenzene (10.9; 1.0), isopropylbenzene (8.6; 0), and tert-butylbenzene (0; 0). The first figure in brackets refers to benzoylation and the second to acetylation.

6. PARTICIPATION OF ASSOCIATIVE AND IONIZATION PROCESSES IN HYDROGEN REPLACEMENT REACTIONS

The selectivity of replacement depends on the polarity of the bond formed in the transition state and is connected with the polarity of the reagent. Metal halides and iodine catalyze bromination [39] (p. 262) precisely because they ionize the bromine molecule, for example:

$$ArH + Br_2 + I_2 \rightleftarrows ArH \cdot \overset{\delta+}{Br} - \overset{\delta-}{Br} \cdot I_2.$$

The molecule is polarized much less in catalysis by $ZnCl_2$, $FeBr_3$, and especially $AlBr_3$.

TABLE 92. Distribution of Isomers (%) in the Alkylation of Toluene

Reagent	ortho-	meta-	para-
CH_3^+	53.8	17.3	28.8
$CH_3CH_2^+$	45	30	25
$(CH_3)_2CH^+$	37.5	29.8	32.7
$(CH_3)_3C^+$	0	7	93

E. A. Shilov and N. I. Konyaev [35] showed that the activity of a brominating agent is stronger the more it is polarized toward the formation of the positive ion Br^+. Therefore, the most powerful brominating agent is the Br^+ ion, the rate of bromination with which is almost 1500 times the rate of bromination with molecular bromine as bromine molecules polarize each other only very weakly (ArH + Br_2 + + $Br_2 \rightleftharpoons ArH \cdot \overset{\delta+}{Br} - \overset{\delta-}{Br} \cdot Br_2$). The polarity of the reagents affects the polarity of the bond formed in the activated complex and the rate of hydrogen replacement increases, while the selectivity decreases (cf., data given on p. 260 and in Table 86). It is evident that in a reaction with an electrophilic reagent, the polarity of the activated complex should also be affected by the degree of nucleophilicity of the substrate. In actual fact, the rate of halogenation (like other electrophilic replacement reactions) increases with an increase in the strength of the hydrocarbon as a base. This was demonstrated conclusively above on the example of methylbenzenes (see p. 166).

The polarity of the bond in the transition state may vary over quite a wide range. Reactions whose points lie on the upper part of the lines in Figs. 31 and 32 proceed by a mechanism which is close to molecular, while reactions whose points lie on the lower part of the lines proceed by a form of ionization mechanism. The fact that all the reactions give points lying on a common line may be interpreted as the result of the existence of intermediate forms between molecular and ionic mechanisms. It is also quite probable that reactions occupying an intermediate position involve complexes with different degrees of polarization and ion pairs.

Brown and his co-workers [41, 42] described phenomena which would be difficult to explain from the point of view of purely ionic or purely molecular mechanisms of replacement reactions. For example, it was found that the partial rate factors of the methylation of toluene catalyzed by $AlBr_3$ change markedly if CH_3I is used as the reagent instead of CH_3Br:

Partial rate factors

Reagent	f_0	f_m	f_p
CH_3Br	6.07	1.96	6.56
CH_3I	7.02	1.73	11.05

Had the reaction involved only the carbonium ions CH_3^+, the nature of the halogen originally attached to the methyl group would have made no difference. Consequently, the results obtained are incompatible with a purely ionic reaction mechanism and agree better with the hypothesis that there is an interaction between the nucleophilic aromatic molecule and a polarized complex of the alkyl halide with aluminum bromide, whose polarity depends on the nature of the halogen atom coordinated with the aluminum. Otherwise, it is difficult to understand how the halogen of the alkyl halide affects the partial rate factors.

Further support for this point of view was provided by a quantitative comparison of alkylations of benzene and toluene by alkyl bromides of different structures. $GaBr_3$ was used as the catalyst [42, 43]. In contrast to $AlBr_3$, in the presence of this catalyst the reaction is not complicated by isomerization. In addition, its lower activity facilitates the measurement of kinetics. If the methylation rate is taken as unity, the relative rates of ethylation, isopropylation, and tert-butylation are given by the following figures: $13.7; 15.9; 3 \cdot 10^5$, and $8 \cdot 10^5$. As regards selectivity factors, as is shown by Table 85, F_S has the minimum value for isopropylation. Judging by this, the reagent has the highest activity in isopropylation. Had carbonium ions participated in all reactions, the selectivity of replacement would have been expected to decrease in the order: tert-butyl > isopropyl > ethyl > methyl as a carbonium ion is more stable, the more branched it is (p. 133). In addition, had the methylation and ethylation proceeded by means of carbonium ions, it would have been difficult to understand why the rate of these reactions is almost five orders less than the rate of isopropylation and tert-butylation.

In the opinion of Brown, these facts confirm the participation of polarized complexes in reactions with low-molecular alkyl halides. This is expressed by the formula

$$Ar \ldots \overset{\overset{\displaystyle H}{\displaystyle |}}{R} - \overset{\delta+}{Hal} \cdot \overset{\delta-}{Ga_2Hal_6} \ldots HalR.$$

The degree of ionic character of the transition state varies, depending on the structure of the alkyl group R, the halogen, the catalyst, the aromatic compounds, and the medium. The reaction mechanism may approach that which Ingold designated by S-1 or S-2. In the first case, the molecule reacts with a previously formed ion, while in the second case, the bonds in the two molecules are broken at the moment of reaction and new bonds are formed. According to Brown, it is also necessary to consider the participation of complexes which are closer in nature to either molecules or ions.

This conclusion is in complete agreement with the deduction made below in connection with an examination of associative and ionization mechanisms of hydrogen exchange (see pp. 281 and 288).

Further confirmation of Brown's views was provided by the work of Olah and his co-workers [44-47]. These authors synthesized complexes with a 1:1 composition $(RF \cdot BF_3)$ by passing boron trifluoride into an alkyl fluoride at a low temperature. The specific electrical conductivity $\varkappa$ of the complexes of boron trifluoride with CH_3F and C_2H_5F was found to be three orders less than the electrical conductivity of analogous complexes with $n-C_3H_7F$ and $tert-C_4H_9F$, while the first two complexes were colorless and the others colored (Table 93).

TABLE 93. Properties of the Complexes $RF \cdot BF_3$

R	M.p., °C	$\varkappa$, ohm^{-1}cm^{-1}	Color	Alkylation product yield, %
CH_3	− 110	$2 \cdot 10^{-6}$	Colorless	62
C_2H_5	− 105	$5 \cdot 10^{-6}$	"	81
C_3H_7	− 95	$4 \cdot 10^{-3}$	Yellow	84
$tert-C_4H_9$	− 80	$4 \cdot 10^{-3}$	"	68

Consequently, the first two complexes have a covalent character and are very weakly ionized in contrast to the others, which decompose to a considerable extent into R^+ and BF_4^- ions. All the complexes smoothly alkylate toluene in high yield.

Olah made a detailed study of the mechanism of alkylation on the example of the reaction between ethyl fluoride and methylbenzenes (toluene and mesitylene) catalyzed by boron trifluoride. If an equivalent amount of the alkylbenzene was added at low temperature to the colorless complex of ethyl fluoride and boron trifluoride, there slowly appeared a bright color and the electrical conductivity increased considerably (to $10^{-3} \sim 10^{-2}$ ohm^{-1}cm^{-1}). A homogeneous readily identifiable molecular compound was obtained. The same compound was formed when BF_3 was added to a mixture of C_2H_5F and the hydrocarbon in equimolecular amounts. Table 94 gives the melting points, the specific electrical conductivity of the complexes in the molten state, and the color of complexes with composition $ArH \times C_2H_5^+$ $\times BF_4^-$.

The authors represented the structure of the complex with mesitylene by the following formula:*

---

* This formula is equivalent in meaning to the formulas given on pp. 6, 15, and 22.

$$\left[\begin{array}{c} \text{CH}_3 \\ \text{H}_3\text{C} - \!\!\!\!\!\diagdown\!\!\!\!\!\diagup - \text{CH}_3 \\ \text{H} \diagup \diagdown \text{C}_2\text{H}_5 \end{array} \right]^{+} \text{BF}_4^{-} \; .$$

Slow thermal decomposition of the homogeneous complex yielded two phases: an upper one containing an organic substance, which, after washing and distillation, was identified as the ethylated hydrocarbon and obtained in high yield, and a lower phase containing a stoichiometric amount of hydrogen fluoride and a small amount of hydrocarbon dissolved in it.

TABLE 94. Properties of the Complexes $\text{ArHC}_2\text{H}_5^{+} \cdot \text{BF}_4^{-}$

Hydrocarbon	M.p., °C	ϰ	Color
Toluene	−80	$2 \cdot 10^{-3}$	Yellow
m-Xylene	−75	$1 \cdot 10^{-3}$	Red
Mesitylene	−15	$2 \cdot 10^{-2}$	Orange

Thus, the authors were able to isolate two individual complexes, which could be regarded as intermediate products* in a reaction of the Friedel-Crafts type. Its mechanism is formulated in the following way:

$$\text{C}_2\text{H}_5\text{F} + \text{BF}_3 \rightleftharpoons \overset{\delta +}{\text{C}_2\text{H}_5} - \overset{\delta -}{\text{F}} \to \text{BF}_3$$

$$\overset{\delta +}{\text{C}_2\text{H}_5} - \overset{\delta -}{\text{F}} \to \text{BF}_3 + \langle \bigcirc \rangle - \text{CH}_3 \rightleftharpoons \left[\begin{array}{c} \text{H} \\ \diagup \diagdown \\ \text{H}_5\text{C}_2 \end{array} \!\!\! \langle \bigcirc \rangle \text{CH}_3 \right]^{+} \text{BF}_4^{-} \rightleftharpoons$$

$$\rightleftharpoons \text{HF} + \text{BF}_3 + \text{H}_5\text{C}_2 \langle \bigcirc \rangle \text{CH}_3 \; .$$

Instead of a polarized complex of the alkyl fluoride and boron trifluoride, a carbonium ion formed by the reaction of BF_3 with the corresponding alkyl fluoride may participate in the alkylation. Consequently, the mechanism of alkylation depends on the structure of the alkyl halide. The structure of the hydrocarbon also is undoubtedly of importance [48, 49]. The considerations put forward on the mechanism of alkylation also apply to other reactions of the electrophilic replacement type, including acid hydrogen exchange in aromatic compounds.

Winstein [50] and Swain [51] consider that there are mechanisms intermediate between S-1 and S-2 in nucleophilic replacement.

---

* The complexes of halogens with aromatic hydrocarbons, which are considered to be intermediate products in halogenation, are discussed on p. 164.

7. SUMMARY

1. Orientation in the replacement of hydrogen in an aromatic ring depends on the polarity, dimensions, and configuration of a substituent present in it, on whether the reagent is an electrophile or nucleophile, the degree of its polarity, and its dimensions (steric factor), and also on the medium in which the reaction occurs.

2. The work of Brown and his school has shown that the orientation rules in electrophilic replacement of hydrogen in aromatic compounds also depend on the chemical "activity" of the reagent. A high activity corresponds to a reduction in the differences in the nonequivalent atoms (ortho-, meta-, and para-positions) with respect to replacement reactions. The selectivity of the replacement is low and the yields of the isomers approach the statistically probable values, i.e., 40% of the ortho-, 40% of the meta-, and 20% of the para-isomer. It was established that the logarithm of the partial rate factor for replacement in the para-position (log f_p) is related linearly to the selectivity factor F_S, which equals the logarithm of the ratio of the partial rate factors of replacement in the para- and meta-positions ($F_S = \log f_p / f_m$).

3. The rules describing the effect of the chemical "activity" of a reagent on the reactivity of nonequivalent atoms in the ring in aromatic replacement are analogous to the rules of acid−base interaction in which there is a reduction in the differences in the strengths of protolytes when a strong acid or a strong base participates in the reaction. The carbon atoms of an aromatic ring, which differ in degree of electronegativity as a result of the presence of a substituent, behave like bases of different strength in a reaction with an acidlike substance.

4. Intermediate products from the addition of the electrophilic reagent to the aromatic compound may be formed in electrophilic replacement of hydrogen in an aromatic ring. Some of these complexes have been isolated.

5. The mechanism of aromatic replacement cannot always be reduced to a molecular (associative) or ionization mechanism. Polarized molecular complexes and ion pairs may participate in reactions together with molecules and ions. This circumstance and the occurrence of a reaction by different mechanisms simultaneously may explain the continuous change in the selectivity factor with a change in the chemical "activity" of the reagent, which is expressed as a linear equation relating the logarithm of the partial rate factor of replacement to the selectivity factor.

LITERATURE CITED

1. A. F. Holleman, Chem. Rev. 1, 187 (1924).
2. K. Le Roi Nelson and H. C. Brown, In the book: The Chemistry of Petroleum Hydrocarbons (Ed., B. T. Brooks, C. E. Boorod, S. S. Kart, and L. Schmerling), Vol. III (New York, 1955), 56 (Aromatic Substitution − Theory and Mechanism).
3. J. Bunnett and R. Zahler, Chem. Rev. 49, 273 (1951).
4. H. C. Brown and L. M. Stock, J. Am. Chem. Soc. 79, 1421 (1957).
5. P. B. D. de la Mare and C. A. Vernon, J. Chem. Soc. (1951) p. 1764.
6. P. B. D. de la Mare, J. Chem. Soc. (1954) p. 4450.
7. H. C. Brown and L. M. Stock, J. Am. Chem. Soc. 79, 5175 (1957).
8. P. B. D. de la Mare and J. T. Harvey, J. Chem. Soc. (1956) p. 36.
9. M. J. S. Dewar and D. S. Urch, J. Chem. Soc. (1958) p. 3079.
10. H. Cohn, E. D. Hughes, M. H. Jones, and M. G. Peckling, Nature 169, 291 (1952).
11. P. R. Robertson, P. B. D. de la Mare, and B. E. Swedlung, J. Chem. Soc. (1953) p. 782.

12. P. B. D. de la Mare and M. Hassan, J. Chem. Soc. (1957) p. 3004.

13. M. J. S. Dewar and T. Mole, J. Chem. Soc. (1957) p. 342.

14. M. J. S. Dewar, T. Mole, and E. W. T. Warford, J. Chem. Soc. (1956) pp. 3576, 3581.

15. N. N. Lebedev, Zhur. Obshchei Khim. 27, 2460 (1957).

16. O. Simamura and Y. Mizuno, Bull. Chem. Soc. Japan 30, 196 (1957).

17. H. C. Brown and K. Le Roi Nelson, J. Am. Chem. Soc. 75, 6292 (1953).

18. K. Le Roi Nelson, J. Org. Chem. 21, 145 (1956).

19. C. K. Ingold and F. R. Shaw, J. Chem. Soc. (1957) p. 2918.

20. H. C. Brown and C. W. McGary, J. Am. Chem. Soc. 77, 2300 (1955).

21. H. C. Brown and C. R. Smoot, J. Am. Chem. Soc. 78, 6255 (1956).

22. H. C. Brown and L. M. Stock, J. Am. Chem. Soc. 79, 5175 (1957).

23. F. E. Condon, J. Am. Chem. Soc. 70, 1963 (1948).

24. P. B. D. de la Mare, J. T. Harvey, M. Hassan, and S. Varma, J. Chem. Soc. (1958) p. 2756.

25. H. C. Brown and C. W. McGary, J. Am. Chem. Soc. 77, 2306 (1955).

26. H. C. Brown and C. W. McGary, J. Am. Chem. Soc. 77, 2310 (1955).

27. H. C. Brown and H. L. Young, J. Org. Chem. 22, 719 (1957).

28. H. C. Brown, B. A. Bolton, and F. R. Jensen, J. Org. Chem. 23, 414 (1958).

29. H. C. Brown and F. R. Jensen, J. Am. Chem. Soc. 80, 2296 (1958).

30. H. C. Brown and H. L. Young, J. Org. Chem. 22, 724 (1957).

30a. L. M. Stock and H. C. Brown, J. Am. Chem. Soc. 81, 5621 (1959); 82, 1942 (1960).

31. E. Berliner and F. J. Bondhus, J. Am. Chem. Soc. 70, 1948 (1948).

32. L. J. Andrews and R. M. Keefer, J. Am. Chem. Soc. 78, 4549 (1956).

33. H. C. Brown and H. Jungk, J. Am. Chem. Soc. 77, 5584 (1955).

33a. L. M. Stock and H. C. Brown, J. Am. Chem. Soc. 81, 3323 (1959).

34. E. L. Mackor, P. J. Smith, and J. H. van der Waals, Trans. Farad. Soc. 53, 1309 (1957).

35. N. I. Konyaev and E. A. Shilov, Doklady Akad. Nauk SSSR 24, 890 (1939).

36. P. B. D. de la Mare, A. D. Ketley, and C. A. Vernon, Research 6, 125 (1953).

37. M. S. Newman (Ed.), Steric Effects in Organic Chemistry, Ch. 3 (New York, 1956) pp. 164-201.

38. H. C. Brown and W. H. Bonner, J. Am. Chem. Soc. 76, 605 (1954).

38a. H. C. Brown and G. Marino, J. Am. Chem. Soc. 81, 5611 (1959).

39. C. S. Price and C. H. Arntzen, J. Am. Chem. Soc. 60, 2835 (1938).

40. H. C. Brown and H. Jungk, J. Am. Chem. Soc. 78, 2182 (1956).

41. H. Jungk, C. R. Smoot, and H. C. Brown, J. Am. Chem. Soc. 78, 2185 (1956).

42. C. R. Smoot and H. C. Brown, J. Am. Chem. Soc. 78, 6245 (1956).

43. C. R. Smoot and H. C. Brown, J. Am. Chem. Soc. 78, 6249 (1956).

44. G. Olah, S. Kuhn, and J. Olah, J. Chem. Soc. (1957) p. 2174.

45. G. Olah and S. Kuhn, Nature 178, 1344 (1956).

46. G. A. Olah, A. E. Pavlath, and J. A. Olah, J. Am. Chem. Soc. 80, 6540 (1958).

47. G. A. Olah and C. J. Kuhn, J. Am. Chem. Soc. 80, 6541 (1958).

48. H. C. Brown, H. W. Pearsall, L. P. Eddy, W. J. Wallace, M. Gryson, and K. Le Roi Nelson, Ind. and Eng. Chem. 45, 1462 (1953).

49. G. A. Russell, J. Am. Chem. Soc. 81, 4834 (1959).

50. S. Winstein, E. Grunwald, and H. W. Jones, J. Am. Chem. Soc. 73, 2700 (1951).

51. C. G. Swain and W. P. Langsdorf, J. Am. Chem. Soc. 73, 2813 (1951).

CHAPTER 3. MECHANISMS OF HYDROGEN EXCHANGE

1. COMPARISON OF RULES OF HYDROGEN EXCHANGE AND OTHER REPLACEMENTS OF HYDROGEN IN AROMATIC COMPOUNDS

The similarity between metallation and hydrogen exchange with strong bases was traced in the first chapter of Section III on hydrocarbons as acids, while the similarity between the reactivity of hydrocarbons in irreversible electrophilic replacement of hydrogen and in isotopic exchange with acids was pointed out in the second chapter of this section. This indicates the common features of the mechanism of these reactions. The same is indicated by the identical effect of substituents in the aromatic ring on the course of each of these reactions. Now that the views on the mechanisms of chemical replacements of hydrogen have been discussed, we will turn to a comparison of their rules with the rules of exchange reactions. This will facilitate the proof of the mechanisms of the latter.

A. Basic Hydrogen Exchange

a. COMPARISON OF HYDROGEN EXCHANGE AND METALLATION

The discussion of the mechanism of metallation in Chapter 1 indicates that these reactions are produced by an acid—base interaction between an organic substance and a metallating reagent. The latter (for example, an organoalkali compound, metal amide, etc.) protonizes the hydrogen of a CH bond of the substance metallated. The reaction rate is limited by the rupture of the CH bond and depends on the degree of polarity of this bond and the strength of the base. Electron-acceptor substituents promote metallation as the electrons of the CH bond are displaced toward them and this facilitates protonization of the hydrogen. These substituents include groups containing electronegative halogen, oxygen, sulfur, and nitrogen atoms (CF_3, OC_6H_5, OCH_3, SCH_3, and $N(CH_3)_2$) or halogen atoms replacing hydrogen in the aromatic ring. Electron-donor substituents, in particular, alkyl groups have the opposite effect and retard metallation of the aromatic ring.

As regards the metallation of the benzene ring when some substituent is present in it, we only have qualitative data at present (with the exception of the case of alkylbenzenes, for which the partial rate factors are known). The following sequence is found:

$$F > C_6H_5O > CH_3O > H > CH_3.$$

This is indicated by the work of Wittig, who established that fluorobenzene is metallated faster than anisole [1] and that diphenyl ether is metallated faster than anisole [2], and also the measurements of Bryce-Smith [3], according to which the partial rate factors of metallation of isopropylbenzene are less than unity, i.e., the alkyl group retards metallation. In addition, Bryce-Smith showed that the metallation of alkylbenzenes in the α-position of the alkyl group by ethylpotassium represents the following percentages of the total replacement of hydrogen by the metal in toluene, ethylbenzene, and isopropylbenzene: 100, 50, and 13%, respectively, i.e., the lability of the hydrogen atoms in the alkyl groups falls in the series $CH_3 > CH_2 > CH$. A similar conclusion was drawn by Morton [4] from determinations of the yields of carboxylic acids obtained by carboxylation of the products from the metallation of alkylbenzenes in the presence of sodium oxide (the yields are given in brackets):

$C_6H_5CH_3$ (87%) > $C_6H_5CH_2CH_3$ (31%) > $C_6H_5CH(CH_3)_2$ (11%). The same sequence of reactivities of the α-atoms was found by measuring the rates of deuterium exchange of monoalkylbenzenes with a solution of potassamide in liquid ammonia [5]. If the rate constants of hydrogen exchange in the aromatic ring of alkylbenzenes are taken as unity, then the following values are obtained for the rate constants of exchange of the α-hydrogen atoms in the alkyl groups of toluene, ethylbenzene, and isopropyl-benzene:

CH_3	CH_2	CH	Ring
280	40	8	1

Thus, among the α-hydrogen atoms in the alkyl groups, the hydrogen atom of the methyne group is most difficult to replace by deuterium and also by an alkali metal. Removal of the proton from it requires a higher consumption of energy and the tertiary carbanion is less stable than a secondary or primary carbanion [6-8].

The same rule also applies to purely aliphatic compounds. For example, in parallel experiments with isobutanes containing deuterium in methyl and methyne groups, isotopic exchange could be produced only in the first compound by heating with a solution of potassamide in liquid ammonia [9]. A study of metallation shows that the acidity of butane isomers decreases with branching of the alkyl group (p. 116).

By a study of the kinetic isotope effect in hydrogen exchange in liquid ammonia it was possible to demonstrate that the rate of exchange with a base, like the rate of metallation, is limited by rupture of the $C-H$ bond. This was demonstrated in work in the Isotopic Reactions Laboratory [10] by measurement of the kinetics of exchange of tritium and deuterium in the methylene group of fluorene for protium in ammonia. It was established that at 25°, the rate of tritium exchange is half that of deuterium exchange ($k = 8 \cdot 10^{-5}$ and $1.5 \cdot 10^{-4}$ sec^{-1}). This is because the zero point energy of the $C-T$ bond is lower than that of the $C-D$ bond as a result of the higher mass of the tritium atom.

The same result was obtained in some other work in the Isotopic Reactions Laboratory [10a] on tritium and deuterium exchange between liquid ammonia and methyl β-naphthyl ketone, containing the heavy isotope in the methyl group. Streitwieser [10b] found that $k_D/k_T = 3.0 \pm 0.3$ for exchange of the heavy isotope in the methylene group of ethylbenzene. The catalyst was lithium cyclohexylamide dissolved in cyclohexylamine (t 49.9°). In the same note, there is a reference to analogous work of Longworthy, who obtained the values of 2.75 and 2.98 for the ratio k_D/k_T for replacement of deuterium and tritium in the methyl group of toluene. As yet it is difficult to explain the differences in the absolute values of the ratio k_D/k_T for different reactions. It may be the result of different polarities of the bond breaking in the transition state.

In hydrogen exchange between a base and aromatic compounds, as in the metallation of the latter, the relative acidity of the $C-H$ bonds is of decisive importance. It is controlled predominantly by the inductive effect of the substituent. This conclusion was drawn from the determination of partial rate factors of deuterium exchange, which are examined below.

b. PARTIAL RATE FACTORS IN BASIC HYDROGEN EXCHANGE

The partial rate factors in hydrogen exchange in aromatic compounds equals the ratio of the rate of deuterium exchange in the corresponding isomer (with respect to the position of the deuterium) to the rate of deuterium exchange in benzene. Partial rate factors for deuterium exchange [11-14] catalyzed by a solution of potassamide in liquid ammonia are summarized in Table 95.

The rate constant for the exchange of a deuterium atom in the ortho-position of fluorobenzene is more than seven orders higher than the rate constants for the exchange of a corresponding deuterium atom in toluene. Substituents producing an inductive displacement of the electrons of CH bonds toward themselves activate hydrogen exchange in the same way as metallation, i.e., $f > 1$. Substituents of electron-donor type, which act in the opposite direction, produce the reverse effect, i.e., $f < 1$.

TABLE 95. Partial Rate Factors of Hydrogen Exchange with a Solution of Potassamide in Liquid Ammonia

Substance	f_o	f_m	f_p	Literature reference
C_6H_4DF	$> 10^6$	10^3	10^2	[11]
$C_6H_4DCF_3$	$> 10^5$	10^4	(10^4)	[11]
$C_6H_4DOC_6H_5$	$2 \cdot 10^4$	$5 \cdot 10^1$	4	[12]
$C_6H_4DOCH_3$	$5 \cdot 10^2$	—	$5 \cdot 10^{-1}$	[12,13]
$C_6H_4DC_6H_5$	4.7	3.3	2.9	[14]
$C_6H_4DN(CH_3)_2$	1.5	$2 \cdot 10^{-1}$	$7 \cdot 10^{-2}$	—
C_6H_5D	1	1	1	—
$C_6H_4DCH_3$	$1 \cdot 10^{-1}$	$3 \cdot 10^{-1}$	—	[14]

The substituents lie in the following series with respect to decreasing f_0:

$$F > CF_3 > OC_6H_5 > OCH_3 > C_6H_5 > N(CH_3)_2 > H > CH_3,$$

which is very similar to that obtained by qualitative comparisons for metallation by organoalkali compounds (p. 275). When a direct comparison of the partial rate factors of deuterium exchange and metallation for the aromatic ring of alkylbenzenes is possible, the factors are found to be very similar [3, 5].

Reaction	f_0	f_m	f_p
Metallation	0.10	0.38	0.43
Deuterium exchange . .	0.14	0.32	—

In accordance with the hypothesis on the influence of the inductive effect on metallation and deuterium exchange in the presence of an electron-donor substituent it is found that $f_0 < f_m < f_p$, while with an electron-acceptor substituent, the exchange rates lie in the opposite order: $f_0 > f_m > f_p$.

In work in the Isotopic Reactions Laboratory [12a] it was shown that a plot of $\log f_0$ against $pK_i^{H_2O}$ [15], where $K_i^{H_2O}$ is the ionization constant of an acetic acid derivative (CH_2XCOOH) with the same substituent X as that in the aromatic ring, the points lie on a straight line (Fig. 35). This is an additional and convincing demonstration of the importance in hydrogen exchange with a base of the relative acidity of the CH bonds in the aromatic ring, which, like the strength of carboxylic acids, depends on the inductive displacement of the electrons toward or away from the substituent.

Together with $pK_i^{H_2O}$, Table 96 gives the values of the inductive parameters σ_I of Taft [15a].

There is a linear relation between $\log f_0$ and σ_I. The data obtained obey the Hammett-Taft equation:

$$\log f_0 = \log \frac{k_0}{k_{C_6H_6}} = \rho\sigma_I = 11\,\sigma_I.$$

c. HYDROGEN EXCHANGE AND THE ALKYLATION OF AROMATIC HYDROCARBONS AND ISOMERIZATION OF OLEFINS

It is not difficult to follow the parallelism between deuterium exchange in hydrocarbons with bases and other chemical reactions (apart from metallation) proceeding through the formation of carbanions. It is not essential that free carbanions should be formed. In the transition state there may be only an electronic displacement toward the formation of $\equiv C^{\ominus}$.

An example of this is the alkylation of aromatic hydrocarbons promoted by organoalkali compounds, which was discovered by Ipatieff and Pines [16] (p. 120). In the alkylation of benzene by ethylene, the yield of ethylbenzene is higher than the yield of isomeric ethylbenzenes in the reaction of ethylene with tert-butylbenzene as the hydrogen atoms in the aromatic ring of the latter have a lower acidity than those in benzene (pp. 93 and 101). Pines and Mark [17] observed that the alkylation of aliphatic-aromatic hydrocarbons always proceeds in the direction corresponding to the higher stability of primary carbanions as compared with secondary and especially tertiary carbanions.

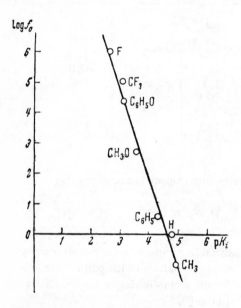

Fig. 35. Relation between the partial rate factors of deuterium exchange with a solution of potassamide in liquid ammonia ($\log f_0$) and the ionization constants $pK_i^{H_2O}$ of acetic acid derivatives with the same substituents.

This rule also applies to the isomerization of olefins with migration of the double bond, catalyzed by potassamide, for which B. A. Kazanskii and I. V. Gostunskaya proposed a carbanion mechanism [18, 19] (p. 116). The isomer with the greatest number of hydrogen atoms in an allyl position always reacts at the highest rate. For example, 2-methyl-1-butene is converted into 2-methyl-2-butene more rapidly than 3-methyl-1-butene. Correspondingly, in alkenes a CH_3 group is metallated in preference to a CH_2 and especially a CH group [20] (p. 95).

By measurement of the kinetics of deuterium exchange in olefins it was also established that there is a fall in the lability of the hydrogen atoms with a change from a CH_3 to a CH_2 and a CH group [21]. Depending on the structure of the hydrocarbon, which determines its degree of acidity, deuterium exchange, metallation, alkylation, and isomerization may occur in the presence of an alkoxide, a caustic alkali, an alkali or alkaline earth metal amide, or an organoalkali compound, i.e., bases of different strengths.

The similarity in the mechanisms of deuterium exchange with bases and metallation led to the idea that hydrogen exchange in hydrocarbons may be induced by carbanions of organoalkali compounds [22, p. 391]. This prediction was soon confirmed by the work of G. A. Razuvaev [23] and Hart [24].

TABLE 96. Values of $pK_i^{H_2O}$ of Acetic Acid Derivatives (25°) and Inductive Parameters σ_I

X	$pK_i^{H_2O}$	σ_I
F	2.66	+ 0.52
CF_3	3.07	+ 0.41
C_6H_5O	3.12	+ 0.38
CH_3O	3.53	+ 0.25
C_6H_5	4.31	+ 0.10
$N(CH_3)_2$	—	+ 0.10
H	4.76	0.00
CH_3	4.88	− 0.05

B. Acid Hydrogen Exchange

a. PARTIAL RATE FACTORS IN ACID HYDROGEN EXCHANGE

Ingold's conclusion that deuteration of aromatic compounds by acids is a typical electrophilic replacement reaction stemmed, in particular, from the actual fact that the normal orientation rules for electrophilic replacements are obeyed in deuteration; for example, deuteration occurs in the ortho- and para-positions in the presence of such substituents as OR and NR_2. Now it is known that the distribution of isomers depends not only on the nature of the substituent, but also on the reagent and some quantitative data are available on the partial rate factors of electrophilic replacement reactions and acid deuterium; Ingold's conclusion can therefore be checked in the light of present knowledge. Table 97 gives partial rate factors for acid exchange of deuterium in different positions of the aromatic ring. (In the case of naphthalene, the corresponding factors f_α and f_β are given in the columns of f_0 and f_m.)

Figure 36 gives a plot of log f_p against F_S for the exchange of deuterium in isomeric monodeuterotoluenes with three acids (liquid HBr, 0.25 M H_2SO_4 in CF_3COOH, and 68% H_2SO_4 in water) and for a series of chemical electrophilic replacements [26].* The points lie close to a common straight line. This obviously indicates that the activated complex has a similar structure in acid hydrogen exchange and in chemical replacements of hydrogen in which acidlike substances participate.

---

* The same comparison was made in the articles of Stock and Brown [L. M. Stock and H. W. Brown, J. Am. Chem. Soc. 81, 3323, 5615 (1959)], which was published when this book had been sent to press [See also the article of these authors published in J. Am. Chem. Soc. 82, 1942 (1960).]

b. SELECTIVITY OF REPLACEMENT AND POLARITY OF BOND

It is extremely probable that Brown was correct in his conclusion, which was based on an analysis of a large amount of material, that both the relative rate of replacement of nonequivalent hydrogen atoms and its selectivity depends on the polarity of the bond formed in the transition state, which, in its turn, is determined by the degree of polarity of the reagent and the dielectric constant of the medium (p. 266).

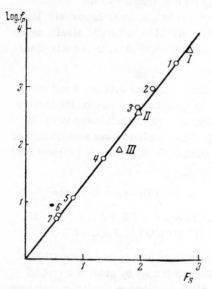

Fig. 36. Relation between partial rate factors and selectivity factors in electrophilic replacement of hydrogen and acid deuterium exchange. 1) Bromination (Br_2); 2) chlorination; 3) chloromethylation; 4) bromination (Br^+) and nitration; 5) methylation; 6) ethylation; 7) isopropylation. I) Deuterium exchange with liquid HBr; II) deuterium exchange with CF_3COOH; III) deuterium exchange with H_2SO_4.

On the one hand (upper part of the line in Fig. 36) there are molecules with covalent bonds (halogens) and on the other (lower part of the line in Fig. 36), cations with a high affinity for an electron such as the carbonium ion $CH(CH_3)_2^+$. The selectivity of replacement gradually changes with an approach to the statistically probable distribution of isomers in a chemical reaction and similar rates of isotopic exchange of all the hydrogen atoms. For example, the selectivity of bromination falls sharply if the reagent is the Br^+ ion instead of the Br_2 molecule ($F_S = 1.3$ and 2.6) or if a catalyst such as $AlBr_3$, which polarizes the bromine molecule, is added (p. 262). The action of the same catalysts is completely analogous in deuterium exchange with liquid HBr as a result of polarization of the HBr molecule:

$$ArD + HBr + AlBr_3 \rightleftharpoons$$
$$\overset{\delta^+}{\rightleftharpoons ArD \cdot H} \overset{\delta^-}{- Br \cdot AlBr_3}.$$

There is a decrease in the energy required to break the H−Br bond, the replacement rate increases, and the replacement loses its selectivity. Thus, the rate constant for deuterium exchange between liquid HBr and m-deuterobiphenyl without a catalyst equals $2 \cdot 10^{-8}$ sec^{-1}, while for the para-isomer k = $1.4 \cdot 10^{-4}$ sec^{-1}. Exchange equilibrium with both isomers is reached after 5 min in the presence of only 0.02 mole of $AlBr_3$ per 1000 g of HBr. A meta-deuterium atom in toluene exchanges equally rapidly when this catalyst is added [14]. Rapid exchange of hydrogen is also observed [29] when the exchange is carried out with liquid deuterium fluoride, which has a very high dielectric constant (84 at 0°) as well as a high acidity ($H_0 = -10.2$). Even such a weak base as toluene is ionized appreciably in liquid hydrogen fluoride (p. 139) and all the hydrogen atoms of the toluene nucleus exchange with this solvent in a few minutes.

TABLE 97. Partial Rate Factors in Acid Hydrogen Exchange (at 25°)

Substance	Acid	f_o and f_α	f_m and f_β	f_p	F_S	Lit. ref.
$C_6H_4DOCH_3$	$HClO_4-H_2O$	$2.3 \cdot 10^4$	$2.5 \cdot 10^{-1}$	$5.5 \cdot 10^4$	5.34	[25]
$C_{10}H_7D$	HBr	$5 \cdot 10^4$	$1 \cdot 10^3$	–	–	[14]
$C_{10}H_7D$	$CF_3CO_2H-H_2SO_4$	$1.8 \cdot 10^2$	$2.3 \cdot 10^1$	–	–	[27]
$C_6H_4DC_6H_5$	HBr	$5 \cdot 10^2$	< 0.5	$3 \cdot 10^3$	> 3.8	[14]
$C_6H_4DC_6H_5$	$CF_3CO_2H-H_2SO_4$	$3.7 \cdot 10^1$	< 0.3	$3.7 \cdot 10^1$	> 2.1	[27]
$C_6H_4DCH_3$	HBr	$1 \cdot 10^3$	5	$4 \cdot 10^3$	2.90	[14]
$C_6H_4DCH_3$	$CF_3CO_2H-H_2SO_4$	$2.5 \cdot 10^2$	3.5	$3.5 \cdot 10^2$	2.00	[27]
$C_6H_4DCH_3$	$CF_3CO_2H-H_2O$ (70°)	$2.5 \cdot 10^2$	3.8	$4.2 \cdot 10^2$	1.82	[27a]
$C_6H_4DCH_3$	$H_2SO_4-H_2O$	$8.3 \cdot 10^1$	1.9	$8.3 \cdot 10^1$	1.64	[28]

A similar reduction in the differences in the reactivities of nonequivalent hydrogen atoms was observed in experiments with diphenyl ether. The meta-atoms in it do not exchange with liquid HBr even after 300 hr, but exchange equilibrium is reached even after 3 hr if $AlBr_3$ is added or the ether is dissolved in liquid HF [30, 31].

Thus, the general conclusions drawn above on the mechanism of electrophilic replacement of hydrogen in aromatic compounds also apply to acid hydrogen exchange. The details of its mechanism are still not clear (p. 64).

c. REACTIVITY OF AROMATIC CH BONDS IN HYDROGEN EXCHANGE WITH A BASE AND WITH AN ACID

Acid hydrogen in an aromatic molecule with a substituent depends mainly on the effect of conjugation of the electrons of the substituent with the π-electron system of the ring, while the inductive effect of the substituent transmitted by the σ-bonds is of prime importance in hydrogen exchange with a base. This difference is caused by the different mechanisms of exchange with an acid and with a base. In the first stage, an electrophilic reagent, i.e., an acid reacts with the π-electrons of the ring, while a nucleophilic reagent, i.e., a base, protonizes the hydrogen of a CH bond.

In both cases there is an acid—base reaction, but in the first case the substance acts as a base and in the second case, it acts as an acid. Different aspects of the intereffect of atoms in the same molecule are therefore revealed.

As an illustration, the sequences of partial rate factors of deuterium exchange with a base and with an acid are compared in Table 98.

This table shows, for example, that methoxyl and phenyl groups promote most rapid exchange in the ortho-position with a base and in the para-position with an acid. Had the same conjugation effect played a decisive part in hydrogen exchange with reagents of both chemical types, in view of the distribution of electron density at the nonequivalent carbon atoms, one would have expected that hydrogen exchange with a base would have occurred more readily at that carbon atom at which the reaction with an acid was most difficult, i.e., the meta-atom. The fact that this does not agree with experiment confirms the view that different intereffects of the atoms are of greatest importance in acid and basic hydrogen exchange.

This hypothesis is not always justified. For example, the α-hydrogen atoms in the naphthalene molecule exchange both with an acid and with a base more rapidly than the β-atoms and the rate of exchange of the latter is higher than for benzene [14]. This is evidently explained by the higher polarizability of the α-CH bond in comparison with the β-CH bond [32] and the higher over-all polarizability of the naphthalene molecule in comparison with that of the benzene molecule (p. 106).

TABLE 98. Sequences of Partial Rate Factors in Deuterium Exchange with a Base and with an Acid

Substituent	Base	Acid
F	$O > M > P \gg 1$	
CF_3	$O > M = P \gg 1$	
C_6H_5O	$O \gg M > P > 1$	
CH_3O	$O \gg M = 1 > \Pi$	$P > O \gg 1 > M$
C_6H_5	$O > M > P > 1$	$P > O \gg 1 > M$
$N(CH_3)_2$	$O > M > P > 1$	$P > (O) \gg 1 > M$
CH_3	$1 > (P) > M > O$	$P > O \gg M > 1$

We should also note that the exchange of hydrogen with a base in the para-position relative to the substituent may depend substantially on the conjugation effect, evidently as a result of the rapid fall-off of the inductive effect with distance from the substituent.

An examination of the results obtained in the Isotopic Reactions Laboratory on deuterium exchange in aromatic compounds with an acid and with a base indicates that there is an inverse relation between the electron-donor and electron-acceptor properties of the same substituents ($N(CH_3)_2$, $N(CH_3)C_6H_5$, OCH_3, OC_6H_5, and C_6H_5), i.e., the more strongly the properties of the substituent as an electron donor (in a reaction with an acid) are expressed, the lower its affinity for an electron (in a reaction with a base).

In contrast to the substituents listed above, an alkyl group is an electron donor regardless of the reagent. As a result of the difference in the mechanisms of deuterium exchange with an acid and with a base and depending on whether the conjugation effect or the inductive effect predominates in the transmission of the effect of the alkyl group to the ring, the partial rate factors in an exchange between toluene and an acid are greater than unity and in an exchange with a base, less than unity (Tables 95 and 97). The differences in the mechanisms of acid and basic exchange appear particularly clearly in experiments with polymethylbenzenes.

Data obtained in the Isotopic Reactions Laboratory on deuterium exchange between methylbenzenes deuterated in the aromatic ring and a solution of potassamide in liquid ammonia are given below. For convenience in comparison with results of Lauer and Stedman's experiments (Table 68, p. 178), the rate constants of deuterium exchange with benzene are taken as equal to 6, while for toluene they are taken as equal to the sum of the partial rate factors.

Deuterium Exchange in Methylbenzenes

Substance	Found	Calculated
Benzene	6	6
Toluene	1.75	1.75
p-Xylene	$6.0 \cdot 10^{-1}$	$4.0 \cdot 10^{-1}$
Mesitylene	$9.0 \cdot 10^{-2}$	$6.2 \cdot 10^{-2}$
Durene	$2.2 \cdot 10^{-2}$	$2.0 \cdot 10^{-2}$

There is good agreement between the rate constants determined experimentally and the values calculated on the principle of additivity, established by Condon (p. 263). The relative rate of the exchange of deuterium in mesitylene with trifluoro-acetic acid is 9 orders higher (Table 68) than the rates of deuterium exchange with an ammonia solution of potassamide. Calculation for exchange with hydrogen bromide, using the partial rate factors for toluene given in Table 97, shows that the differences in the relative rate constants reach 12 orders. A more graphic demonstration of the differences in the mechanisms of hydrogen exchange in aromatic compounds with acids and with bases can hardly be required.

2. MECHANISMS OF HYDROGEN EXCHANGE

A. A. I. Brodskii's Classification

a. FAST AND SLOW TYPES OF EXCHANGE

In 1949 A. I. Brodskii [33] made an important generalization on hydrogen exchange in solutions. It was necessary to explain why the rate of hydrogen exchange in the CH bonds of organic compounds with heavy water differs so sharply from the rate of exchange of hydrogen in OH, NH, SH, and HaLH bonds (see pp. xv, 21, and 66). On analyzing the reasons for this, A. I. Brodskii came to the conclusion that it cannot be the result of differences in any physical parameters of the CH bond (energy, length, force constant, and polarizability). In his opinion, the structure of the electron cloud of the atom attached to the hydrogen is of decisive importance. Oxygen, nitrogen, sulfur, and halogen atoms in electrically neutral molecules have a free pair of electrons. A deuteron may add to this simultaneously with the elimination of a proton. This synchronous process occurs with an appropriate disposition of the reacting molecules. The activation energy required for their orientation is small; the hydrogen therefore exchanges at a high rate:

$$D^+ \rightarrow : X : H \rightarrow .$$

There are no free electrons in the electron cloud of a carbon atom of an organic compound. Therefore, in contrast to the case just examined, the addition of a deuteron is only possible when the bond between the carbon and hydrogen atoms has been broken.

The rupture of the C−H bond and the elimination of the proton requires a considerable activation energy. Therefore, the hydrogen in a CH bond exchanges slowly.

Thus, according to A. I. Brodskii, fast and slow types of hydrogen exchange differ qualitatively in mechanism. The simplest scheme corresponding to fast exchange is represented by the equation:

$$:X-H + :Y-D \rightarrow \begin{matrix} X - H \\ \overset{\cdot\cdot}{\downarrow} \quad \uparrow \\ D - \overset{\cdot\cdot}{Y} \end{matrix} \rightarrow :X-D + :Y-H.$$

In actual fact, the reaction may proceed not in a four-membered complex, but, for example, in a six-membered cyclic transition complex formed by hydrogen bonds [34].

A. I. Brodskii's idea of the participation of a free electron pair in hydrogen exchange was found to be very fruitful. It not only helped in the generalization of many disconnected observations on hydrogen exchange in different substances with amphoteric solvents, which had accumulated in the first and most important of their investigation, but also made it possible to predict new facts of fundamental importance.

b. EXPERIMENTAL CHECK OF A. I. BRODSKII'S RULE

A. I. Brodskii's conclusion represented a generalization from a vast number of facts on hydrogen exchange with heavy water and heavy alcohol. For an additional check and substantiation of his idea on the relation between the exchange rate and the structure of the electron cloud of the atom of the element to which the hydrogen atom is connected, A. I. Brodskii undertook a series of experimental studies [35-45]. They are widely known and therefore we will only mention them briefly.

As was anticipated in accordance with the absence of a free pair of electrons from the silicon atom in organosilicon compounds, no deuterium exchange was detected between heavy water and silanes of the type R_3SiH, where R is C_2H_5O or C_6H_5 [42, 43].

In phosphorous acid, the last hydrogen atom could not be exchanged for the deuterium of heavy water because it is connected to a pentavalent phosphorus atom which has no electron pair:

$$H-\underset{\underset{OH}{|}}{\overset{\overset{O}{\|}}{P}}-OH.$$

At the same time, all the hydrogen atoms in hypophosphorous acid exchange comparatively rapidly. This is explained [45] by the equilibrium between tautomeric molecules with penta- and trivalent phosphorus:

$$\underset{H}{\overset{H}{>}}P\underset{OH}{\overset{O}{<}} \rightleftarrows H-P\underset{OH}{\overset{HO}{<}} .$$

It was possible to demonstrate the retardation of exchange of hydrogen in an ammonium salt when a concentrated acid was added [13]. The acid suppresses hydrolysis of the ammonium salt and reduces the ammonia concentration to a minimum. Exchange is retarded in the ammonium ion itself because the atom in it does not have a free pair of electrons as the pair at the nitrogen atom is saturated as a result of the addition of a proton to it.

This interesting result, which stimulated A. I. Brodskii's theory, was confirmed and developed by other authors [46-49] with alkylammonium salts. For example, the half-time of exchange between triethylammonium chloride and ethylene glycol in ethylenic glycol solution containing 0.1 mole of HCl equals one hour at 6° [46], i.e., the exchange is actually retarded. The kinetics of hydrogen exchange between ammonium salts and sulfuric acid of various concentrations were measured recently [49a]:

$$
\begin{array}{lccc}
H_2SO_4, \% & 49.1 & 58.4 & 69.4 \\
k \cdot 10^3 \text{ sec}^{-1} & 6.6 & 0.6 & 0.01
\end{array}
$$

At the same time, new reports were published on the rapid hydrogen exchange between water and alcohols [50, 51], ethylamine and tert-amyl alcohol and between amines [52], which completely agrees with A. I. Brodskii's rule.

However, there are also deviations from A. I. Brodskii's rule. A. I. Brodskii has listed a series of these facts [39-41]. Thus, despite the presence of a free pair of electrons at the phosphorus atom in the phosphine molecule ($: PH_3$), hydrogen exchange is retarded and this cannot be explained by diffusion processes [53]. The rate of exchange between hydrogen sulfide and alcohol [54], and hydrogen sulfide and thiophenol [55] is comparatively low. The exchange equilibrium between deuterium oxide and normal water is established slowly [56]. A. I. Brodskii partly explains the deviations from the rule he formulated by the fact that the electron pairs are blocked as a result of associative processes. G. P. Miklukhin [55] was inclined to ascribe them to the difference in the polarity of the bonds with hydrogen, for example, OH and SH. However, it is evident that the slow exchange between two isotopic forms of water cannot be explained in this way (cf., p. 71).

On the other hand, examples were found of deuterium exchange in CH bonds which proceeds quite rapidly in water or alcohol without a catalyst. Hydrogen exchange in thiazolonium salts [57], pyrone [58], and fluoradene [59] has been mentioned above (p. 96).

The differences in the rates of hydrogen exchange in bonds with atoms with different structures of the electron cloud, which are normally quite marked when the reagent is an amphoteric solvent, are most often obliterated completely if the solvents and catalysts used are strong bases or strong acids such as a solution of KND_2 in liquid ND_3, $AlBr_3$ in liquid DB3, or BF_3 in liquid DF (see Section II). According to A. I. Brodskii [38, 39, 41] it is not the exchange rate itself that is important in the classification of exchange reactions, but whether or not this rate is dependent on the structure of the substrate (for example, a substituent) and the experimental conditions (catalysis). In the first case we have the type of exchange which is conditionally called slow and in the second case, exchange reactions of the fast type. At the same time, it is assumed that the activation energy of an exchange of the fast type does not exceed 8-10 kcal/mole. However, this is not always the case. For example, the activation energy of exchange of hydrogen in the CH bonds of acetophenone and fluorene is of the order of 11-12 kcal/mole (p. 98), while the activation energy of hydrogen exchange between phosphine and water reaches 18 kcal/mole, though the first reaction should be regarded as a slow exchange in accordance with the structure of the electron cloud of the carbon atom, while the second is a fast exchange according to the structure of the electron cloud of the phosphorus cloud. The effect of a change in the structure of the substrate produced, for example, by the introduction of some substituent may actually be observed in fast exchange also and, in particular, it is observed in the exchange of hydrogen between different alcohols (p. 71). A. I. Brodskii [39] pointed out that this was probable. The existence of acid—base catalysis is not only characteristic of slow exchange, but is also a feature common to hydrogen exchange in bonds of hydrogen with any element (p. 66).

The extensive factual material presented in this book shows that it is expedient and accurate to classify hydrogen exchange from the point of view of the protolytic function of the substrate and reagent [22, 60], (p. 21).

B. Associative and Ionization Mechanisms of Hydrogen Exchange

In the examination of the mechanisms of reactions in solution, associative and ionization mechanisms are often distinguished by determining whether the reaction occurs between associated molecules or through an ionic state. These mechanisms also appear in the classification of hydrogen exchange reactions. A. I. Brodskii [36] considers that fast exchange always proceeds by an associative mechanism in the form of synchronous breakage and formation of bonds between atoms in a cyclic complex, which consists of several molecules. Slow exchange (exchange in CH bonds) proceeds either by an associative or by an ionization mechanism. In the opinion of A. I. Brodskii, catalysis by bases is more characteristic of the latter, while he regards reactions of a substrate with acids as typical of the associative mechanism of slow exchange and identifies them with electrophilic replacements, following Ingold's scheme:

$$\text{>C—H} + \text{D—OSO}_3\text{D} \longrightarrow \text{>C}\overset{\delta-}{:}\begin{matrix}\overset{\delta+}{\text{H}}\\ \\ \overset{\delta+}{\text{D}}\end{matrix}\overset{\delta-}{:}\text{OSO}_3\text{D} \longrightarrow \text{>C—D} + \text{H—OSO}_3\text{D} .$$

However, this view is not always justified [61, p. 706; 62, p. 62]. It is easy to provide examples of hydrogen exchange in an aromatic ring under ionizing conditions (for example, exchange of the meso-hydrogen atoms of anthracene with D_2SO_4 [63] (see also p. 170). On the other hand, it is unlikely that hydrogen exchange between a solution of a strong base such as potassamide in liquid ammonia and heptane [64], for example, is accompanied by appreciable ionization of the hydrocarbon. Associative and ionization processes occur in hydrogen exchange both with bases and with acids. To understand the mechanism of an exchange it is of prime importance to know the chemical function of the substance during the reaction and only then, the state in which it is.

Gold and Satchell [65] also considered the acidic or basic function of the substance. It is true that the terminology which they used can easily be misunderstood. They consider as electrophilic those mechanisms in which two-stage exchange is produced by the elimination from or addition to the substrate of a proton. Exchange is therefore closely connected with acid—base catalysis. In contrast to A. I. Brodskii, Gold and Satchell classify ionization and not associative mechanisms as electrophilic. They distinguish between two ionization mechanisms: dissociative (in which the substrate is an acid) and additive (in which the substrate is a base). These terms are superficially similar to the terms dissociative (ionization) and associative, but they are used in a completely different sense. What A. I. Brodskii called the ionization mechanism of slow exchange, Gold and Satchell denote as an electrophilic mechanism. The mechanism of the type proposed by Ingold to explain deuterium exchange with sulfuric acid, i.e., associative according to A. I. Brodskii, is called synchronous by Gold and Satchell.

In the article cited, which was published in 1955, Gold and Satchell considered the ionization of an acid and a base in the light of Brønsted's simple scheme, which has become classical. The complexity of an acid—base interaction is now well known from many studies, which were discussed in Section IV. N. A. Izmailov substantiated experimentally a different scheme for the ionization of acids. According to this scheme, the first stage is the formation of a complex between the acid and the solvent, i.e., their association (p. 206). Consequently, according to N. A. Izmailov, ionization and associative processes are connected inseparably.

In a solvent with a low dielectric constant, the ions are connected in ion pairs or even in more complicated complexes. Noncoulombic as well as electrostatic forces may be involved in their formation. This was observed long ago [66, 67, p. 412] and now there is no doubt about it. It is sufficient to read the latest articles of Kraus [68], who is the greatest authority on the theory of solutions (see also p. 200).

In a review in 1958, Gold [69] considered that ionization is a multistage process which he wrote in the form of the scheme:

$$\overbrace{RX \rightleftarrows R^+X^-}^{\text{Ionization}} \rightleftarrows \overbrace{R^+ \| X^- \rightleftarrows R^+ + X^-}^{\text{Dissociation}}$$

in which RX is an electrolyte which decomposes into the ions R^+ and X^-.

The scheme provides for the existence of ion pairs of different types, "intimately" and "loosely" bound. In the latter, the ions are separated by a solvation envelope, which is represented by two lines in the scheme. Consequently, in addition to ions and molecules, structures of an intermediate type exist. They may be closer to either molecules or free ions. Therefore, the participation of ionic complexes in reactions (including hydrogen exchange) is by no means equivalent to the participation of molecules and ions in them.

A. I. Brodskii pointed out in discussion [39, p. 17] [70] that there is a difference between ionization and associative mechanisms which is approximately equivalent to the difference between S-1 and S-2 mechanisms according to Ingold and that this classification is as well founded as the division of chemical bonds into ionic and covalent. This is true only if we consider limiting cases as A. I. Brodskii also agrees with the fact that depending on the degree of polarization of the bonds in the activated complex and on the medium, there may be any transitional form between mechanisms [39, p. 16].

Ionic and covalent bonds and ionic and molecular reactions are antipodes in classification. In dealing with real substances and actual chemical reactions, one has transitional forms most often. Features approaching one of the limiting types may be expressed more or less clearly in them. This is shown, for example, by data on the mechanism of electrophilic replacement. Therefore, it is incorrect to merely contrast ionization and associative mechanisms without considering transitional forms [62, 71, p. 925] (see also [72, p. 243]). Particular attention should be paid to the latter to derive a classification which reflects reality more fully.

With a knowledge of the rules of protolytic reactions and the chemistry and physical chemistry of solutions, it is often possible to predict which processes, associative or ionization, will be predominant in any case. Special and not always simple investigations are required to establish the degree of their participation in an actual reaction.

Several examples are given to provide a clearer idea of the change from a clearly molecular to a clearly ionization mechanism in hydrogen exchanges of the same type.

First example. It was shown above (p. 280) that if the logarithm of the partial rate factor f_p is plotted against the selectivity factor F_S (Fig. 36), then a common straight line passes through the points obtained by measuring the rate of hydrogen exchange between monodeuterotoluene isomers and various acids (HBr, CF_3CO_2H, and H_2SO_4). The selectivity of the replacement, i.e., its selectivity with respect to nonequivalent hydrogen atoms, depends on the degree of polarity of the bond formed in the transition state. Points corresponding to reactions whose mechanisms approach a molecular mechanism lie on the upper part of the straight line, while points corresponding to reactions whose mechanisms are close to an ionization mechanism lie on the lower part of the line. The continuity of the transition (points lie along the whole line) may be interpreted as an indication of the participation in the reaction of polar molecules, complexes of various polarities, ion pairs, and ions. Particles of the activated complex may also vary (p. 281).

Second example. Let us examine acid hydrogen exchange in an aromatic ring. According to the data of G. M. Panchenkov [73], the hydrogen in the aromatic ring of cumene is exchanged for deuterium comparatively readily when the vapor is in contact with a solid deuterated aluminosilicate catalyst. Measurements of the infrared spectra of an aromatic carbon absorbed on a similar catalyst, made under the direction of A. N. Terenin [74], indicate that a hydrogen bond is formed between the hydroxyl group of the catalyst and the aromatic hydrocarbon (π-complex). Appreciable dissociation of a monoalkylbenzene and, even more so, benzene does not occur when the hydrocarbon is dissolved in liquid deuterium bromide, whose dielectric constant is low ($DC_{25°} = 4$). No appreciable increase in the electrical conductivity of this solvent is observed when even a polymethylbenzene is dissolved in it. Nonetheless, exchange occurs at room temperature between benzene and liquid deuterium bromide (p. 168): $k_{20°} = 5 \cdot 10^{-8} \sec^{-1}$. It is most likely that the reaction proceeds by an associative mechanism in this case. The rate of hydrogen exchange can be increased to different extents by adding electrophilic acidlike substances of different electron-acceptor powers: $TiBr_4 < BBr_3 < AlBr_3$. The acceleration of the exchange is explained by the formation of complexes in which the H−Br bond is polarized as a result of the tendency of the halogen ion to add to the coordinationally unsaturated central atom of the bromide (p. 42). The different degrees of acceleration of the exchange may be explained by the different polarities of the complexes formed in the ternary system aromatic hydrocarbon-hydrogen bromide-bromide. For example, in the presence of even a very low concentration of aluminum bromide, the exchange between benzene and deuterium bromide proceeds so rapidly that its rate is difficult to measure at room temperature (p. 168) and the solution conducts a current.

Sharp acceleration of the reaction is also observed in experiments on hydrogen exchange in benzene and alkylbenzene dissolved in liquid deuterium fluoride (p. 174).

As a result of the high protogenicity and very high dielectric constant ($DC_{0°}$ = 84) of this solvent, aromatic hydrocarbons are ionized in it (p. 132). The rate constant for exchange of the hydrogen in benzene $k_{25°}$ = $1 \cdot 10^{-3} sec^{-1}$ (p. 168), while the dissociation constant K = $1 \cdot 10^{-8}$. Mackor [75] considers that even in the system benzene-hydrogen fluoride, associative processes still continue to play a substantial part. Exchange may occur in the complex, which is given in brackets in the equation:

$$ArD + HF \rightleftarrows (ArD.HF) \rightleftarrows (ArH.DF) \rightleftarrows ArH + DF.$$

Finally, if boron trifluoride is added to a solution of benzene (and even more so, an alkylbenzene), the hydrocarbon is completely converted into ions, for example, benzene is present in the solution in the form of $C_6H_7^+$ ions (p. 141). The exchange acquires a clearly expressed ionic character.

Thus, it is possible to effect acid exchange of hydrogen in the same hydrocarbon by any desired mechanism from associative to ionization, including intermediate forms where complexes of different degrees of polarity participate in the exchange.

Third example. We will now consider basic hydrogen exchange. Deuterium exchange may be effected between a hydrocarbon such as benzene, in the vapor phase and solid deuterated potassium or calcium amide [19]. It proceeds even more readily if the benzene and the potassamide are dissolved in liquid deuteroammonia (p. 104). The rate of hydrogen exchange in the methyl group of toluene is 70 times higher than that in the benzene ring, but measurement of the infrared spectrum of a solution of toluene and potassamide in liquid ammonia does not show any ionization of the CH bonds of the methyl group [76]. Exchange occurs more rapidly in the methyl group of methylnaphthalene. Ionization of methylnaphthalene and the formation of a carbanion in the presence of potassamide are indicated by the electronic spectrum of the ammonia solution [77].

Hydrogen exchange in the methyl group of quinaldine dissolved in liquid ammonia is a very slow reaction ($k_{120°}$ = $7 \cdot 10^{-8} sec^{-1}$). The solution does not contain ions in a concentration that can be determined by spectral measurements [77]. If 0.02 mole of potassamide per mole of quinaldine is added to the solution, the rate constant of the exchange even at $-31°$ is increased by two and a half orders ($k_{-31°}$ = $2 \times 10^{-5} sec^{-1}$) (p. 24). Measurement of the absorption spectrum shows that quinaldine is converted into a carbanion by ionization of a CH bond of the methyl group.

Fourth example. This example again concerns hydrogen exchange in the methyl group of quinaldine, but with alcohols in this case (p. 35). When nitrogen bases are dissolved in alcohols, a hydrogen bond is formed between the nitrogen atom and the hydrogen of the hydroxyl group of the alcohol. This is indicated by infrared spectral measurements. Measurements of the dipole moments of alcohols in the presence of organic bases show that the polarity of the O−H bond changes differently, depending on the acidity of the alcohol (p. 228). The rate of hydrogen exchange in the methyl group of quinaldine changes in parallel. The highest rate constants correspond to hydrogen exchange with the most acid of the alcohols compared: isopropanol < ethanol < methanol < ethylene glycol. This is explained by the fact that then there is the most clearly expressed effect of the change in the valence of the nitrogen atom, which may be completed by its conversion into a tetravalent ion with positively charged nitrogen by the formation of a salt with a strong mineral acid. In actual fact, when 0.2 mole of hydrogen chloride per mole of quinaldine is added, i.e., when part of the latter is converted into the hydrochloride, the reaction rate increases considerably ($k_{120°} = 5 \cdot 10^{-5}$ instead of $2 \cdot 10^{-6}$ sec^{-1} in the absence of acid).

Thus, the lability of the hydrogen atoms of the methyl group of quinaldine estimated quantitatively by means of hydrogen exchange, depends on the valence state of the nitrogen of the heterocycle. It may change not in a jump, as is required by Brønsted's theory, but gradually, for example, with a gradual increase in the degree of polarity of the hydrogen bond and, finally, conversion of the heterocycle into a quaternary ammonium salt.

All these examples show quite clearly and convincingly that a gradual change in the degree of the acid−base interaction from a very weak one (formation of a hydrogen bond) right up to complete transfer of a proton and ionization of the protolyte molecules may affect the rate, selectivity, and, in the end, the mechanism of hydrogen replacement without changing the characteristics of the hydrogen exchange, which are determined by the protolytic function of the substrate. The stronger the acid−base interaction, the more rapid the exchange normally is. The examples given confirm the conclusion drawn previously that hydrogen exchange is accelerated by an increase in the polarity of the reaction complex, the limiting polarization state of which is − ionization [62, p. 6]. The polarity of the intermediate reaction complex, if one is formed, and the polarity of the transition state are evidently symbatic.

The most important conclusion from all that has been presented is the fact that the classification of hydrogen exchanges in solution (like other heterolytic reactions) should not be limited to merely associative and ionization mechanisms. Not only molecules and ions, but also polarized molecular complexes, ion pairs, and their aggregates of various types may participate in reactions; the mechanism of the reactions depends on this.

3. SUMMARY

1. The rules of basic deuterium exchange (for example, with a solution of potassamide in liquid deuteroammonia) and of metallation, alkylation of hydrocarbons, and isomerization and dimerization of olefins catalyzed by bases are similar as all these reactions proceed by a carbanion protophilic mechanism with the hydrocarbon fulfilling the function of an acid; the rates of metallation and deuterium exchange are limited by the rupture of the CH bond.

2. The rules of acid hydrogen exchange with aromatic compounds and electrophilic replacements of hydrogen in them are analogous because both types of reaction are caused by an interaction between an aromatic ring, which is a base (electron donor), and an electrophilic reagent (an acid or an acidlike substance).

3. As a result of the different mechanisms of hydrogen exchange with an acid and with a base, mainly different aspects of the intereffect of atoms in the aromatic compound (conjugation effect and inductive effect) are shown.

4. The most important feature for classifying heterolytic hydrogen exchange in solutions is the protolytic function of the substrate and reagent. In discussing the mechanism of hydrogen exchange it is necessary to consider new ideas on the mechanisms of acid−base interaction, which provide for processes that are not accompanied by complete transfer of a proton and ionization of electrically neutral molecules.

5. Together with associative and ionization mechanisms of hydrogen exchange it is necessary to consider intermediate types of exchange mechanisms.

LITERATURE CITED

1. G. Wittig, Angew. Chem. 66, 10 (1954).
2. G. Wittig, Angew. Chem. 69, 245 (1957); Uspekhi Khim. 27, 298 (1958).
3. D. Bryce-Smith, J. Chem. Soc. (1954) p. 1079.
4. C. E. Claff and A. A. Morton, J. Org. Chem. 20, 981 (1955).
5. A. I. Shatenshtein and E. A. Izrailevich, Zhur. Fiz. Khim. 32, 2711 (1958).
6. P. D. Bartlett, S. Friedman, and M. Stiles, J. Am. Chem. Soc. 75, 1771 (1953).
7. H. Pines and V. Mark, J. Am. Chem. Soc. 79, 4316 (1956).
8. J. E. Leffler, The Reactive Intermediates of Organic Chemistry (New York, 1956).
9. Yu.G.Dubinskii,E.A.Yakovleva,and A.I.Shatenshtein,Zhur.Obshchei Khim.(in press).
10. F. S. Yakushin, Yu. G. Dubinskii, E. A. Yakovleva, and A. I. Shatenshtein, Zhur. Fiz. Khim. 33, 2820 (1959).
10a. F. S. Yakushin and A. I. Shatenshtein, Kinetika i Kataliz 1, 489 (1960).
10b. A. Streitwieser, E. E. van Sickle, and L. Reif, J. Am. Chem. Soc. 82, 1513 (1960).
11. G. E. Hall, R. Piccolini, and J. D. Roberts, J. Am. Chem. Soc. 77, 4540 (1955).
12. A. I. Shatenshtein, A. N. Talanov, and Yu. I. Ranneva, Zhur. Obshchei Khim. 30, No. 2 (1960).
12a. A. I. Shatenshtein, Collection: Structure and Reactivity of Organic Compounds, Scientific Conference, October, 1959 [in Russian] (Goskhimizdat, Leningrad, 1959) p. 4.
13. A. I. Shatenshtein and A. V. Vedeneev, Zhur. Obshchei Khim. 28, 2644 (1958).
14. E. N. Yurygina, P. P. Alikhanov, E. A. Izrailevich, P. N. Manochkina, and A. I. Shatenshtein, Zhur. Fiz. Khim. 34, No. 3 (1960).
15. H. C. Brown, D. H. McDaniel, and O. Häfliger, in the book: Determination of Organic Structures by Physical Methods (Ed., E. A. Braude, and F. C. Nachod) (New York, 1955) p. 577.
15a. R. W. Taft, J. Am. Chem. Soc. 75, 4231 (1953); 79, 1045 (1957); 80, 2436 (1958).
16. H. Pines, J. A. Vesely, and V. N. Ipatieff, J. Am. Chem. Soc. 77, 554 (1955).
17. H. Pines and V. Mark, J. Am. Chem. Soc. 78, 4316, 5946 (1956).
18. I. V. Gostunskaya, N. I. Tyun'kina, and B. A. Kazanskii, Doklady Akad. Nauk SSSR 108, 473 (1956).
19. A. I. Shatenshtein, Yu. G. Dubinskii, E. A. Yakovleva, I. V. Gostunskaya, and B. A. Kazanskii, Izvest. Akad. Nauk SSSR, Otdel. Khim. Nauk (1958) p. 104.
20. A. A. Morton, M. L. Brown, M. E. T. Holden, B. L. Letsinger, and E. E. Magat, J. Am. Chem. Soc. 67, 2224 (1945).
21. A. I. Shatenshtein and L. N. Vasil'eva, Doklady Akad. Nauk SSSR 95, 115 (1954).
22. A. I. Shatenshtein, Uspekhi Khimii 24, 377 (1955).
23. G. A. Razuvaev, G. G. Petukhov, M. M. Shubenko, and V. A. Voitovich, Ukr. Khim. Zhur. 22, 45 (1956).
24. H. Hart, J. Am. Chem. Soc. 78, 2619 (1956).
25. D. P. N. Satchell, J. Chem. Soc. (1956) p. 3911.
26. A. I. Shatenshtein, Eighth Mendeleev Conference, Radiochemistry and Chemistry of Isotopes Section [in Russian] (Izd. AN SSSR, Moscow, 1958) p. 43; Zhur. Fiz. Khim. 34, No. 3 (1960).
27. G. Dallinga, A. A. Verrijn Stuart, P. J. Smit and E. L. Mackor, Z. Elektrochem. 61, 1019 (1957).

27a. W. M. Lauer, G. W. Matson, and G. Stedman, J. Am. Chem. Soc. 80, 6433(1958).
28. V. Gold and D. P. N. Satchell, J. Chem. Soc. (1956) p. 2743.
29. Ya. M. Varshavskii, M. G. Lozhkina, and A. I. Shatenshtein, Zhur. Fiz. Khim. 31, 1377 (1957).
30. A. V. Vedeneev, Dissertation [in Russian] (L. Ya. Karpov Physicochemical Institute, Moscow, 1955).
31. A. I. Shatenshtein, A. V. Vedeneev, and P. P. Alikhanov, Zhur. Obshchei Khim. 28, 2638 (1958).
32. P. H. Hermans, Introduction to Theoretical Organic Chemistry (Amsterdam, London, 1954) p. 410.
33. A. I. Brodskii, Izvest. Akad. Nauk SSSR, Otdel. Khim. Nauk (1949) p. 3.
34. Ya. K. Syrkin, Doklady Akad. Nauk SSSR 105, 1018 (1955).
35. A. I. Brodskii, Doklady Akad. Nauk SSSR 83, 847 (1953); Zhur. Obshchei Khim. 24, 413 (1954).
36. A. I. Brodskii, Chemistry of Isotopes [in Russian] (Izd. AN SSSR, Moscow, 1952) p. 198.
37. A. I. Brodskii, Collection: Problems in Chemical Kinetics, Catalysis, and Reactivity [in Russian] (Izd. AN SSSR, Moscow, 1955) p. 18.
38. A. I. Brodskii, Collection of Transactions of the Session of the Academy of Sciences of the USSR on the Peaceful Use of Atomic Energy. Meeting of the Division of Chemical Sciences [in Russian] (Izd. AN SSSR, Moscow, 1955) p. 210.
39. A. I. Brodskii, Ukr. Khim. Zhur. 22, 11 (1956).
40. A. I. Brodskii, Chemistry of Isotopes [in Russian] (Izd. AN SSSR, Moscow, 1957) p. 287.
41. A. I. Brodskii, J. chim. phys. (1958) p. 26.
42. A. I. Brodskii and I. G. Khaskin, Doklady Akad. Nauk SSSR 74, 229 (1950).
43. I. G. Khaskin, Zhur. Obshchei Khim. 23, 32 (1953).
44. A. I. Brodskii and L. V. Sulima, Doklady Akad. Nauk SSSR 74, 513 (1950).
45. A. I. Brodskii and L. V. Sulima, Doklady Akad. Nauk SSSR 85, 1277 (1952).
46. L. Kaplan and K. E. Wilzbach, J. Am. Chem. Soc. 76, 2593 (1954).
47. C. G. Swain, J. T. McKnight, M. M. Labes, and V. P. Kreiter, J. Am. Chem. Soc. 76, 4243 (1954).
48. C. G. Swain and M. M. Labes, J. Am. Chem. Soc. 79, 1084 (1957).
49. C. G. Swain, J. T. McKnight, and V. P. Kreiter, J. Am. Chem. Soc. 79, 1088 (1957).
49a. M. T. Emerson, E. Grunwald, M. L. Kaplan, and R. A. Kromhout, J. Am. Chem. Soc. 82, 6307 (1960).
50. J. Hine and C. H. Thomas, J. Am. Chem. Soc. 75, 739 (1953).
51. H. Kwart, L. P. Kuhn, and E. L. Bannister, J. Am. Chem. Soc. 76, 5998 (1954).
52. J. Hine and C. H. Thomas, J. Am. Chem. Soc. 76, 612 (1954).
53. R. E. Weston and J. Bigeleisen, J. Am. Chem. Soc. 76, 3074 (1954).
54. K. H. Geib, Z. Elektrochem. 45, 648 (1939).
55. G. P. Miklukhin and A. F. Rekasheva, Ukr. Khim. Zhur. 22, 59 (1956).
56. R. E. Mardaleishvili, G. K. Lavrovskaya, and V. V. Voevodskii, Zhur. Fiz. Khim. 28, 2195 (1954).
57. R. Breslow, J. Am. Chem. Soc. 79, 1762 (1957).
58. R. C. Lord and W. D. Philips, J. Am. Chem. Soc. 74, 2429 (1952).
59. H. Rapoport and G. Smolinsky, J. Am. Chem. Soc. 80, 2910 (1958).

60. A. I. Shatenshtein, Uspekhi Khimii 28, 3 (1959).
61. A. I. Shatenshtein, Collection: Problems in Chemical Kinetics, Catalysis, and Reactivity [in Russian] (Izd. AN SSSR, Moscow, 1955) p. 699.
62. A. I. Shatenshtein, Ukr. Khim. Zhur. 22, 3, 62 (1956).
63. V. Gold and F. A. Long, J. Am. Chem. Soc. 75, 4543 (1953).
64. A. I. Shatenshtein, L. N. Vasil'eva, N. M. Dykhno, and E. A. Izrailevich, Doklady Akad. Nauk SSSR 85, 381 (1952).
65. V. Gold and D. P. Satchell, Quart. Rev. 9, 51 (1955).
66. A. I. Shatenshtein, Acta physicochim. URSS 3, 37 (1935); Zhur. Fiz. Khim. 6, 33 (1936).
67. A. I. Shatenshtein, Uspekhi Khimii 24, 377 (1955).
68. C. Kraus, J. Phys. Chem. 60, 129 (1956); J. Chem. Educ. 35, 324 (1958).
69. D. Bethell and V. Gold, Quart. Rev. 12, 173 (1958).
70. A. I. Brodskii, Ukr. Khim. Zhur. 22, 63 (1956).
71. A. I. Shatenshtein, Uspekhi Khimii 21, 919 (1952).
72. S. Z. Roginskii, Theoretical Principles of Isotopic Methods of Studying Chemical Reactions [in Russian] (Izd. AN SSSR, Moscow, 1956).
73. G. M. Panchenkov, Z. V. Gryaznova, V. M. Emel'yanova, and L. Ganichenko, Doklady Akad. Nauk SSSR 109, 325, 546 (1956).
74. A. N. Terenin, Collection: Surface Chemical Compounds and Their Role in Adsorption Phenomena [in Russian] (Izd. MGU, Moscow, 1957) pp. 206 and 273.
75. E. L. Mackor, P. G. Smit, and J. H. van der Waals, Trans. Farad. Soc. 53, 1309 (1957).
76. E. A. Izrailevich, D. N. Shigorin, I. V. Astaf'ev, and A. I. Shatenshtein, Doklady Akad. Nauk SSSR 111, 617 (1957).
77. I. V. Astaf'ev and A. I. Shatenshtein, Zhur. Optika i Spektroskopiya 6, 631 (1959).

CONCLUSION

We have now reached the end of the book. It contains an account of many facts on hydrogen exchange and other reactions involving replacement and migration of hydrogen in organic compounds. It has been shown that the similarity between the reactions examined is connected with the similarity in their mechanisms, which include a stage where the substrate is an acid, an acidlike compound, or a base.

Throughout the whole book, the idea has been developed that the chemical function of a substance is perceived only during its conversion. If the nature of the reagents or medium is changed, the same substance is often able to react as a donor or acceptor of electrons or as a base or an acid. It is sometimes possible to observe the chemical functions of substances which remain unchanged and this makes it possible to guide the process in the desired direction. A clear example is hydrogen exchange in hydrocarbons, when they react as protolytes with very strong bases or acids.

A change in the chemical nature of the reagent and the properties of the medium also makes it possible to reveal different aspects of the intereffect of atoms in the molecule of the same substance. Thus, the conjugation effect normally prevails in electrophilic replacements of hydrogen in aromatic compounds, while the effect of the inductive displacement of electrons is most important in protophilic replacements of hydrogen. If an electrophilic reagent is replaced by a nucleophilic reagent, there is a reversal of the orientation rules of hydrogen replacement in an aromatic ring; and not only this, they even change when there is a sharp increase in the chemical activity of reagents of a given type.

Thus, the reactivity of substances depends to a large extent on not only their structure, but also the characteristics of the reaction medium.

The limitations of existing generalizations become clear as knowledge increases. This refers, in particular, to scientific classifications. They should be reexamined continuously and brought into line with the fuller knowledge of nature. For example, the classification which distinguishes between only associative and ionization mechanisms of chemical reactions is inadequate with the present level of science. The classification in which the proton of the acid is transferred to the base and electrically neutral molecules are converted into ions has also been found to be incomplete. There is now no doubt that an acid–base interaction is a much more complex process than is normally considered.

The concept of acids and bases arose long ago and has occupied the minds of chemists for several centuries. In the first pages of this book, the stages in the development of the theory of acids and bases were followed. At the end of the book, we note that opinions on acids and bases have continued to develop up to now. The theory of acids and bases sheds light on many chemical phenomena, in particular, it is fruitful in the discussion of hydrogen exchange and the replacement of hydrogen in organic compounds.

APPENDIX

PREPARATION OF DEUTERATED ORGANIC COMPOUNDS

1. INTRODUCTION

The production and separation of isotopes made possible the origination and development of a new section of synthetic chemistry, namely, the synthesis of substances differing in isotopic composition.

Most elements have several isotopes. In principle, each of the atoms in any compound may be replaced by its isotopes. The amount and relative position of isotopically replaced atoms may be varied. The more complex the composition and structure of the substance, the greater is the number of variants. The synthesis of isotopic compounds is therefore an infinite field.

With the aid of compounds labeled with isotopes it is possible to follow processes and conversions involving chemically identical substances, to establish which of identical atoms participates in a reaction, to determine its true mechanism, etc. [1-3].

Compounds labeled with isotopes play an important part in biological investigations [4, 5]. They make it possible to determine the routes of distribution of substances in an organism, to demonstrate the interrelation between coexisting species, to check the vital activity of separate organs, etc. Compounds containing radioactive isotopes may be used as radiation sources to treat a particular part of an organism.

Substances labeled with isotopes have found wide application in many branches of technology.

The study of the properties of isotopic modifications of substances and the differences in their chemical behavior (isotope effect) is of importance in itself. Isotope effects become particularly noticeable when protium is replaced by tritium and deuterium as a result of the relatively large differences in their masses [1, 2]. In particular, measurements of vibration spectra of deuterated compounds are of great scientific value. The replacement of hydrogen by one of its isotopes changes the vibration frequencies of the bonds without changing the force constants. A comparison of the spectra of deuterated and normal molecules makes it possible to obtain the additional parameters required to determine the force constants characterizing the structure of the molecule and this, in its turn, makes it possible to calculate theoretically vibration spectra, including those of substances which cannot be obtained in a pure form for measurements. These considerations stimulated cooperative work on the preparation of deuterated hydrocarbons and the measurement of their vibration spectra in the laboratory of Academician G. S. Landsberg (Institute of Physics, Academy of Sciences, USSR) and in the Isotopic Reactions Laboratory (L. Ya. Karpov Physicochemical Institute) [6-9].

The measurements of the electronic spectra of deuterated hydrocarbons at liquid hydrogen temperature (20°K) in the laboratory of A. F. Prikhot'ko (Institute of Physics, Academy of Sciences, USSR) and the Isotopic Reactions Laboratory were carried out to determine the changes in the structure of molecules and crystals which accompany isotopic replacement [10].

These investigations are examples from the wide field of application of deuterated organic compounds in the solution of concrete scientific problems. The new methods of preparing deuterated compounds by hydrogen exchange in nonaqueous solutions described below were used.

The general principles of the preparation of substances of abnormal isotopic composition were discussed in S. Z. Roginskii's report [11]. The literature contains reviews of the synthesis of such substances, in particular, organic compounds with heavy hydrogen [12-18].

2. METHODS OF PREPARING DEUTERATED ORGANIC COMPOUNDS

Either organic syntheses with deuterium and its compounds or hydrogen exchange are used to prepare deuterated organic compounds. The latter method is more general as deuterium exchange can be used to prepare many deuterated substances from normal organic compounds by means of the same deuterated reagents. In addition, it is possible to obtain purer substances by exchange than by chemical syntheses in which there are often formed by-products that are not always readily separable. It is true that exchange reactions sometimes are also accompanied by chemical conversions which contaminate the substance.

If one of several equivalent hydrogen atoms in the molecule of a substance is to be replaced by deuterium it is necessary to use synthetic methods as all similar hydrogen atoms will participate in an exchange. As a rule, a larger amount of deuterating reagent (for example, heavy water) is consumed in the preparation of deuterium compounds by isotopic exchange than in synthesis as the deuterium concentration is reduced due to the protium in the starting substance. Therefore, to prepare a substance of the highest possible isotopic purity it is necessary to repeat deuterium exchange several times with the fullest possible removal of the reagent used in the previous operation. This is not always simple to do. A homogeneous system provides the optimal conditions for isotopic exchange, but it is often possible to deuterate difficultly soluble substances.

Clusius [19] derived a formula for calculating the deuterium concentration in a substance after the establishment of exchange equilibrium, whose equilibrium constant equals K; n_{sb} and n_{sl} are the numbers of moles of the substance and solvent participating in the exchange reaction:

$$n_{sb} = \frac{P_{sb}}{M_{sb}} N_{sb}, \quad n_{sl} = \frac{P_{sl}}{M_{sl}} N_{sl},$$

where P, M, and N are the weight, molecular weight, and number of exchangeable hydrogen atoms of the substance (subscript "sb") and solvent (subscript "sl"),

$$\% D = \frac{K}{2n_{sb}(K-1)} \left[n_{sb} + n_{sl} - \sqrt{(n_{sb} - n_{sl})^2 + 4n_{sb}n_{sl}/K} \right]$$

when K = 1, the expression is simplified:

$$\% D = \frac{n_{sl}}{n_{sb} + n_{sl}}.$$

It is assumed that all the exchangeable atoms are replaced by deuterium in the solvent used in the reaction.

3. SYNTHESIS OF DEUTERATED ORGANIC COMPOUNDS

We will limit ourselves to a comparatively small number of the simplest typical examples of methods of synthesizing deuterated organic compounds. These are reduction with deuterium, hydrolysis and hydration with heavy water, etc. The substances obtained in this way may be used to synthesize more complex compounds. In recent years, systematic work has been carried out in some laboratories on the development of methods for synthesizing deuterium compounds [see, for example, the series of articles of L. C. Leitch and his co-workers in the Canadian Journal of Chemistry].

A. Reduction

Most reductions with deuterium are catalyzed with platinum or nickel:

$$CO + 3D_2 \rightarrow CD_4 + D_2O;$$
$$CO_2 + 4D_2 \rightarrow CD_4 + D_2O;$$
$$HC \equiv CH + D_2 \rightarrow HDC = CHD;$$
$$DHC \equiv CHD + D_2 \rightarrow D_2HC - CHD_2;$$
$$C_6H_6 + 3D_2 \rightarrow C_6H_6D_6.$$

Ethylene, ethane, and cyclohexane with complete isotopic replacement have been obtained by reduction of deuteroacetylene and deuterobenzene, prepared as described below.

Lithium aluminum hydride is used increasingly nowadays for the reduction of organic compounds. We will consider work [20] in which $LiAlD_4$ and $LiAlH_4$ were used to prepare isobutanol with deuterium in the α-, β-, and γ-positions. Isobutyraldehyde was reduced to prepare the α-isomer:

$$(CH_3)_2CHCHO \xrightarrow{D_2} (CH_3)_2CHCDOD.$$

The β-isomer was synthesized by decarboxylation of dimethylmalonic acid, deuterated in the carboxyl group:

$$(CH_3)_2C(COOD)_2 \xrightarrow{-CO_2} (CH_3)_2CDCOOD.$$

The monocarboxylic acid was reduced to $(CH_3)_2CDCH_2OD$. Finally, the γ-isomer was prepared by the following sequence of reactions: deuteroacetone, which was prepared from normal acetone by deuterium exchange with a solution of alkali in D_2O, was reduced with lithium aluminum hydride:

$$(CD_3)_2\,CO \xrightarrow{\text{LiAlD}_4} (CD_3)_2\,CHOH;$$

$$(CD_3)\,CHOH \xrightarrow{\text{PBr}_3} (CD_3)_2\,CHBr;$$

$$(CD_3)_2\,CHBr \xrightarrow{\text{Mg}} (CD_3)_2\,CHMgBr;$$

$$(CD_3)_2\,CHMgBr \xrightarrow{\text{CO}_2} (CD_3)_2\,CHCOOH;$$

$$(CD_3)_2\,CHCOOH \xrightarrow{\text{LiAlH}_4} (CD_3)_2\,CHCH_2OH.$$

B. Hydration and Hydrolysis

a. HYDROLYSIS OF INORGANIC SUBSTANCES

Carbides and carbon suboxide* react with D_2O according to the following equations:

$$CaC_2 + 2D_2O \rightarrow DC \equiv CD + Ca\,(OD)_2;$$
$$Al_4C_3 + 12D_2O \rightarrow 3CD_4 + 4Al\,(OD)_3;$$
$$C_3O_2 + 2D_2O \rightarrow D_2C\,(COOD)_2.$$

b. HYDRATION OF ORGANIC SUBSTANCES

The addition of D_2O to acid anhydrides yields acids deuterated in the carboxyl group:

$$(CH_3CO)_2O + D_2O \rightarrow 2CH_3COOD.$$

The hydration of acetylene to yield heavy acetaldehyde is catalyzed by Hg_2SO_4 and H_3PO_4:

$$C_2D_2 + D_2O \rightarrow CD_3CDO.$$

c. HYDROLYSIS OF HALOGEN COMPOUNDS

In the hydrolysis of halogen compounds by heavy water, the halogen is replaced by the OD group. For example, hydrolysis of acetyl chloride yields CH_3COOD:

$$CH_3COCl + D_2O \rightarrow CH_3COOD + DCl.$$

---

*Carbon suboxide is obtained by dehydration of malonic ester with phosphorus pentoxide:

$$CH_2(COOC_2H_5)_2 \xrightarrow{\text{P}_2\text{O}_5} 2C_2H_4 + 2H_2O + C_3O_2.$$

d. HYDROLYSIS OF ORGANOMAGNESIUM HALIDES

To introduce deuterium into a given position it is possible to replace the hydrogen by a halogen, apply the Grignard reaction to prepare an organomagnesium halide, and hydrolyze the latter with heavy water [21-23].

For example:

$$C_6H_5CH_3 + Br_2 \rightarrow p\text{-}C_6H_4BrCH_3 + HBr;$$
$$p\text{-}C_6H_4BrCH_3 + Mg \rightarrow p\text{-}C_6H_4MgBrCH_3;$$
$$p\text{-}C_6H_4MgBrCH_3 + D_2O \rightarrow p\text{-}C_6H_4DCH_3 + Mg(OD)Br.$$

For the preparation of pure dideuterobenzene it is recommended [22] that the monomagnesium halide derivative is first synthesized from a dihalogen compound. After decomposition of this with deuterium oxide, a Grignard reagent is again prepared and hydrolyzed to give a dideuterobenzene. If sufficient magnesium is used to form the disubstituted organomagnesium compound, hydrolysis of the latter forms a mixture of mono- and dideuterobenzenes.

1,2,4,5-Tetradeuterobenzene [23] was synthesized by replacement of the four hydrogen atoms in p-dibromobenzene by exchange with deuterosulfuric acid (p. 300) with subsequent replacement of the bromine atoms by hydrogen by hydrolysis of the Grignard reagent with normal water. Pentadeuterobenzene was obtained [23] by bromination of hexadeuterobenzene and replacement of the bromine by light hydrogen.

If DCl is used instead of D_2O for decomposition of the Grignard reagent, it is possible to obtain the required deuterated product from polyhalogen compounds in one step without risk of the formation of a mixture of substances with different degrees of deuteration (see [15]). A method of replacing a halogen by deuterium without preparation of a Grignard reagent has been reported [24-26]. The halogen compound is boiled with zinc dust in deuteroacetic acid. This method was used to prepare isomers of deuteropyridine and thiophene and also toluene and xylene deuterated in the methyl group.

e. HYDROLYSIS OF ORGANOALKALI COMPOUNDS

Hydrogen may be replaced by deuterium through the metallation of an organic compound. For example:

$$2C_2H_5OH + 2Na \rightarrow C_2H_5ONa + H_2;$$
$$C_2H_5ONa + D_2O \rightarrow C_2H_5OD + NaOD;$$
$$2\,(C_6H_5)_3\,CH + 2K \rightarrow 2\,(C_6H_5)_3\,CK + H_2;$$
$$(C_6H_5)_3\,CK + D_2O \rightarrow (C_6H_5)_3\,CD + KOD;$$
$$C_6H_5Li + D_2O \rightarrow C_6H_5D + LiOD.$$

C. Ammonolysis

Ammonolysis with deuteroammonia has not been used up to now for the preparation of deuterated compounds. This method may be used to synthesize amines, amides, and amidines (see [28]):

$$RI + 2ND_3 \rightarrow RND_2 + ND_4I;$$
$$RCOCl + 2ND_3 \rightarrow RCOND_2 + ND_4Cl;$$
$$RCN + ND_3 \rightarrow RCNDND_2.$$

D. Condensation

An example of condensation is the protolytic polymerization of deuteroacetylene over tellurium oxide at 650°, which leads to the formation of a mixture of aromatic hydrocarbons: hexadeuterobenzene, octadeuterotoluene, octadeuteroindene, deca-deuterofluorene, and decadeuteropyrene [29-31]. Individual hydrocarbons may be isolated from this mixture of substances in the form of picrates. Deuterobenzene and deuteronaphthalene have been prepared by this method [32]. In the latter case, the reactions conditions were changed and the condensation temperature was 1025°. The naphthalene yield reached 19%.

E. Decarboxylation

When dibasic acids are heated to temperatures below 200°, carbon dioxide is eliminated to form a monobasic acid, for example, deuterooxalic gives deuterofor-mic acid, deuteromalonic gives deuteroacetic acid, and deuteromethylmalonic acid gives deuteropropionic acid:

$$(COOD)_2 \rightarrow DCOOD + CO_2;$$
$$D_2C(COOD)_2 \rightarrow CD_3COOD + CO_2;$$
$$CH_3CD(COOD)_2 \rightarrow CH_3CD_2COOD + CO_2.$$

The decarboxylation of calcium salts of aromatic acids mixed with $Ca(OD)_2$ lib-erates calcium carbonate and forms benzene in which a deuterium atom occupies the position of the carboxyl group. This reaction occurs at 300-500°. Decarboxylation of the calcium salt of benzoic acid gives monodeuterobenzene [33], phthalic acid gives dideuterobenzene, trimesic acid gives trideuterobenzene, pyromellitic acid gives tet-radeuterobenzene [34], and finally, the calcium salt of mellitic acid gives hexadeu-terobenzene [35].

Two side reactions produce excess deuteration in comparison with the expected amount: exchange between $Ca(OD)_2$ and the calcium salt of the acid and exchange between $Ca(OD)_2$ and the partially deuterated benzene obtained by decarboxylation [36]. This observation was later used [37] to prepare octadeuteronaphthalene by heat-ing a mixture of normal naphthalene, $Ca(OD)_2$, and D_2O for 24 hr at 400°. This pro-cedure was repeated seven times with the naphthalene distilled each time.

4. ISOTOPIC EXCHANGE OF HYDROGEN

A. Hydrogen Exchange with Heavy Water and Deuteroalcohol

Heavy water or deuteroalcohol are usually used for hydrogen exchange. The latter has the advantage that organic substances are more soluble in it than in water.

The hydrogen in O−H, N−H, S−H, and Hal−H bonds exchanges instantaneously with heavy water and alcohol. The exchange of hydrogen in CH bonds of organic com-pounds is much more difficult. The hydrogen of CH bonds exchanges with water and alcohol in only a few substances.

In the case of hydrocarbons, the hydrogen in acetylene exchanges with aqueous alkali comparatively readily. This is the result of the polarity of the CH bonds, which is caused by the sp-hybridization of the bond between the carbon atoms. In ethylenic, aromatic, and more so, saturated hydrocarbons there is no exchange of hydrogen with heavy water, even when bases and acids are used as catalysts. However, isotopic exchange of hydrogen between heavy water and benzene occurs at room temperature under the action of a high-frequency current in the presence of platinum black [38]. The latter evidently plays a decisive part as it was possible to prepare C_6D_6 from C_6H_6 and D_2O in the presence of it at 110° [39]. After four exchanges, the protium content was less than 1%.

The presence of polar groups in the molecule of an organic substance facilitates exchange with water or alcohol. Exchange catalyzed by alkali is normally promoted by the presence of an electronegative substituent in the molecule of the substance (NO_2, SO_3H, COOH, CHO, and CN) (see [12, 40, 41]), while electropositive substituents (OH, OCH_3, NH_2, and $N(CH_3)_2$)* promote exchange catalyzed by acid [41]. For example, aniline is deuterated by sulfuric acid with the deuterium replacing the ortho- and para-hydrogen atoms. This was used to prepare 1,3,5-trideuterobenzene [23]. The amino group was eliminated from the deuteroaniline by diazotization and treatment of the solution of the diazo compound with sodium stannite.

Some exchanges proceed more readily in the gas phase with heterogeneous catalysis than in solution [42]. D_2 or D_2O is used as a source of deuterium. Taylor and his co-workers [43] considered that if a mixture of benzene vapor and deuterium oxide was passed at 200° over nickel on kieselguhr, pure deuterobenzene was obtained. A check [44] showed that even at 150°, hydrogen and carbon dioxide was formed. The benzene was contaminated by higher-molecular products, mainly, biphenyl. As a rule, hydrocarbons undergo partial cracking on metal and oxide catalysts.

B. Hydrogen Exchange with Deuterosulfuric Acid

Exchange of hydrogen in ethylene, benzene, and even saturated hydrocarbons with a tertiary carbon atom occurs if the deuterium donor is deuterosulfuric acid. Ingold and his co-workers [45] used it for preparative purposes. Benzene was shaken for 10 days at room temperature with deuterosulfuric acid with a concentration of 51-52 mol. %. This operation was repeated several times. Later, naphthalene was deuterated by shaking with 50% D_2SO_4 (for 100 hr at 120°) [46]. Octadeuteronaphthalene was also prepared [37] by boiling and stirring a solution of naphthalene in CCl_4 with 65% deuterosulfuric acid for 200 hr with the acid replaced by a fresh portion every 24 hr.

C. Hydrogen Exchange with DCl + AlCl₃

Klit and Langseth [47] prepared hexadeuterobenzene by exchange between C_6H_6 and gaseous deuterium chloride in the presence of aluminum chloride. An apparatus for the preparation of considerable amounts of C_6D_6 by this method was described subsequently [48]. The hydrogen chloride was obtained by hydrolyzing thionyl chloride ($SOCl_2$) with heavy water. The reagents used for the reaction were 50 g of C_6H_6, 5 g of $AlCl_3$, 200 g of D_2O, (99.5%), and 1200 g of $SOCl_2$. After 51 hr, 36 ml of heavy benzene containing 99.3% of D was obtained.

---

* See footnote on p. 9.

D. Hydrogen Exchange with Liquefied Gases

Let us now examine the methods of preparing deuterated compounds which arose as a result of the study in the Isotopic Reactions Laboratory of the L. Ya. Karpov Physicochemical Institute of hydrogen exchange with liquefied gases, namely, liquid deuteroammonia and liquid deuterium bromide and fluoride (see pp. 98-116, 168-187 [6-8, 10, 49, 50]). By means of those solvents it is possible to exchange the hydrogen of CH bonds of many organic substances for deuterium, especially if a catalyst is used (potassamide, aluminum bromide, and boron trifluoride, respectively). An advantage of these solvents is their high volatility, as a result of which they are readily removed after an exchange.

It is very important that under the action of an acidic or basic solvent, the exchange of nonequivalent hydrogen atoms normally proceeds selectively, i.e., in different ways and at different rates. This may be used for the preparation of partially deuterated substances. Successive treatments of organic substances with several solvents (deuterated and normal) considerably extends the range of partially deuterated compounds.

a. HYDROGEN EXCHANGE WITH LIQUID DEUTEROAMMONIA

Liquid deuteroammonia is obtained by the reaction between magnesium nitride and heavy water:

$$Mg_3N_2 + 3D_2O \rightarrow 2ND_3 + 3Mg(DO)_2.$$

Some details of the method of preparing deuteroammonia are given in [6].

By the use of liquid deuteroammonia without a catalyst it is possible to replace by deuterium only sufficiently labile hydrogen atoms such as, for example, three hydrogen atoms in the indene molecule, three hydrogen atoms in the methyl groups of acetophenone,* the hydrogen atoms of the methylene group of fluorene,* etc.

With catalysis by potassamide in liquid deuteroammonia there is complete exchange of hydrogen in ethylenic hydrocarbons with a straight chain of carbon atoms from ethylene to cetene. The exchange is accompanied by isomerization, for example, 1-pentene is converted to 2-pentene. Therefore, this method can be used to prepare a mixture of isomers enriched in deuterated 2-pentene, but a pure substance cannot be obtained.

Measurement of the exchange kinetics has shown that the rates of exchange of nonequivalent hydrogen atoms in alkenes sometimes differ considerably. For example, the hydrogen of the methyne group of propylene exchanges much more slowly than the other hydrogen atoms. In principle, this may be used to prepare deuteropropylene with a hydrogen atom in the methyne group or, on the other hand, propylene with five hydrogen atoms and a deuterium atom in the methyne group (reverse exchange between C_3D_6 and $NH_3 + NH_2^-$).

In the presence of potassamide, liquid deuteroammonia exchanges with the hydrogen of aromatic rings of hydrocarbons and those heterocyclic compounds which do not react chemically with the amide. This method may be uesd to prepare fully deuterated benzene,* naphthalene,* phenanthrene,* biphenyl,* triphenylmethane, diphenylmethane, bibenzyl, pyridine, etc.

---

* Substances whose deuteroderivatives have been prepared in the Isotopic Reactions Laboratory of the Karpov Physicochemical Institute are marked with an asterisk.

In the presence of potassamide, there is very rapid replacement by deuterium of hydrogen in aliphatic CH bonds, whose carbon is attached to an aromatic ring (σ, π-conjugation effect), for example, in the methyne groups of triphenylmethane* and isopropylbenzene, in the methylene groups of indene, fluorene, diphenylmethane, and bibenzyl, at the α-carbon atoms of the hydrogenated rings of tetralin, in the methyl groups of toluene,* methylnaphthalene, picoline, quinaldine, etc. This may be used for partial deuteration of the substances listed in brief experiments. With longer experiments, replacement occurs in the aromatic ring. Samples of toluene,* p- and m-xylene,* mesitylene,* durene,* and hexamethylbenzene* in which all the hydrogen atoms were replaced by deuterium were prepared [10].

Under drastic conditions (a high potassamide concentration and a temperature of 120°), deuterium replaces hydrogen at the β-carbon atoms of alkyl groups in alkylbenzenes. For example, the complete deuteration of ethylbenzene and isopropylbenzene is possible.

It should be noted that with catalysis by potassamide in liquid deuteroammonia, there is hydrogen exchange not only in methyl groups attached directly to an aromatic ring, but also in groups attached to it through an oxygen or nitrogen atom, for example, in anisole,* methoxynaphthalene, dimethylaniline,* and dimethylnaphthylamine.

b. HYDROGEN EXCHANGE WITH LIQUID DEUTERIUM BROMIDE
Liquid deuterium bromide is obtained by synthesis from the elements:

$$D_2 + Br_2 \rightarrow 2DBr.$$

The details of the synthesis method are given in [51].

In exchange reactions with liquid deuterium bromide there is complete exchange of the hydrogen in the following aromatic hydrocarbons: naphthalene,* anthracene, phenanthrene, and pyrene. The hydrogen of benzene exchanges with liquid deuterium bromide very slowly, but in the presence of aluminum bromide, the exchange is instantaneous.

The nonequivalence of hydrogen atoms in the molecule of an organic substance appears in hydrogen exchange with liquid deuterium bromide to a much greater extent than in liquid deuteroammonia, catalyzed by potassamide. By using the differences in the exchange rates of nonequivalent hydrogen atoms, it is possible to prepare substances in which some of the hydrogen atoms in definite positions have been replaced by deuterium atoms. For example, when six hydrogen atoms in the biphenyl molecule have exchange, no further exchange is observed even after 300 hr. 2,4,6, 2',4',6'-Hexadeuterobiphenyl* is obtained. Analogously, 10 hydrogen atoms are replaced in p-terphenyl. All the hydrogen atoms in biphenyl exchange very rapidly in the presence of AlBr$_3$; decadeuterobiphenyl* has been prepared by this method.

The α-hydrogen atoms in the naphthalene molecule exchange with liquid deuterium bromide 40-50 times as fast as the β-atoms [52]. Exchange of the α-atoms is practically complete in 30 min, while the β-atoms cannot be replaced in this time. This makes it possible to prepare almost pure tetra-α-deuteronaphthalene* [7].

---

* See footnote on p. 302.

In alkylbenzene molecules, the rate of exchange of the ortho- and para-hydrogen atoms of the aromatic ring is two to three orders higher than the rate of exchange of the meta-atoms. Trisubstituted deuteroderivatives therefore can be prepared.

In the presence of aluminum bromide, all five hydrogen atoms of the ring of alkylbenzenes exchange instantaneously, but even under these conditions, the hydrogen atoms of the alkyl group do not participate in the exchange. Consequently, it is possible to prepare alkylbenzenes deuterated in the ring with unsubstituted alkyl groups, for example, $C_6D_5CH_3$.*

In polymethylbenzenes, all the hydrogen atoms in the ring are replaced by deuterium rapidly without a catatlyst. In this way it was possible to prepare $C_6D_4(CH_3)_2$,* $C_6D_3(CH_3)_3$,* and $C_6D_2(CH_3)_4$* with a high degree of isotopic purity (for spectroscopic measurements).

c. COMBINED EXCHANGE

By combining forward and reverse exchange in different solvents it is possible to prepare many substances in which the deuterium is in a definite position. For example, if octadeuteronaphthalene is treated for a short time with liquid hydrogen bromide, the deuterium is removed from the α-positions and tetra-β-deuteronaphthalene* is obtained.

Similarly, when decadeuterobiphenyl, prepared by the action of $ND_3 + ND_2^-$ on normal biphenyl, was treated with liquid hydrogen bromide, the deuterium was removed from the ortho- and para-positions to leave 3,5,3',5'-tetradeuterobiphenyl* [7].

The number of possible combinations is increased still further if liquid hydrogen fluoride is used for reverse exchange as a result of the characteristics of exchange with this solvent. Table 99 gives 15 formulas of ethylbenzene, which differ in the number and positions of the deuterium atoms. The possible methods of preparing these substances are indicated under the formulas. Table 100 gives methods of preparing anisole in which the hydrogen atoms have been replaced by deuterium completely or partially. It is obvious that the number of combinations which make it possible to prepare different deuterated organic compounds by the methods developed in the Isotopic Reactions Laboratory is exceptionally great.

5. COMPARISON OF METHODS OF PREPARING DEUTERIUM COMPOUNDS

Many deuterated compounds which are prepared comparatively readily by isotopic exchange with DBr, $ND_3 + KND_2$, etc., are very difficult to synthesize. This is illustrated by examples given above. 2,2',4,4',6,6'-Hexadeuterobiphenyl, which was obtained by a multistage synthesis [52], is readily obtained by treatment of normal biphenyl several times with liquid DBr. The number of examples may be multiplied readily. The same deuterated organic compounds may be prepared by different methods. Substances can often be obtained most readily by means of liquefied gases. At the present time, octadeuteronaphthalene can be synthesized from acetylene (p. 300) and prepared by two methods of isotopic exchange with deuterosulfuric acid (p. 301), with $Ca(OD)_2$ (p. 300), with liquid deuterium bromide (p. 303), and with deuteroammonia with catalysis by potassamide (p. 302). In the latter case, exchange equilibrium is reached (with C_{KND_2} = 0.1 N) after 10 min at room temperature. The exchange reaction with liquid deuterium bromide continues for about a day.

---

* See footnote on p. 302.

TABLE 99. Preparation of Deuterated Ethylbenzenes

CD_2CH_3 (I) ring: D, D, D, D, D	CD_2CD_3 (II) ring: D, D, D, D, D	CH_2CH_3 (III) ring: D, D, H, H, D
$ND_3 + ND_2^-$ Room temp.	$ND_3 + ND_2^-$ 120°	DBr
CH_2CH_3 (IV) ring: D, D, D, D, D	CH_2CH_3 (V) ring: H, H, D, D, H	CD_2CH_3 (VI) ring: H, H, H, H, H
$DBr + AlBr_3$	IV + HBr	I + HF
CD_2CH_3 (VII) ring: H, H, D, D, H	CD_2CH_3 (VIII) ring: D, D, D, D, D	CH_2CD_3 (IX) ring: H, H, H, H, H
I + HBr	VI + DBr	$II + NH_3 + NH_2^-$
CH_2CD_3 (X) ring: D, D, D, D, D	CH_2CD_3 (XI) ring: D, D, H, H, D	CH_2CD_3 (XII) ring: H, H, D, D, H
$IX + DBr + AlBr_3$	IX + DBr	X + HBr
CD_2CD_3 (XIII) ring: H, H, H, H, H	CD_2CD_3 (XIV) ring: H, H, D, D, H	CD_2CD_3 (XV) ring: D, D, H, H, D
II + HF or $II + HBr + AlBr_3$	II + HBr	XII + DBr

G. V. Peregudov [7], measured the Raman spectra of the preparations obtained by the last two methods and showed that within the limits of experimental error, they coincide with spectra of octadeuteronaphthalene preparations according to literature data.

The exchange of protium for deuterium in some organic substances can be achieved with both $ND_3 + KND_2$ and DBr (or $DBr + AlBr_3$). Deuterium bromide has advantages.

In the laboratory, it is much simpler to prepare DBr with a high degree of isotopic purity than ND_3. If industrial gaseous deuterium ($\sim$ 99.5 at. %) is used, then the DBr contains the same percentage of deuterium.

TABLE 100. Preparation of Deuterated Anisoles

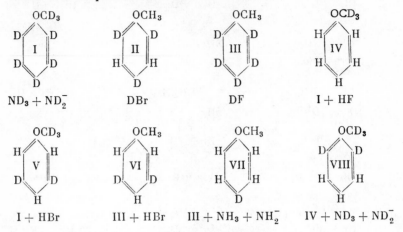

Note. The different rates of exchange of the o-, m-, and p-atoms with a solution of potassamide in ammonia (of deuteroammonia) may be used to prepare VII and VIII.

If ND_3 is parepared from deuterium oxide (99.5 mol. %) by the reaction with magnesium nitride without special precautions, the deuterium content does not exceed 95 at. %. If the same operations with magnesium nitride are carried out in a dry box and the nitride is charged into metal ampoules, pumped out with heating, treated with a small amount of ND_3, and again pumped out, it is possible to increase the deuterium concentrations in the ammonia to 98.6 at. % [54].

The same amount of heavy water yields 25 times as much DBr as ND_3. As already mentioned above, with the same concentration of deuterium in the two solvents, with an equal number of exchanges with DBr it is possible to reach a higher concentration of deuterium in the organic substance as a result of the high deuterium distribution coefficient [55]. Because of this, the number of repeat deuterations is reduced and the requirements for the isotopic purity of the deuterating reagent are lower.

Naturally, the same method [56] may be used to obtain preparations labeled with tritium if water enriched in tritium is used instead of deuterium oxide (and the solvents prepared from it).

The method of preparing organic substances with the hydrogen replaced by its isotopes developed in the Isotopic Reactions Laboratory is now being used to supply the needs of investigators in the State Institute of Applied Chemistry (Leningrad) [57].

LITERATURE CITED

1. A. I. Brodskii, Chemistry of Isotopes [in Russian] (Izd. AN SSSR, Moscow, 1957).
2. S. Z. Roginskii, Theoretical Principles of Isotopic Methods of Studying Chemical Reactions [in Russian] (Izd. AN SSSR, Moscow, 1956).

3. L. Ya. Margolis, Labeled Atoms in Catalysis [in Russian] (Izd. AN SSSR, Moscow, 1958).

4. G. Hevesy, Radioactive Indicators, Their Application in Biochemistry, Normal Physiology, and Pathological Physiology of Man and Animals [Russian translation] (IL, Moscow, 1950).

5. M. Kamen, Radioactive Tracers in Biology [Russian translation] (IL, Moscow, 1948).

6. G. S. Landsberg, A. I. Shatenshtein, V. Peregudov, E. A. Izrailevich, and L. A. Novikova, Izvest. Akad. Nauk SSSR, Ser. Fiz. (1954) p. 669; Zhur. Optika i Spektroskopiya 1, 34 (1956).

6a. G. S. Landsberg, Selected Works [in Russian] (Izd. AN SSSR, Moscow, 1958) p. 448.

7. A. I. Shatenshtein, G. V. Peregudov, E. A. Izrailevich, and V. R. Kalinachenko, Zhur. Fiz. Khim. 32, 146 (1958).

8. A. I. Shatenshtein and E. A. Izrailevich, Collection: Isotopes in Catalysis [in Russian] (Izd. AN SSSR, Moscow, 1957) p. 430.

9. M. A. Kovner and G. V. Peregudov, Zhur. Optika i Spektroskopiya 5, 134 (1958).

10. V. L. Braude, E. A. Izrailevich, A. L. Liberman, M. I. Onoprienko, O. S. Pakhomova, A. F. Prikhot'ko, and A. I. Shatenshtein, Zhur. Optika i Spektroskopiya 5, 113 (1958).

11. S. Z. Roginskii, Collection: Isotopes in Catalysis [in Russian] (Izd. AN SSSR, Moscow, 1957) p. 411.

12. G. P. Miklukhin, Exchange Reactions of Hydrogen Isotopes, Uspekhi Khimii 17, 663 (1948).

13. F. Adickes, Angew Chem. 51, 89 (1938); Uspekhi Khimii 7, 1052 (1938).

14. H. Erlenmeyer, Z. Elektrochem. 44, 8 (1938).

15. M. A. Langseth, Le Isotope, Rapports et discussions (Inst. Internat de chimie Solvay, 1948) p. 242.

16. S. L. Thomas and H. S. Turner, Quart. Rev. 7, 407 (1953).

17. A. Murray and D. L. Williams, Organic Syntheses with Isotopes, Vol. II (Interscience, New York, 1958).

18. G. P. Miklukhin, Isotopes in Organic Chemistry [in Russian] (Izd. AN Ukr.SSR, Kiev, 1961).

19. K. Clusius and H. Knopf, Z. Naturforsch 2b, 169 (1947).

20. F. E. Condon, J. Am. Chem. Soc. 73, 4675 (1951); Canad. J. Chem. 34, 75(1956).

21. O. Redlich and W. Stricks, Monatsh. Chem. 67, 213 (1936).

22. L. H. P. Weldon and C. L. Wilson, J. Chem. Soc. (1946) p. 235.

23. A. P. Best and C. L. Wilson, J. Chem. Soc. (1946) p. 239.

24. B. Bak, J. Org. Chem. 21, 797 (1956).

25. R. Renaud and L. C. Leitch, Canad. J. Chem. 34, 98 (1956).

26. M. E. Leblanc, A. T. Morse, and L. C. Leitch, Canad. J. Chem. 34, 354 (1956).

27. W. M. Lauer and W. E. Noland, J. Am. Chem. Soc. 75, 3689 (1953).

28. W. C. Fernelius and G. W. Bowman, Chem. Rev. 26, 3 (1940).

29. J. W. Murray, C. F. Squire, and D. H. Andrews, J. Chem. Phys. 2, 714 (1934).

30. G. R. Clemo, A. McQuillen, and A. C. Robson, J. Chem. Soc. (1935) p. 851; (1939) p. 429.

31. G. R. Clemo and A. McQuillen, J. Chem. Soc. (1935) p. 1325.

32. J. Gubeau, H. Luther, K. Feldman, and G. Brandes, Ber. 86, 214 (1953).

33. N. Morite and T. Titani, Bull. Chem. Soc. Japan 10, 557 (1935).

34. O. Redlich and W. Stricks, Monatsh. Chem. 68, 374 (1937).

35. H. Erlenmeyer, H. Lobeck, H. Gärtner, and A. Epprecht, Helv. Chim. Acta 18, 1464 (1935); 19, 336 (1936).

36. L. H. P. Weldon, and C. L. Wilson, J. Chem. Soc. (1946) p. 244.

37. E. R. Lippincott, and E. J. O. Reilly, J. Chem. Phys. 23, 238 (1955).

38. J. Horiuti and T. Koyano, Bull. Chem. Soc. Japan 10, 601 (1935).

39. L. C. Leitch, Canad. J. Chem. 32, 813 (1954).

40. K. F. Bonhöffer, K. H. Geib, and O. Reitz, J. Chem. Phys. 7, 664 (1939).

41. M. S. Kharasch, W. G. Brown, J. McNab, and W. R. Sprowls, J. Org. Chem. 2, 36 (1937); 4, 442 (1939).

42. H. S. Taylor, Uspekhi Khimii 15, 359 (1946).

43. P. J. Bowman, W. S. Benedict, and H. S. Taylor, J. Am. Chem. Soc. 57, 960 (1935).

44. J. A. Dixson and R. W. Schiessler, J. Am. Chem. Soc. 76, 2197 (1954).

45. C. K. Ingold, C. G. Raysin, C. L. Wilson, C. H. Bailay, and B. Topley, J. Chem. Soc. (1936) pp. 915, 1637.

46. W. P. Pearson, G. C. Pimental, and O. Schnepp, J. Chem. Phys. 23, 230 (1955).

47. A. Klit and A. Langseth, Z. Phys. Chem. 176, 65 (1936).

48. M. Brüllmann, H. J. Gerber, and D. Meyer, Helv. Chim. Acta 41, 1831 (1958).

49. A. I. Shatenshtein and A. V. Vedeneev, Zhur. Obshchei Khim. 28, 2644 (1958).

50. A. I. Shatenshtein, A. N. Talanov, and Yu. I. Ranneva, Zhur. Obshchei Khim. 30, No. 2 (1960).

51. V. R. Kalinachenko, Ya. M. Varshavskii, and A. I. Shatenshtein, Zhur. Fiz. Khim. 30, 1140 (1956).

52. A. I. Korolev, A. I. Shatenshtein, E. N. Yurygina, V. R. Kalinachenko, and P. P. Alikhanov, Zhur. Obshchei Khim. 26, 1661 (1956).

53. R. J. Akavie, J. M. Searborough, and J. G. Burr, NAA-SR-2144, 1.12.1957. See Nucl. Sci. Abstr. No. 3618 (1958); J. Org. Chem. 24, 1946 (1959); 26, 243 (1961).

54. A. I. Shatenshtein and Yu. I. Antonchik, Zhur. Anal. Khim. 14, 9 (1959).

55. Ya. M. Varshavskii, V. R. Kalinachenko, S. E. Vaisberg, and A. I. Shatenshtein, Zhur. Fiz. Khim. 30, 1647 (1956).

56. A. I. Shatenshtein, E. A. Izrailevich, and V. R. Kalinachenko, Methods of preparing fully and partially deuterated organic substances, State registration No. 5483, January 22, 1957.

57. R. S. Bardasova, E. G. Komarova, I. F. Tupitsyn, and A. I. Shatenshtein, Collected Works of the State Institute of Applied Chemistry (Leningrad, 1960) No. 45, p. 111.